Computer Analysis of Structural Frameworks

James A. D. Balfour
BSc, PhD, CEng, MICE

COLLINS
8 Grafton Street, London W1

Collins Professional and Technical Books
William Collins Sons & Co. Ltd
8 Grafton Street, London W1X 3LA

First published in Great Britain by
Collins Professional and Technical Books 1986

Copyright © James A. D. Balfour 1986

British Library Cataloguing in Publication Data
Balfour, James A. D.
  Computer analysis of structural frameworks.
  1. Structures, Theory of—Data processing
  I. Title
  624.1'71'0285   TA647

ISBN 0 00 383057 8

Printed and bound in Great Britain by
Robert Hartnoll (1985) Ltd, Bodmin, Cornwall

All rights reserved. No part of this publication may
be reproduced, stored in a retrieval system or
transmitted, in any form, or by any means, electronic,
mechanical, photocopying, recording or otherwise,
without the prior permission of the publishers.

> While efforts have been made to ensure the accuracy of the programs in this book the publishers and author disclaim all responsibility for their use for any purpose whatsoever. The programs are intended for purposes of illustration only and should not be regarded as fully validated commercial software.

Contents

PREFACE   vi

1 INTRODUCTION   1
   1.1 Structural analysis in perspective   1
   1.2 The computer revolution   3

2 FUNDAMENTALS OF STRUCTURAL BEHAVIOUR   6
   2.1   Introduction   6
   2.2   Definition of a structure   6
   2.3   Stable and unstable structures   6
   2.4   The need to idealise   7
   2.5   Idealisation of the structural loading   8
   2.6   Idealisation within the structure   10
   2.7   Idealisation at the structural boundaries 12
   2.8   Equilibrium   15
   2.9   Deformation of structures   23
   2.10 Statically indeterminate structures   25
   2.11 Superposition   29
   2.12 Simple beam behaviour   30

3 INTRODUCTION TO THE STIFFNESS METHOD   37
   3.1 Classification of framed structures   37
   3.2 Degree of freedom   38
   3.3 Matrix methods of structural analysis   42
   3.4 Stiffness coefficients   44
   3.5 Direct application of the stiffness method 49
   3.6 Axes systems   55
   3.7 Properties of the structure stiffness matrix 57

4 SIMULTANEOUS LINEAR EQUATIONS   62
   4.1 Introduction   62

CONTENTS

4.2	Gauss-Jordan elimination - Program GAUSSJ.EQ	63
4.3	Multiple right-hand sides	67
4.4	Gauss elimination - Program GAUSS.EQ	69
4.5	Triangular decomposition - Program UDU.EQ	75
4.6	Matrix inversion using Gauss-Jordan elimination	83
4.7	Ill-conditioned equations - Program INVERT.EQ	85

5 PLANE TRUSSES   95

5.1	Introduction	95
5.2	The element stiffness matrix in the member axes system	95
5.3	Transformation of force and displacement	97
5.4	The element stiffness matrix in the global axes system	100
5.5	Nodal equilibrium	102
5.6	The initial structure stiffness matrix - Programs ESTIFF.PT and ISTIFF.PT	105
5.7	Application of the boundary conditions	118
5.8	The final structure stiffness matrix - Program FSTIFF.PT	122
5.9	Member end forces and reactions - Program MFORCE.PT	131
5.10	Putting it all together - Program PTRUSS.PT	136
5.11	A data preprocessor - Program PRE.MA/PRE.01/PRE.02	147

6 SPACE TRUSSES   165

6.1	Introduction	165
6.2	The element stiffness matrix in the member axes system	165
6.3	Transformation of force and displacement	166
6.4	The element stiffness matrix in the global axes system	170
6.5	The structure stiffness matrix	173
6.6	Member end forces and reactions	176
6.7	Space truss program - STRUSS.ST	177

7 PLANE FRAMES   192

7.1	Introduction	192
7.2	The element stiffness matrix in the member axes system	192
7.3	Transformation of force and displacement	194
7.4	The element stiffness matrix in the global axes system	197
7.5	Nodal equilibrium	199
7.6	The initial structure stiffness matrix - Programs ESTIFF.PF and ISTIFF.PF	201
7.7	Application of the boundary conditions	214
7.8	The final structure stiffness matrix - Program FSTIFF.PF	216
7.9	Member end forces and reactions - Program MFORCE.PF	224

8 PLANE FRAMES - FURTHER TOPICS   231

8.1	Introduction	231

CONTENTS

8.2	Member loads, temperature effects, settlement, and and lack of fit - Program FIX.PF	231
8.3	Inclined supports	255
8.4	Elastic supports	265
8.5	Elements with pins	270
8.6	Non-prismatic members	285
8.7	Putting some of it together - Program PFRAME.PF	287

9	GRILLAGES	305
9.1	Introduction	305
9.2	The element stiffness matrix in the member axes system	306
9.3	Transformation of force and displacement	307
9.4	The solution algorithm	309
9.5	Grillage program - GRID.GD	314

10	SPACE FRAMES	327
10.1	Introduction	327
10.2	The element stiffness matrix in the member axes system	327
10.3	Transformation of force and displacement	329
10.4	The element stiffness matrix in the global axes system - Program TNODE.SF	334
10.5	Member end forces and reactions	345
10.6	Space frame program - SFRAME.SF	346

11	BUYING AND USING COMMERCIAL FRAME ANALYSIS PROGRAMS	363
11.1	Introduction	363
11.2	Buying a frame analysis program	364
11.3	Using frame analysis programs	366

APPENDIX A - Definition of Program Variables	375
APPENDIX B - Summary of Key Formulae and Matrices	380
INDEX	391

Preface

Practicing engineers now use computers to solve all non-trivial problems of structural analysis. Almost certainly the computer program used will be based upon the stiffness method, and equally certainly it will not have been written by the engineer using it. To obtain a solution to a structural analysis problem the engineer first selects a suitable computer program and then prepares data describing the problem in accordance with the user manual. Say, for instance, a static analysis is required, then the data will consist of member properties and locations, restraints, material properties, and details of the loading. Using the supplied data the computer then performs the structural analysis and outputs the displacements and stresses (or stress resultants). If a buckling, a plastic or dynamic analysis is required then the input and output will, of course, be different.

The twin facts that the stiffness method is extremely general and computer programs are expensive to develop has divided the engineering profession into two groups: a small group of specialist program writers, and a much larger group of practising engineers who use the programs. Today's engineers have at their fingertips an analysis capability undreamt of by previous generations of engineers, and it is possible to wield that power without knowing anything of its source. In general if the input is correct then so too is the output. This raises a thorny issue; if the practising engineer *needs* know nothing about structural analysis programs in order to use them, then how much *should* he know? Opinions on this matter vary. I am convinced, however, that the practising engineer should know enough about structural analysis programs to enable him to :-

1. Idealise structures in a sensible manner.

2. Prepare input data knowing, at least in general

PREFACE

      terms, how it will be used by the computer.

3. Interpret the computer output.

4. Know when, and why, a structure is beyond the capabilities of a particular computer program.

5. Select, from a number of different computer programs the one best suited to the problem in hand.

6. Devise cross checks when problems arise.

This book is intended to give engineering students and practising engineers the background knowledge necessary to make safe and efficient use of commercial frame analysis programs. If, after working through the text, the reader feels that he has mastered the six skills listed above then the author's efforts will have been worth while. This book is not intended for the expert structural analyst, but it should serve as an introductory text to more advanced studies as the methodology and notation have been chosen to make easy the transition to finite element theory.

    In my experience applying newly learned concepts to practical problems is an essential part of the learning process. This makes the teaching of computer methods of structural analysis difficult because computer methods are intended for use on a computer. When computer methods are applied by hand the volume of arithmetic generated by even the simplest problem is likely to dull the enthusiasm of the keenest student. One way out of this difficulty is to quickly move from hand calculation to solving problems using a standard structural analysis program. While this will familiarise the student with the mechanics of using a standard program it is unlikely to give him much insight into how the program works. Unfortunately standard structural analysis programs tend to appear as 'black boxes' to their users with input going in one end and solution appearing magically at the other.

    This book tries to avoid these difficulties by using short computer programs to perform each of the standard procedures used in commercial structural analysis programs. Each program is complete in itself, and all were written with the sole aim of clarity (efficiency, compactness and elegance were non-considerations). By transferring the burden of arithmetic calculation to the computer these programs allow the analysis of reasonably complex structures to be undertaken, yet leave control of the analysis procedure in the hands of the student. Also presented are automatic structural analysis programs for a number of different framework types. These programs have been constructed by splicing together the programs for individual procedures.

The programs are written in BASIC and are designed to run on any computer from a desktop microcomputer to a mainframe machine. Unfortunately there are many versions of BASIC in common use. The programs in this book are written in the popular Microsoft BASIC (MBASIC), and the programs may need some minor conversion work before they will run under other versions of BASIC. The author will make the programs (as ASCII text files) available on disk for a number of different microcomputers including the BBC Microcomputer, the DEC Rainbow, the APPLE IIe, and the IBM PC (disks will be prepared for other machines depending upon the demand). For futher details contact

> Dr James A.D. Balfour,
> Department of Civil Engineering,
> Heriot-Watt University,
> Riccarton,
> Edinburgh EH14 4AS   Tel 031 449 5111

The reader of this book should have an elementary knowledge of statics, strength of materials, matrix algebra, and computer programming in BASIC. If deficient in any of these subjects then I suggest that the reader should make good these deficiencies before proceeding.

I am indebted to my colleague, Professor A.E. Edwards, for many helpful suggestions made during the preparation of this book. I would like to conclude by recognising the debt owed by myself and the engineering profession to the many researchers who have brought the computer analysis of structures to its current level of sophistication. By their efforts the arithmetic tedium has been taken out of structural analysis, and the designer has been freed to ask the question - *what if* ?

1 Introduction

1.1 STRUCTURAL ANALYSIS IN PERSPECTIVE

Civil engineering structures are many and varied, including tall buildings, airports, factory buildings, roads, harbours, railways, water treatment works, sewage disposal works, offshore structures, dams, pipelines, and bridges. To design and build such structures is a long and complex process where the time from conception to completion is measured in years rather than in months. Say, for instance, a manufacturing company requires a new factory to expand its operations. Having acquired a suitable site the company then commissions a firm of consulting engineers to design the factory and supervise its construction. Once the consulting engineer has established the requirements of the client, he then sets about the process of designing the structure.

The first stage in the design process is to produce a preliminary scheme or schemes. This phase requires decisions to be made regarding the type of structure, its layout, and the materials to be used in its construction. In other words the preliminary scheme determines the structure's appearance, cost, functionality and to a large extent its ultimate success or failure. Here the designer uses a blend of art and science to produce a design that is close to the optimum. To do this he must integrate his understanding of engineering materials, building costs, structural behaviour, construction methods, and site conditions. If successful then the structure will be economical, functional, durable, safe, and pleasing to the eye. It must be remembered, however, that the preliminary design for some structures, such as tall buildings, is the responsibility of the architect, who may or may not involve a structural engineer at the preliminary design stage.

Structural analysis is conducted during the preliminary design to ensure that the proposed scheme is not structurally impractical. The calculations are usually kept short by oversimplifying the assumed structural behaviour. At this stage the designer is only interested in approximate values of member sizes, settlements, deflections, etc.

Once the outline design is complete the structure is analysed in detail to ensure that it has adequate strength and stiffness. The strength is checked to ensure that the maximum expected loads can be carried without collapse, and the stiffness is checked to avoid excessive deflection under everyday (service load) conditions. Other aspects of structural behaviour such as vibration, cracking in concrete, and foundation settlement are also assessed during this phase. From the detailed structural analysis come final member sizes, reinforcement layouts, connection details, and foundation dimensions. Codes of practice are extensively used during the detailed analysis and design of a structure. These codes relate to different structural types and construction materials. They give guidance on many aspects of analysis and design, including loadings, material specifications, stresses and deflections.

Structural analysis is, therefore, inextricably bound up with structural design, being one of the tools that the designer uses to ensure the economy and safety of the final structure. It is important to remember, however, that structural analysis is a means to an end and not an end in itself. In essence it is a check, and it is salutary to remember that engineers of the past built highly successful structures without the aid of structural analysis.

Of course, the structural analysis must be competently executed. As structures are invariably idealised prior to analysis care must be taken to ensure that the idealised structure incorporates the principal characteristics of the actual structure. The loads must be properly assessed and the calculations must be carried out accurately. However, given a preliminary design, any two competent engineers will compute deflections and stresses that are essentially the same. Hence structural analysis, while a necessary part of the design process, does not affect the ultimate success of the structure (gross errors excepted).

The creative part of the structural analysis phase occurs when the analysis shows that the preliminary design is unsatisfactory. Parts of the structure may be overstressed making the structure unsafe, while other parts may be grossly understressed making the structure uneconomical. In addition, vibration, deflection, ground pressures, etc., may be found to be unacceptable. Seldom will a preliminary design emerge as a final design without modification. If the detailed structural analysis shows the structure to be unacceptable then the designer must decide what changes are required in order to satisfy the structural requirements. It is here that the engineer

INTRODUCTION

can use his experience and understanding of structural behaviour to good effect. It takes a certain skill to select, from the infinite number of possible structural modifications, one that will change an unsatisfactory structure into a satisfactory one.

Once the design has been finalised the contract documents are prepared. These documents, which include working drawings, specifications, conditions of contract, and bills of quantity, are then issued to civil engineering contractors who tender competitively for the work. The contract is normally awarded to the contractor making the lowest bid. The construction is monitored by the resident engineer to ensure that the structure is built according to the contract documents, and that the contractor is fairly rewarded for the work done.

1.2 THE COMPUTER REVOLUTION

Today, computers play an integral part in the analysis of civil engineering structures. Hand calculation is limited to simple structures, and initial member sizing (either during the preliminary design or prior to computer analysis). Prudent engineers also estimate, by hand, the expected deflections and forces at selected points of a structure about to undergo computer analysis. This can provide a useful check that there is nothing seriously wrong with the input data, and gives the engineer a feel for how the structure will work. When compared with the analysis procedures of only a few years ago it is evident that a revolution has taken place. There follows a brief account of the events leading up to that revolution. The background to a revolution is always interesting.

By the end of the nineteenth century the principles governing the elastic behaviour of structures had been laid down by men such as Mohr, Rankine, Euler, Maxwell, Castigliano and Muller-Breslau. These principles had the virtues of generality and mathematical elegance, but although they gave engineers a better understanding of structural behaviour, they proved to be of limited practical value due to the volume of arithmetic associated with their application. In many cases to find a solution entailed solving a system of simultaneous equations. Consequently the first half of this century saw the appearance of a myriad of structural analysis techniques designed to reduce the arithmetic to manageable proportions. Many of these techniques were problem orientated, being concerned with only one type of structure, and yet, in most cases, these techniques were based upon the same underlying principles. Of the value to practising engineers of techniques such as moment distribution there can be no doubt, yet it is apparent that engineers were required to grapple with a huge range of apparently different analysis techniques.

During the early 1940s a machine called ENIAC (Electronic Numerical Integrator and Calculator) was built at the University of Pennsylvania. It was housed in a room approximately 20 metres by 10 metres, it contained 18,000 valves and weighed about 30 tonnes. ENIAC is generally recognised as being the world's first digital computer. The use of a large number of valves, each having a relatively short lifespan, made keeping ENIAC operational a major problem. Typically the machine would run for only a few minutes before valve failure would halt operations.

The reliability problem was largely solved by the invention of the transistor in 1948, and by the mid 1960s a second generation of computers based upon solid state technology were being commercially produced. At about the same time manufacturing techniques were being developed that allowed an integrated circuit (IC) consisting of a number of transistors to be produced on a single silicon wafer. Since that major technological breakthrough computers have become steadily smaller, cheaper, and more powerful. Today's microcomputers are small enough to sit on a desk top. They offer computing power that is vastly superior to that of ENIAC and they are extremely reliable. The low cost of these machines, coupled with a growing computer awareness in society, has led to the widespread use of computers in schools, offices, homes, and factories.

Developments in computer software have tended to parallel those in computer hardware. New software has been required to take advantage of the improved facilities offered by each new generation of computers. Each year sees an improvement in the quality and scope of commercially available software. In contrast to the cost of computer hardware, which has been falling, the cost of software has shown no corresponding decrease. The net effect of these two factors is that in many computer installations the capital investment in software is similar to that in hardware.

Of the many different ways in which computer technology has affected the engineering profession, it is in the field of structural analysis that the impact has been most profound (although some engineers claim that word processing has had a greater impact on their daily work). Early computers were used primarily for numerical calculation, and engineers were quick to realise that the computational speed and accuracy offered by these machines might herald a new era in structural analysis. Much of the pioneering work was done by aircraft engineers because at that time aircraft design was severely hampered by a lack of sufficiently accurate analysis techniques. Aircraft structures are extremely complex and the type

INTRODUCTION

of analysis that can be undertaken by hand will give only a poor representation of the true structural behaviour. Some attempts were made to computerise hand methods of calculation, but it was soon realised that a more general approach stemming directly from the fundamental principles of structural analysis would be better suited to computer implementation. The computer's ability to handle vast amounts of arithmetic with speed and accuracy has made viable these computationally intensive methods.

By 1953 engineers were writing stiffness equations in matrix notation and solving the resulting simultaneous equations using digital computers. Matrix notation is used to describe modern structural analysis techniques because it is efficient, compact and is well suited to computer implementation. Of the different matrix methods that have been successfully translated into computer code, the stiffness method has proved to be by far the most generally useful. The principal advantage of the stiffness method is that it is easy to automate and hence a single program can be used to analyse a structure of any geometry. Another advantage of the stiffness method is that it is not limited to linearly elastic problems of statics, having been successfully extended to problems involving buckling, plasticity, and dynamics.

The development cost of structural analysis programs is high. A simple single element, static analysis program consisting of about 1000 statements is likely to take a number of man-months to develop, test, and document. At the other end of the scale a large general purpose finite element program with pre- and post-processors may consist of 250,000 statements and will take hundreds of man-years to develop. This has led to the situation where a small number of establishments have made a large investment in the development of structural analysis software. These programs are then marketed to the engineering community, thus dividing engineers into a small group of specialist program writers, and a large group of practising engineers who use the programs.

Many civil engineering structures are supported by frameworks and consequently frame analysis programs are extensively used by civil engineers. To the user a frame analysis program appears as a "black box" which will produce results at one end if fed with input data at the other. This book is concerned with the structural theory behind these programs and with how that theory is turned into useful computer programs.

2 Fundamentals of Structural Behaviour

2.1 INTRODUCTION

The theory of structures is a wide and sometimes complex subject. Fortunately the structural theory used in this book is limited to simple beam theory, Castigliano's First Theorem and the fundamental concepts of equilibrium, compatibility, and superposition. This chapter presents all of these topics with the exception of Castigliano's First Theorem, which is developed in Chapter 3. Should the reader require a more general introduction to the theory of structures then he should refer to one of the many excellent textbooks on the subject. This chapter begins by discussing the way in which the highly complex behaviour of real structures is made tractable by the process of structural idealisation.

2.2 DEFINITION OF A STRUCTURE

The function of a structure is to transmit force. The trunk of a tree transmits wind and gravity forces from the upper parts of the tree into the roots and hence safely into the ground. Floor beams are used to transmit floor loadings to the building walls. The walls then carry the load from the beams to the foundations and the foundations transmit the load into the ground below. From these simple examples it is evident that the world is full of both man-made and natural structures, and that force can be transmitted in a variety of ways.

2.3 STABLE AND UNSTABLE STRUCTURES

A structure is said to be stable if it can support an arbitary system of infinitesimal loads. No part of a stable structure is a mechanism, nor can the structure, as a whole, undergo rigid body

FUNDAMENTALS OF STRUCTURAL BEHAVIOUR

motion. Figure 2.1 shows a structure that is a mechanism. As can be seen the structure can support the particular system of loads shown in the left hand diagram, but not the system of loads shown in the right hand diagram.

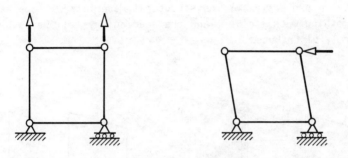

FIGURE 2.1   EXAMPLE OF A MECHANISM

Figure 2.2 shows a structure that is inadequately restrained and is therefore capable of rigid body motion. Again a particular system of loads can be supported, but not a general system.

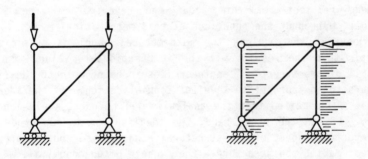

FIGURE 2.2   EXAMPLE OF AN INADEQUATELY RESTRAINED STRUCTURE

Civil engineers are concerned to produce stable structures. If a structure is unstable then the structural displacements caused by the loads become indeterminate, and no meaningful solution can be found.

2.4   THE NEED TO IDEALISE

The actual behaviour of even a simple structure is extremely complex. Take for example the building shown in fig 2.3. Obviously the structure is three-dimensional. Of course the roof and cladding will act together with the frame of the structure to resist loads. We know that there will be a complex interaction between the foundations and the subsoil. The loading will be highly variable and impossible to exactly predict in advance. Yet more likely than not, for analysis purposes, the structure will be idealised as a two-dimensional series of interconnected line elements encastré at ground level as shown on the right of the diagram. Moreover, the complex stress-strain properties of the structural materials will be assumed to be something much simpler, and the loading may well be assumed to be uniformly distributed.

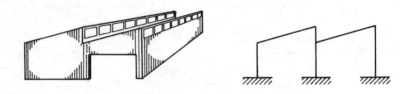

FIGURE 2.3   STRUCTURE IDEALISATION

The structural engineer does not therefore analyse the real structure, but an idealised version of the structure. Any results obtained will not tell him about the behaviour of the actual structure, but about the behaviour of this mathematical model. Hence it is obvious that unless the mathematical model is sufficiently realistic then the results will be of questionable value. Part of the skill of the structural engineer lies in being able to idealise a structure to an appropriate level of sophistication. If the idealisation is too detailed then the analysis will be excessively long and costly. If, on the other hand, the idealisation is too crude then some important aspect of the structural behaviour may not be properly modelled, making the structure either unsafe or uneconomical.

2.5 IDEALISATION OF THE STRUCTURAL LOADING

The principal objective of structural analysis is to ensure that all anticipated loadings (design loads) are carried safely and economically. Loads are usually classified as dead or live. Dead loads are permanent loads. In practice dead load is the self weight of the structure, which can be subdivided into load carrying components (such as a building frame) and non-load carrying components (such as cladding and partition walls). Live loads are non-permanent loads such as snow, people, furniture, machinery, vehicles, and wind. While some loads, such as the weight of water in a water tower, can be assessed with some degree of confidence, most

FUNDAMENTALS OF STRUCTURAL BEHAVIOUR

design loadings are but crude approximations to the actual loads the structure will carry. For instance, the calculation of the wind pressures experienced by a building during a once in 100-year storm is a calculation fraught with uncertainty. The following load types are usually used to simplify the actual structural loadings:

1. Uniformly distributed load
2. Uniformly varying load
3. Point load
4. Line load.

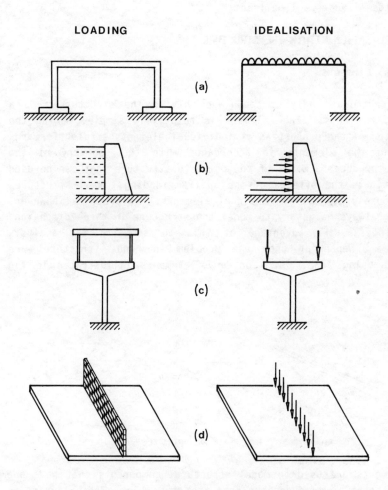

FIGURE 2.4   LOAD IDEALISATION

A high intensity uniformly distributed load that acts over a relatively small area is sometimes called a patch load. Figure 2.4 gives examples of all these load types:

(a) Shows how the self weight of a concrete bridge deck might be considered as a uniformly distributed load.
(b) Shows how water pressure, which varies linearly with depth, can be regarded as a uniformly varying load.
(c) Shows how the loads from the upper level of a two level roadway might be idealised as point loads.
(d) Shows how a partition wall crossing a slab might be idealised as a line load.

2.6 IDEALISATION WITHIN THE STRUCTURE

Structural Elements

Modern structural analysis techniques divide the structure into a number of elements. Each element is relatively simple, making the calculation of the properties of individual elements straightforward. Assembling the elements in accordance with the geometry of the structure produces a model of the complete structure that can be used to investigate the effects of the applied loading. Real structures consist purely of three-dimensional elements. In practice, however, it is not always necessary to model the structure in three dimensions and engineers take advantage of this to simplify the analysis procedure. Depending upon the problem in hand, structures are idealised using *line*, *plate* or *brick* elements (illustrated in fig 2.5).

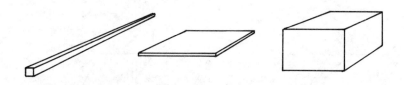

FIGURE 2.5 LINE, PLATE AND BRICK ELEMENTS

Line elements are used to model structural components that have one dimension that is much greater than the other two. Typical examples are truss members, beams, columns and slender arches. Structures consisting entirely of line elements are often referred to as

FUNDAMENTALS OF STRUCTURAL BEHAVIOUR

skeletal, or frame, structures. Plate elements are used to model structural components, such as slabs and shells, of which two dimensions are much greater than the third. Plate elements are also used where, although the structure is obviously three-dimensional, the structural action is essentially two-dimensional (e.g. cases of plane strain). Brick elements are used to model structures such as thick slabs and machine components where the structural action is truly three-dimensional. As this text deals exclusively with structural frameworks, only line elements are of concern.

Structural Materials

The science of engineering materials is a large and often complex subject. When designing structures, however, the engineer invariably adopts a simplified model of material behaviour.

The framed structures encountered in civil engineering are usually constructed from steel or reinforced concrete. Figure 2.6 shows typical *idealised* stress-strain diagrams for steel reinforcement and structural concrete.

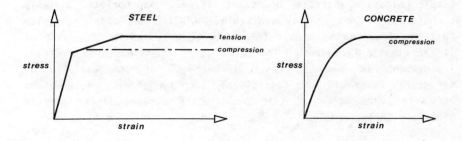

FIGURE 2.6   TYPICAL IDEALISED STRESS-STRAIN DIAGRAMS

Both stress-strain curves have a section where strain increases yet the stress remains constant (i.e. the horizontal portion of the curve). Deformations associated with that portion of the curve are plastic. Plastic deformation is non-recoverable deformation that remains after the removal of the applied stress, and it should be noted that the onset of plastic deformation occurs before the stress reaches the level of the horizontal portion of the curve. Steel will sustain large plastic strains prior to failure and is therefore described as a *ductile* material. Concrete, by contrast, fails shortly after the onset of plasticity and is therefore described as a *brittle* material.

Engineers make extensive use of codes of practice when designing structures. Most of these codes are based upon a limit state design philosophy which demands that the structure never reaches the limit states of serviceability or collapse.

The serviceability limit state is reached when the structure begins to lose utility or to cause public concern through causes such as excessive cracking in concrete, large deflections, or unacceptable levels of vibration. The engineer must ensure that the structure will remain serviceable under the service loads as laid down in the relevant codes of practice. Service loading represents a severe working load and the structure is invariably assumed to be linearly elastic under service conditions.

The ultimate limit state is reached when the structure is on the point of collapse. Ultimate loads represent the extreme loading for the structure and the structure must be able to carry these loads without collapse. Structures close to collapse are usually behaving inelastically. In spite of this the forces within the structure are usually determined using linear elastic theory. The individual members are then designed on the basis of inelastic material behaviour. Linear elastic analysis under ultimate loads is used because, with the exception of yield line analysis of slabs and simple plastic analysis of steel frames, appropriate analysis techniques are not readily available. Limit state codes recognise this by allowing the member forces to be assessed on the basis of linear elastic behaviour. In recent years there has been a steady improvement in computer based techniques for the analysis of structures under ultimate conditions. Such techniques are, however, not within the scope of this book, which assumes linear elastic behaviour throughout.

2.7 IDEALISATION AT THE STRUCTURAL BOUNDARIES

In section 2.3 the need to restrain the structure against rigid body motion was discussed. To facilitate analysis the engineer makes assumptions about the displacement conditions pertaining at the points of restraint. As these points are at the boundary of the structure they are known as *boundary conditions*.

Take for example the bridge shown in fig 2.7(a). PTFE, a very low friction material, has been incorporated in the bearings at the abutments and on the top of the columns. The bridge deck is dowelled to the left-hand abutment and will slide freely on the PTFE bearings. Consequently the boundary conditions assumed during the analysis of the bridge deck are likely to be those illustrated in fig 2.7(b). These boundary conditions imply zero displacement in the directions shown in fig 2.7(c).

FUNDAMENTALS OF STRUCTURAL BEHAVIOUR

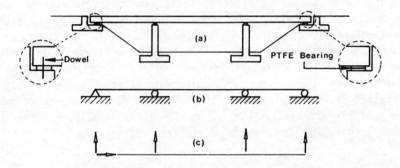

FIGURE 2.7   NON-MONOLITHIC BRIDGE STRUCTURE

The bridge deck might, however, be constructed as monolithic with the columns as shown in fig 2.8(a). In this case the bridge deck should not be analysed independently from the columns. Figure 2.8(b) shows the boundary conditions that might be assumed, and fig 2.8(c) indicates the locations and directions of zero displacement. Note that although rotation has been restrained at the base of the columns, the degree of restraint will depend upon the nature of the connection between the columns and the foundations.

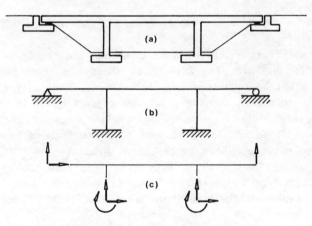

FIGURE 2.8   MONOLITHIC BRIDGE STRUCTURE

To make assumptions about the boundary conditions is mathematically convenient, yet perfect boundary conditions never occur in practice. For, instance pinned connections always have some moment carrying capacity, and a fully fixed connection will always shed some moment due to the rotation of the connection itself. Consider the boundary conditions shown in fig 2.7(a). The PTFE bearings are assumed to offer no resistance to horizontal movement. In practice they will offer some resistance. Also the vertical displacement at the top of the columns and at the abutments is assumed to be zero, although it is evident that stress in the columns and the subsoil will result in some vertical displacement. As the boundary conditions have a significant effect on the structural behaviour it is important that the boundary conditions assumed during the analysis are realistic. Figure 2.9 gives examples of connection details that might be considered as fixed and pinned.

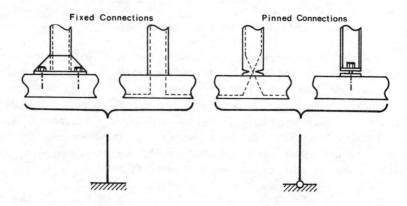

FIGURE 2.9   FIXED AND PINNED BASE CONNECTIONS

A further complication arises when more than one member is involved. Consider, for example, fig 2.10 which shows how three members of a framework might be connected to a foundation. Obviously the members are rigidly connected together, yet the connection to the foundation is effectively pinned. Hence, for the purposes of analysis, the boundary condition should be assumed to be that shown in the centre diagram and not that shown in the right hand diagram. The boundary condition shown in the right hand diagram allows the ends of the members to rotate by different amounts and does not, therefore, adequately represent the behaviour of the actual structure.

FIGURE 2.10   BOUNDARY CONDITIONS

FUNDAMENTALS OF STRUCTURAL BEHAVIOUR

Other common boundary conditions include roller supports, elastic supports and semi-rigid supports. In addition it may be necessary to consider the effect of settlements at the foundations as part of the analysis.

2.8 EQUILIBRIUM

Equilibrium is the most important concept in structural analysis. Engineers talk about two types of equilibrium, static and dynamic, although it can be argued that static equilibrium is a special case of dynamic equilibrium. This section will consider only static equilibrium, dynamic equilibrium being beyond the scope of this book.

FIGURE 2.11   VECTOR COMPONENTS IN TWO AND THREE DIMENSIONS

Static equilibrium exists if all parts of a structure can be considered static (i.e. motionless) under a particular system of loads. As all elements of the structure are stationary there is no net force acting on any element (otherwise acceleration would occur). In two dimensions force can be completely described by two mutually perpendicular linear components and one rotational component, as illustrated in fig 2.11.

Also shown in fig 2.11 are the three mutually perpendicular linear components and the three rotational components that completely describe force in three dimensions. This enables the well-known equations of static equilibrium to be written:

$\sum P_x = 0$ Two- $\sum P_x = 0$ $\sum M_x = 0$ Three-
$\sum P_y = 0$ dimensional $\sum P_y = 0$ $\sum M_y = 0$ dimensional
$\sum M = 0$ case $\sum P_z = 0$ $\sum M_z = 0$ case

Internal and External Forces

Equilibrium is to do with forces, and forces can be regarded as being either internal or external. External forces consist of applied forces and reactions. The applied forces can be regarded as active forces, and the reactions as passive forces generated in response to the applied loads. The applied loads usually have preset values, whereas the reactions assume values that will maintain the equilibrium of the structure. The engineer can distinguish between applied loads and reactions, but the structure cannot, with reactions and applied loads appearing simply as external forces. Internal forces are the forces generated within the structure in response to the applied loads. In essence structural action is the transmission of the applied loads to the reactions via internal member forces.

Stress Resultants

A better name for the forces generated within a framework in response to the applied loads is stress resultants. In fact internal forces do not actually exist, as force is transmitted through the structure by stress within the members. Internal forces (stress resultants) are simply useful vector quantities obtained by integrating stress, or moment of stress, over the cross-section of the member.

Freebody Diagrams

Stemming directly from the principle of equilibrium is the concept of freebody diagrams. The basic argument is this:

1. If a structure as a whole is in static equilibrium then every part (or element) of the structure must be in static equilibrium.

2. Hence the internal and external forces (and moments) acting on every part of the structure must balance. If they did not, then the out-of-balance force would cause that part of the structure to accelerate.

3. If a part of the structure is "cut" from the rest of the structure, then the stresses which were formerly present at the cuts would be released, causing a change in the structural behaviour. If, however, stresses (or stress resultants) identical to those which existed prior to cutting are applied to the cut faces, then the structure and the severed part will be unaware that they are no longer connected.

This reasoning holds because the only communication between the elements of a structure is the stress present at the inter-element boundaries. To illustrate this concept consider the simple structure shown in fig 2.12.

As can be seen two cuts have been made, dividing the original structure into three elements. Freebody diagrams are obtained by applying stress resultants to the cut faces as shown. Due to the nature of the loading the structure is subjected to only horizontal forces. Hence only one equation of static equilibrium is of value (i.e. $\sum P_x = 0$), the other two equations being automatically satisfied.

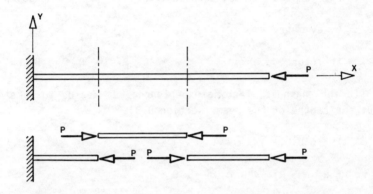

FIGURE 2.12 A SIMPLE EXAMPLE OF FREEBODY DIAGRAMS

In the case of the beam shown in fig 2.13(a) the three equations of static equilibrium, $\sum P_x = 0$, $\sum P_y = 0$, and $\sum M = 0$, can be used to find the shearing force and bending moment at any section along the beam. Say, for instance, the shearing force and bending moment at sections A-A and B-B are required to be found.

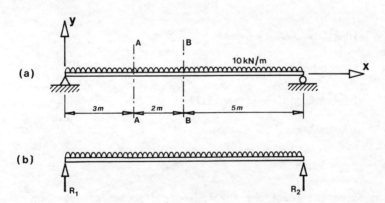

FIGURE 2.13 A SIMPLE BEAM STRUCTURE

To find the reactions the freebody diagram shown in fig 2.13(b) is used. The right-hand support is a roller, therefore the reaction must be vertical, and consequently there can be no horizontal component of force at the left-hand reaction. Having established that both reactions are vertical, the equations of equilibrium $\sum P_y = 0$ and $\sum M = 0$ are used to find the value of the reactions.

These equations can be applied in an infinite number of ways. The engineer's reasoning would be:

For vertical forces to balance $R_1 + R_2$ = 100 kN

by symmetry $R_1 = R_2$

therefore $R_1 = R_2$ = 50 kN

The student engineer, lacking experience, may decide to take moments about the centre of the beam (section B-B).

$$\sum M = 0 \quad +\circlearrowright$$

$$-5 R_1 + 5 R_2 + 0 \times 100 = 0 \tag{2.1}$$

Vertical forces must sum to zero $(\sum P_y = 0) \;+\uparrow$

$$R_1 + R_2 - 100 = 0 \tag{2.2}$$

Equation (2.1) - 5 × equation (2.2) gives

$$-5 R_1 + 5 R_2 - 5 R_1 - 5 R_2 + 500 = 0$$

therefore

$$-10 R_1 = -500$$

hence

$$R_1 = 50 \text{ kN}$$

Substitute for R_1 in equation (2.1)

$$-5 \times 50 + 5 R_2 = 0$$

hence

$$R_2 = 50 \text{ kN}$$

Having found the reactions the freebody diagrams produced by cuts A-A and B-B can be sketched as shown in fig 2.14.

FUNDAMENTALS OF STRUCTURAL BEHAVIOUR

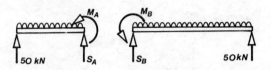

FIGURE 2.14   FREEBODY DIAGRAMS FOR BEAM ENDS

To find the required bending moments and shearing forces consider first the left-hand portion of the beam.

i.e.
$$\sum P_y = 0 \quad +\uparrow$$
$$50 - 3 \times 10 + S_A = 0$$
hence
$$S_A = -20 \text{ kN}$$

i.e.
$$\sum M = 0 \quad +\circlearrowright \quad \text{about section A-A}$$
$$-50 \times 3 + (3 \times 10) \times 1.5 + M_A = 0$$
hence
$$M_A = 105 \text{ kN m}$$

Now consider the right-hand end.

i.e.
$$\sum P_y = 0 \quad +\uparrow$$
$$50 - 5 \times 10 + S_B = 0$$
hence
$$S_B = 0 \text{ kN}$$

i.e.
$$\sum M = 0 \quad +\circlearrowright \quad \text{about section B-B}$$
$$50 \times 5 - (5 \times 10) \times 2.5 + M_B = 0$$
hence
$$M_B = -125 \text{ kN m}$$

The final freebody diagrams are shown on fig 2.15, and the reader should verify that the centre portion of the beam is in equilibrium.

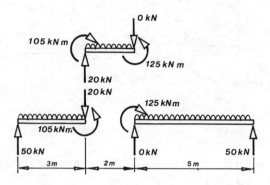

FIGURE 2.15   FINAL FREEBODY DIAGRAMS

Now consider the use of freebody diagrams in the analysis of the frame shown in fig 2.16(a).

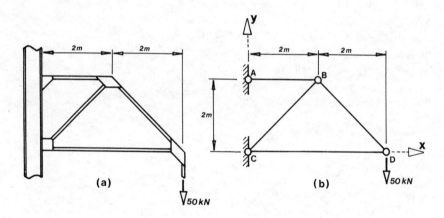

FIGURE 2.16   A SIMPLE FRAMEWORK

Triangulated frameworks such as that shown in fig 2.16(a) transmit load primarily by axial force in the members. Consequently to idealise the joints as perfect pins as shown in fig 2.16(b) is reasonable. In order to find the reactions the freebody diagram shown in fig 2.17(a) is used.

There are four components of reaction (h_a, v_a, h_c, v_c): one of which can be found by consideration of the freebody diagram for joint A (shown in fig 2.17(b)).

FUNDAMENTALS OF STRUCTURAL BEHAVIOUR

i.e.
$$\sum P_y = 0 \quad +\uparrow$$
$$v_a = 0$$

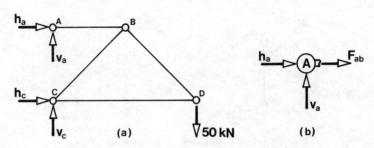

FIGURE 2.17   FREEBODY DIAGRAM FOR THE REACTIONS

Now consider the equilibrium of the whole frame (fig 2.17(a)).

i.e.
$$\sum M = 0 \quad +\curvearrowleft \quad \text{about C}$$
$$-2 \times h_a - 4 \times 50 = 0$$
hence
$$h_a = -100 \text{ kN}$$

i.e.
$$\sum P_x = 0 \quad +\rightarrow$$
$$h_a + h_c = 0$$
hence
$$h_c = -h_a = 100 \text{ kN}$$

i.e.
$$\sum P_y = 0 \quad +\uparrow$$
$$v_a + v_c - 50 = 0$$
hence
$$v_c = 50 \text{ kN} \quad \text{as} \quad v_a = 0$$

If the forces in members BD and CD are of interest, then the best way forward is to consider the freebody diagram for joint D as shown in fig 2.18(a). Replace the force in member BD by its components as shown in fig 2.18(b) and then consider the equilibrium of the joint.

i.e.
$$\sum P_y = 0 \quad +\uparrow$$
$$v_{bd} - 50 = 0$$

hence

v_{bd} = 50 kN

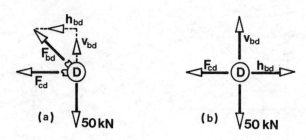

FIGURE 2.18   FREEBODY DIAGRAM FOR JOINT D

As BD is at 45 degrees,

h_{bd} = v_{bd} = 50 kN

and

$F_{bd} = \sqrt{(v_{bd}^2 + h_{bd}^2)}$ = 50$\sqrt{2}$ kN

$\sum P_x = 0$ $+\longrightarrow$

i.e.

$-h_{bd} - F_{cd} = 0$

hence

F_{cd} = -50 kN

The technique that has been used here is, of course, the well-known method of joints.

The force in the remaining members can be found by considering the equilibrium of the freebody diagram produced by cutting the frame on section E-E, as shown in fig 2.19.

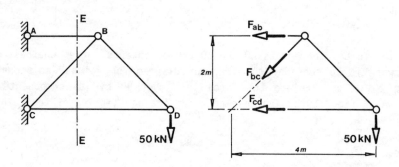

FIGURE 2.19   FREEBODY DIAGRAM TO FIND THE FORCES IN MEMBERS AB AND BC

FUNDAMENTALS OF STRUCTURAL BEHAVIOUR

i.e.
$$\sum M = 0 \quad +\circlearrowright \quad \text{about C}$$
$$-50 \times 4 + 2 \times F_{ab} = 0$$

hence
$$F_{ab} = 100 \text{ kN}$$

i.e.
$$\sum P_y = 0 \quad +\uparrow$$
$$-50 - F_{bc}/\sqrt{2} = 0$$

hence
$$F_{bc} = -50\sqrt{2} \text{ kN}$$

The foregoing calculations are in fact an example of the method of sections commonly used in the analysis of pin jointed frames.

2.9 DEFORMATION OF STRUCTURES

Structures support applied loads by developing internal stresses within their members. As stress cannot exist without strain, it follows that a structure cannot resist loads without deformation.

Compatibility

The principle of compatibility is assumed to apply to all of the structures encountered in this book. Compatibility is concerned with deformation. If compatibility is assumed then geometric fit is implied (i.e. if a joint of a structure moves, then the ends of the members connected to that joint move by the same amount, consistent with the nature of the connection).

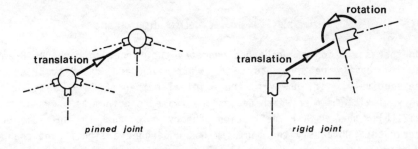

FIGURE 2.20   COMPATIBILITY AND JOINT DISPLACEMENTS

In the case of a pin jointed frame, compatibility means that the ends of the members meeting at a joint undergo equal translation. If the framework is rigidly jointed then, in addition to equal translation, the rotation of the ends of the members meeting at a joint must also be equal, as shown in fig 2.20.

Small Deflection Theory

As all structures deform when loaded it follows that the equations of equilibrium should be based upon the deformed shape of the structure. In most civil engineering structures, however, the deformations caused by the loading are extremely small in comparison to the size of the structure. Consequently the assumption of unchanged geometry can (usually) be safely made, thereby greatly simplifying the analysis procedure. Small deflection theory assumes that the equilibrium equations for the loaded structure can be safely based upon the unloaded geometry. Consider for example the simple frame shown in fig 2.21(a).

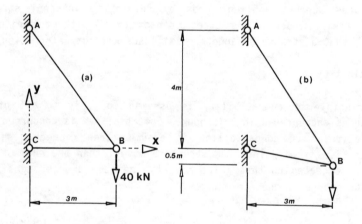

FIGURE 2.21   AN EXAMPLE OF LARGE DISPLACEMENTS

On the assumption of unchanged geometry the forces in members AB and BC can easily be shown to be 50 kN tension and 30 kN compression respectively. If, however, the original member AB is replaced by a very flexible member which causes the frame to deform as shown in fig 2.21(b), then small deflection theory may not be sufficiently accurate. Obviously the geometry has undergone a significant change during loading, and the equilibrium equations for joint B are as follows:

$$\sum P_y = 0 \quad + \uparrow$$

hence

$$F_{ab} \times 4.5/5.408 - F_{cb} \times 0.5/3.041 - 40 = 0$$

$$\sum P_x = 0 +$$

hence

$$F_{cb} \times 3.0/3.041 - F_{ab} \times 3.0/5.408 = 0$$

Solution of these equations yields F_{ab} = 54.08 kN tension, and F_{cb} = 30.41 kN compression.

Fortunately in civil engineering structures the change in geometry under loading is a second order effect that can usually be safely ignored by the engineer. Only when the deformations are large does the use of small deflection theory become suspect. Large deformations sometimes occur in flexible structures such as suspension bridges, slender arches, and tall buildings. Large deformations are also associated with structures that are approaching collapse. In such cases the deformations are usually large enough to warrant basing the equations of equilibrium on the loaded geometry.

2.10 STATICALLY INDETERMINATE STRUCTURES

All of the structures encountered so far have been *statically determinate*. In other words the forces in the members can be determined by the application of the equations of equilibrium alone. Structural deformation will, of course, be dependent on the materials and cross-sections used. However, provided that the structure is linearly elastic and small deflection theory remains valid, then the member forces will be independent of the materials and cross-sections used.

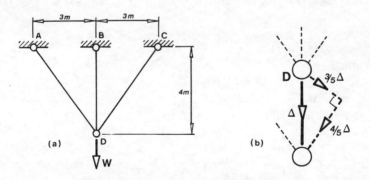

FIGURE 2.22   A STATICALLY INDETERMINATE STRUCTURE

Structures that cannot be analysed using only the equations of equilibrium are said to be *statically indeterminate*. The analysis of indeterminate structures demands a knowledge of the structural properties of the members, and in general the force carried by each member will depend upon its material and cross-section. To illustrate, consider the frame shown in fig 2.22(a) which is constructed using a linearly elastic material. The cross-sectional area of member BD is A_1 and the cross-sectional area of members AD and CD is A_2. Due to the symmetry there will be no horizontal movement of joint D which will displace as shown in fig 2.22(b).

If D displaces downwards by Δ then the change in length of members AD and CD will be $4\Delta/5$.

Strain in BD

$$\varepsilon_{bd} = \delta L/L = \Delta/4$$

Strain in AD and CD

$$\varepsilon_{ad} = \varepsilon_{cd} = (4\Delta/5)/5 = 4\Delta/25$$

Hence if the stress in BD is σ_{bd} then, as stress is proportional to strain, the stress in AD and CD will be $16\sigma_{bd}/25$.

Force = stress x cross-sectional area

Therefore $F_{bd} = \sigma_{bd} A_1$
and
$$F_{ad} = F_{cd} = 16\sigma_{bd} A_2/25$$

Vertical equilibrium of joint D gives $4(F_{ad} + F_{cd})/5 + F_{bd} - W = 0$

Hence
$$128\sigma_{bd} A_2/125 + \sigma_{bd} A_1 = W$$
therefore
$$\sigma_{bd} = W/(128 A_2/125 + A_1)$$

If $A_2 = A_1$

then
$$\sigma_{bd} = 125W/253 A_1$$
$$F_{bd} = 125W/253$$
$$F_{ad} = F_{cd} = 80W/253$$

FUNDAMENTALS OF STRUCTURAL BEHAVIOUR

If $\quad A_2 = A_1/10$

then
$$\sigma_{bd} = 1250W/1378A_1$$
$$F_{bd} = 1250W/1378$$
$$F_{ad} = F_{cd} = 80W/1378$$

If $\quad A_2 = 10A_1$

then
$$\sigma_{bd} = 125W/1405A_1$$
$$F_{bd} = 125W/1405$$
$$F_{ad} = F_{cd} = 800W/1405$$

These results are illustrated in fig 2.23, and as might be expected the stiffer the member the more load it tends to attract, showing clearly how the member properties influence the force distribution in a statically indeterminate structure.

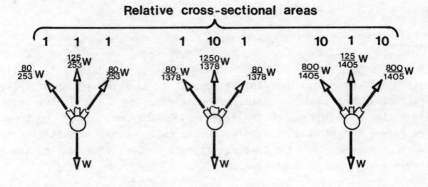

FIGURE 2.23   MEMBER FORCES

Redundancy

If a structure is statically indeterminate then it has at least one reactive or member force that can be released without the structure becoming a mechanism. Such forces are known as redundants, and the total number of redundants is known as the *degree of indeterminacy* of the structure.

Figure 2.24 shows three statically determinate structures. Such structures have zero degree of indeterminacy, implying that to remove any internal or reactive component of force will produce a mechanism.

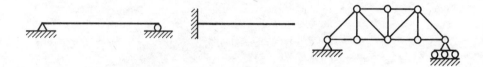

FIGURE 2.24   EXAMPLES OF STATICALLY DETERMINATE STRUCTURES

By introducing an additional support to each of the beams, and two additional members to the truss as shown in fig 2.25, the beams are made indeterminate to the first degree, and the truss is made indeterminate to the second degree.

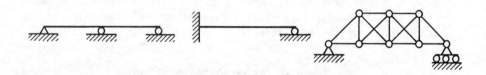

FIGURE 2.25   EXAMPLES OF STATICALLY INDETERMINATE STRUCTURES

It should be noted that the beams can be rendered statically determinate by releasing either a support or an internal moment. The internal moment could be released by introducing a pin at any point along the beam (except at the free ends). In the case of the truss the reader should note that only the two centre panels are statically indeterminate which means that releasing a member force outside these panels would produce a mechanism.

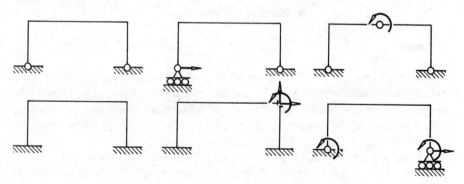

FIGURE 2.26   PORTAL FRAMES WITH PINNED AND FIXED FEET

FUNDAMENTALS OF STRUCTURAL BEHAVIOUR

The left-hand diagrams in fig 2.26 show portals with pinned and fixed feet. The portal with pinned feet is indeterminate to the first degree as it can be rendered determinate by releasing a single component of force. The portal with fixed feet is indeterminate to the third degree. The other diagrams in fig 2.26 illustrate how components of force equal in number to the degree of indeterminacy can be released without the structure becoming unstable. The circular arrows indicate points where a moment has been released, and the linear arrows incidate where a linear force has been released.

2.11 SUPERPOSITION

If a structure is constructed from a linearly elastic material, and small deflection is adopted then the structure behaves linearly. Hence deflections, reactions, and member forces are directly proportional to the applied loads. Consequently the effect of a given load system on a linear structure will be independent of the order in which the loads are applied, making valid the principle of superposition. Superposition is most commonly used to investigate the combined effect of two load systems. If, for instance, a structure has been independently analysed for imposed load and for wind load, then the combined effect of imposed and wind load can be found by simply superimposing the individual effects of the two load systems. The propped cantilever shown in fig 2.27 illustrates the principle of superposition.

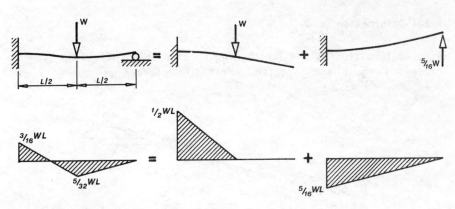

FIGURE 2.27   AN EXAMPLE OF SUPERPOSITION

The diagram shows that, if the effect of the propping force (i.e. the right-hand reaction) is superimposed on the effect of the applied load on the unpropped cantilever, then the true results are obtained.

Here the propping force and the applied load have been treated as two independent load systems and then their effects have been superimposed.

Superposition is not valid when the structural behaviour is non-linear. There are two sources of non-linearity in structures. The first is when the displacements are large, making small deflection theory inapplicable. The second is when the stiffness of the members of the structure is non-linear due to non-linear material behaviour. Non-linear behaviour is beyond the scope of this book, with linear behaviour being assumed throughout.

2.12 SIMPLE BEAM BEHAVIOUR

Structural frameworks consist of interconnected line elements. To apply the stiffness method to such structures requires an understanding of how line elements behave under load. This section is concerned with the relationship between the deformed shape of a *prismatic* line element and the loads applied to its ends. In general, a line element can undergo the following three types of deformation:

1. Axial deformation

2. Torsional deformation

3. Flexural deformation.

Axial Deformation

Axial deformation is the lengthening or shortening of a line element caused by equal and opposite longitudinal forces as shown in fig 2.28.

FIGURE 2.28   AXIAL DEFORMATION

The axial deformation of the element is governed by equation (2.3).

FUNDAMENTALS OF STRUCTURAL BEHAVIOUR

$$\delta = PL/EA \tag{2.3}$$

where
- δ is the axial deformation
- P is the axial load
- L is the element length
- E is the elastic constant (Young's modulus)
- A is the cross-sectional area.

Torsional Deformation

Torsion is the twisting of a line element caused by equal and opposite moments at its ends, as shown in fig 2.29.

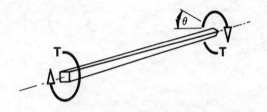

FIGURE 2.29   TORSIONAL DEFORMATION

The angle of twist as shown in fig 2.29 is given by equation (2.4).

$$\theta = TL/GJ \tag{2.4}$$

where
- θ is the relative rotation between the ends of the element
- T is the applied torsional moment
- L is the element length
- G is the modulus of rigidity
- J is the torsional constant.

The modulus of rigidity, G, is sometimes referred to as the modulus of rigidity in shear, or simply the shear modulus. The modulus of rigidity, the elastic constant, and Poisson's ratio are related by the following formula:

$$G = E/\{2(1 + \gamma)\}$$

The torsional constant, J, is a property of the cross-section and has units of second moment of area. The torsional constant cannot be evaluated exactly for most cross-sections and must be calculated from approximate formulae.

Flexural Deformation

Flexural deformation (or bending) of a line element arises from transverse forces and moments as illustrated in fig 2.30. These forces cause translation and rotation of the ends of the members as shown.

The remainder of this section is devoted to finding the relationship between the member end forces and the flexural deformation. Rotational and translational displacements at the member ends will be treated independently as any composite deformation and force system can be found by superposition.

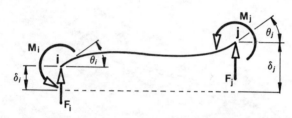

FIGURE 2.30   FLEXURAL DEFORMATION

If a line element is loaded only at its ends then the shear force along the length of the element is constant.

$$SF = k$$

but

$$SF = \text{rate of change of bending moment}$$

i.e

$$= dM/dx = k$$

therefore, by integration

$$M = kx + l \quad \text{(where "l" is a constant of integration)}$$

Simple beam theory assumes that bending moment is proportional to curvature, with EI as the constant of proportionality.

$$M = EI\, d^2y/dx^2 = kx + l$$

Incorporate EI into the constants on the right-hand side and integrate twice:

$$d^2y/dx^2 = kx/EI + l/EI$$

FUNDAMENTALS OF STRUCTURAL BEHAVIOUR

$$dy/dx = kx^2/2EI + lx/EI + c$$

$$y = kx^3/6EI + lx^2/2EI + cx + d$$

or

$$y = ax^3 + bx^2 + cx + d \tag{2.5}$$

proving that the flexural deformation of a prismatic line element loaded only at its ends conforms to a cubic polynomial. Knowing the rotation or slope (dy/dx) and the transverse displacement (y) at both ends of the element gives a total of four boundary conditions, and these enable the four coefficients (a, b, c, d) of the polynomial to be evaluated. An expression that algebraically describes the deformed shape of an element is known as a *shape function*.

Now consider the force system associated with rotation at end "i" of the line element "i,j" shown in fig 2.31.

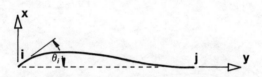

FIGURE 2.31   ROTATION AT END "i"

The shape function and its derivatives are:

$$y = ax^3 + bx^2 + cx + d$$

$$dy/dx = 3ax^2 + 2bx + c$$

$$d^2y/dx^2 = 6ax + 2b$$

The boundary conditions are:

at x = 0

y	=	0	therefore	$0 =$	$0 + 0 + 0 + d$
dy/dx	=	θ_i	therefore	$\theta_i =$	$0 + 0 + c$

at x = L

y	=	0	therefore	$0 =$	$aL^3 + bL^2 + cL + d$
dy/dx	=	0	therefore	$0 =$	$3aL^2 + 2bL + c$

33

Hence
$$d = 0, \quad c = \Theta_i, \quad a = \Theta_i/L^2, \quad b = -2\Theta_i/L$$

giving
$$y = x^3\Theta_i/L^2 - 2x^2\Theta_i/L + x\Theta_i \tag{2.6}$$

but
$$M = EI\, d^2y/dx^2 \quad \text{(i.e. bending moment = EI × curvature)}$$

Substitution of the second differential of equation (2.6) into the above equation yields

$$M = EI\Theta_i(6x/L^2 - 4/L)$$

hence
$$M = -4EI\Theta_i/L \quad \text{when} \quad x = 0$$
$$M = 2EI\Theta_i/L \quad \text{when} \quad x = L$$

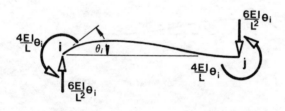

FIGURE 2.32   FORCE SYSTEM ASSOCIATED WITH ROTATION AT "i"

The signs mean that the applied bending moments causes negative curvature at the left-hand end of the beam, and positive curvature at the right-hand end of the beam. Hence the force system associated with rotation at "i" is as shown in fig 2.32. Note that the forces $6EI\Theta_i/L^2$ are necessary to maintain the equilibrium of the element.

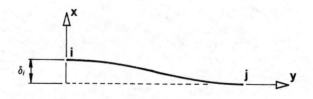

FIGURE 2.33   TRANSVERSE TRANSLATION AT NODE "i"

FUNDAMENTALS OF STRUCTURAL BEHAVIOUR

Now consider the force system associated with transverse translation at end "i" as illustrated in fig 2.33.

The boundary conditions are:

at $x = 0$

$y = \delta_i$ therefore $\delta_i = 0 + 0 + 0 + d$
$dy/dx = 0$ therefore $0 = 0 + 0 + c$

at $x = L$

$y = 0$ therefore $0 = aL^3 + bL^2 + cL + d$
$dy/dx = 0$ therefore $0 = 3aL^2 + 2bL + c$

Hence

$d = \delta_i$, $c = 0$, $b = -3\delta_i/L^2$, $a = 2\delta_i/L^3$

giving

$$y = 2x^3\delta_i/L^3 - 3x^2\delta_i/L^2 + \delta_i \qquad (2.7)$$

but

$M = EI\, d^2y/dx^2$ (i.e. bending moment = EI x curvature)

Substitution of the second differential of equation (2.7) into the above equation gives

$$M = EI\delta_i(12x/L^3 - 6/L^2)$$

hence
$M = -6EI\delta_i/L^2$ when $x = 0$
$M = 6EI\delta_i/L^2$ when $x = L$

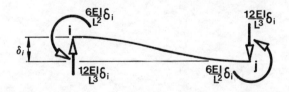

FIGURE 2.34 FORCE SYSTEM ASSOCIATED WITH TRANSLATION AT "i"

The signs mean that the applied bending moments causes negative curvature at the left-hand end of the beam, and positive curvature at the right-hand end of the beam. Hence the force system associated with translation at "i" is as shown in fig 2.34.

3 Introduction to the Stiffness Method

3.1 CLASSIFICATION OF FRAMED STRUCTURES

Framed structures are structures that can be satisfactorily idealised using line elements. Such structures are often referred to as skeletal structures. Usually the members of the structure are assumed to be connected either by frictionless pins or by rigid joints. Framed structures are usually idealised as one of the five types of skeletal structures shown in fig 3.1.

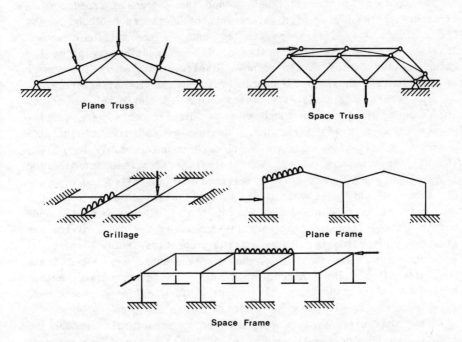

FIGURE 3.1   STRUCTURAL FRAMEWORKS

FRAMEWORK TYPE	JOINT TYPE	JOINT LOADS	MEMBER LOADS	LOAD DIRECTIONS
1. Plane trusses	Pinned	Yes	No	In plane
2. Space trusses	Pinned	Yes	No	Any direction
3. Plane frames	Rigid	Yes	Yes	In plane
4. Grillages	Rigid	Yes	Yes	Normal to plane
5. Space frames	Rigid	Yes	Yes	Any direction

Trusses are usually loaded only at their joints and, because the joints cannot transmit bending moment, trusses must be triangulated to avoid mechanism formation. Plane trusses are loaded in their own plane, whereas the joints of a space truss can be loaded from any direction.

Rigidly jointed frames are often loaded along their members as well as at their joints. Plane frames, like plane trusses, are loaded only in their own plane. In contrast, grillages are always loaded normal to the structure. Space frames can, of course, be loaded in any plane.

3.2 DEGREE OF FREEDOM

In the context of the stiffness method the degree of freedom of a structure is the number of displacement components to be found during the analysis. Finding these displacements by the stiffness method involves the solution of a system of linear equations relating the known applied forces to the unknown displacements, where the number of equations is equal to the degree of freedom of the structure.

In general, elastic structures have an infinite number of independent displacement components as displacements vary continuously throughout the structure. Obviously, to solve an infinite number of simultaneous equations will take an unacceptably long time. The stiffness method renders the analysis tractable by considering displacements and forces only at selected points, called *nodes*.

When the stiffness method is applied to problems in continuum mechanics it is commonly known as the finite element method. Node points are selected and imaginary lines between the nodes divide the structure into elements. To assess the properties of the elements an assumption is made about how the displacements vary along these imaginary lines. The finite element method is an approximate method because of this assumption. In general, using a finer element mesh (i.e. more nodes) will produce a more accurate solution.

When the stiffness method is applied to structural frameworks a node is usually located at each joint of the structure. The analysis determines the nodal displacements caused by the applied loading.

INTRODUCTION TO THE STIFFNESS METHOD

The number of possible displacement components at each node is known as the *nodal degree of freedom*, and the nodal degree of freedom for different framework types is as shown in the following table:

	FRAMEWORK TYPE	NODAL DEGREE OF FREEDOM
1.	Plane trusses	2
2.	Space trusses	3
3.	Plane frames	3
4.	Grillages	3
5.	Space frames	6

Figure 3.2 illustrates the nodal freedoms for different types of structural frameworks.

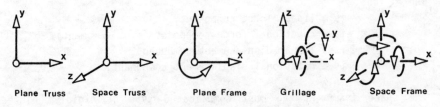

FIGURE 3.2   NODAL DEGREES OF FREEDOM

The stiffness method is concerned with the relationship between nodal (for the time being nodes equate to joints) forces and displacements. The components of force and displacement at a node are conveniently stored in one dimensional arrays known as the *nodal force and displacement vectors*. The number of elements in each vector is equal to the nodal degree of freedom as shown below. In general the displacement vector contains rotations as well as translations, and the force vector contains moments and forces. Consequently these vectors are sometimes referred to as *generalised* force and displacement vectors.

STRUCTURE TYPE	NODAL FORCE VECTOR	NODAL DISPLACEMENT VECTOR
Plane truss	$\begin{bmatrix} P_x \\ P_y \end{bmatrix}$	$\begin{bmatrix} \Delta_x \\ \Delta_y \end{bmatrix}$
Space truss	$\begin{bmatrix} P_x \\ P_y \\ P_z \end{bmatrix}$	$\begin{bmatrix} \Delta_x \\ \Delta_y \\ \Delta_z \end{bmatrix}$

Plane frame $\begin{bmatrix} P_x \\ P_y \\ M_z \end{bmatrix}$ $\begin{bmatrix} \Delta_x \\ \Delta_y \\ \Theta_z \end{bmatrix}$

Grillage $\begin{bmatrix} P_z \\ M_x \\ M_y \end{bmatrix}$ $\begin{bmatrix} \Delta_z \\ \Theta_x \\ \Theta_y \end{bmatrix}$

Space frame $\begin{bmatrix} P_x \\ P_y \\ P_z \\ M_x \\ M_y \\ M_z \end{bmatrix}$ $\begin{bmatrix} \Delta_x \\ \Delta_y \\ \Delta_z \\ \Theta_x \\ \Theta_y \\ \Theta_z \end{bmatrix}$

where

P is a linear force component.
M is a rotational force component (i.e. a moment).
Δ is a linear displacement component.
Θ is a rotational displacement component.

The space frame is the most complicated type of rigidly jointed framework. Each member of a space frame can undergo axial deformation, torsional deformation, and flexural deformation (in two planes). Axial and torsional deformations of a prismatic line element loaded only at its ends vary linearly along the length of the element. Hence it is a simple matter to determine the axial and torsional deformation at any point along the member from a knowledge of the end displacements. Section 2.12 showed that the cubic polynomial defining the flexural deformation of a line element loaded at its ends can be determined from the transverse displacements and the rotations at its ends. It follows that member end displacements define flexural deformation at all points along the member. Hence the deformed shape of a prismatic space frame member loaded only at its ends is uniquely defined by its end displacements.

All other types of rigidly jointed frameworks are special cases of space frame structures. Hence the deformed shape of any type of structural framework, loaded only at its joints, is completely defined by the nodal displacements. Consequently the stiffness method is exact when applied to frameworks loaded only at their joints. That only joint loads have been considered here is not, in fact, a restriction. Member loads, which have been excluded in the interests of clarity, can easily be included in the analysis through the use of equivalent joint forces as described in Chapter 8.

Boundary Conditions

For stability a structure is always restrained at one or more nodes (see section 2.3). These restraints are known as boundary conditions and they have the effect of reducing the degree of freedom of the restrained node.

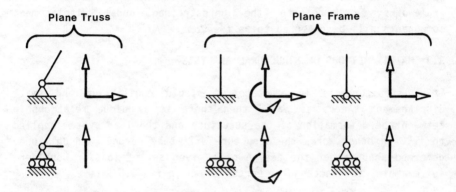

FIGURE 3.3   BOUNDARY CONDITIONS FOR PLANE TRUSSES AND PLANE FRAMES

Figure 3.3 shows typical boundary conditions and restraint directions for plane trusses and plane frames. The boundary conditions for other framework types are similar in concept.

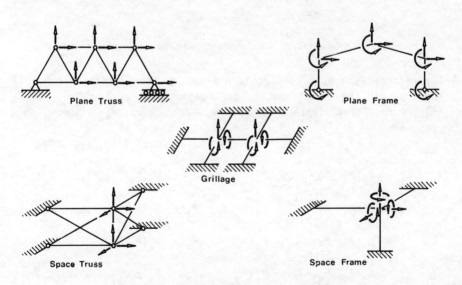

FIGURE 3.4   UNRESTRAINED NODAL DISPLACEMENTS

COMPUTER ANALYSIS OF STRUCTURAL FRAMEWORKS

The determination of the degree of freedom of a framed structure is a straightforward process, requiring only a knowledge of the nodal degree of freedom, and an understanding of the boundary conditions. The degree of freedom of the structure is simply the product of the number of nodes and the nodal degree of freedom, less the number of restrained displacement components. Figure 3.4 shows the unrestrained nodal displacement components for some simple structural frameworks. Henceforth the unrestrained nodal displacement components will be referred to as *freedoms*.

3.3 MATRIX METHODS OF STRUCTURAL ANALYSIS

If a structure is in a state of stable equilibrium and small displacement theory is valid then there is a unique relationship between the deformation of the structure and the load system applied to it. In other words the structure will take up one, and only one, deformed shape under the action of a given set of loads. Consider, for example, the rectangular portal shown in fig 3.5.

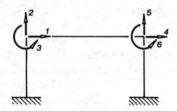

FIGURE 3.5   UNRESTRAINED NODAL DISPLACEMENTS FOR A RECTANGULAR PORTAL

The deformed shape is completely defined by the freedoms shown. If the loads in the directions of the freedoms are P_1, P_2, P_3, ... P_6 and the resulting displacements are $\Delta_1, \Delta_2, \Delta_3, ... \Delta_6$ as shown, then

$$\Delta_1 = G_1(P_1, P_2, ... P_6) \quad \text{or} \quad P_1 = H_1(\Delta_1, \Delta_2, ... \Delta_6)$$

$$\Delta_2 = G_2(P_1, P_2, ... P_6) \quad \quad P_2 = H_2(\Delta_1, \Delta_2, ... \Delta_6)$$

$$\Delta_3 = G_3(P_1, P_2, ... P_6) \quad \quad P_3 = H_3(\Delta_1, \Delta_2, ... \Delta_6)$$

$$\Delta_4 = G_4(P_1, P_2, ... P_6) \quad \quad P_4 = H_4(\Delta_1, \Delta_2, ... \Delta_6)$$

$$\Delta_5 = G_5(P_1, P_2, ... P_6) \quad \quad P_5 = H_5(\Delta_1, \Delta_2, ... \Delta_6)$$

$$\Delta_6 = G_6(P_1, P_2, ... P_6) \quad \quad P_6 = H_6(\Delta_1, \Delta_2, ... \Delta_6)$$

INTRODUCTION TO THE STIFFNESS METHOD

The equations on the left-hand side state that the unknown nodal displacements are a function of the applied loads. Of course, the expressions can be rewritten to give the applied loads as a function of the unknown nodal displacements, as shown by the equations on the right-hand side.

Now, if the structure is assumed to be linearly elastic then there is a linear relationship between the loads and the resulting displacements. That relationship can be expressed with either the loads or the displacements as the independent variables

i.e.
$$\Delta_1 = F_{11}P_1 + F_{12}P_2 + F_{13}P_3 + \ldots\ldots + F_{16}P_6$$
$$\Delta_2 = F_{21}P_1 + F_{22}P_2 + F_{23}P_3 + \ldots\ldots + F_{26}P_6$$
$$\vdots$$
$$\Delta_6 = F_{61}P_1 + F_{62}P_2 + F_{63}P_3 + \ldots\ldots + F_{66}P_6$$

or
$$P_1 = k_{11}\Delta_1 + k_{12}\Delta_2 + k_{13}\Delta_3 + \ldots\ldots + k_{16}\Delta_6$$
$$P_2 = k_{21}\Delta_1 + k_{22}\Delta_2 + k_{23}\Delta_3 + \ldots\ldots + k_{26}\Delta_6$$
$$\vdots$$
$$P_6 = k_{61}\Delta_1 + k_{62}\Delta_2 + k_{63}\Delta_3 + \ldots\ldots + k_{66}\Delta_6$$

Rewriting these equations in matrix form gives:

$$\begin{bmatrix} \Delta_1 \\ \Delta_2 \\ \Delta_3 \\ \Delta_4 \\ \Delta_5 \\ \Delta_6 \end{bmatrix} = \begin{bmatrix} F_{11} & F_{12} & F_{13} & F_{14} & F_{15} & F_{16} \\ F_{21} & F_{22} & F_{23} & F_{24} & F_{25} & F_{26} \\ F_{31} & F_{32} & F_{33} & F_{34} & F_{35} & F_{36} \\ F_{41} & F_{42} & F_{43} & F_{44} & F_{45} & F_{46} \\ F_{51} & F_{52} & F_{53} & F_{54} & F_{55} & F_{56} \\ F_{61} & F_{62} & F_{63} & F_{64} & F_{65} & F_{66} \end{bmatrix} \begin{bmatrix} P_1 \\ P_2 \\ P_3 \\ P_4 \\ P_5 \\ P_6 \end{bmatrix}$$

$$\Delta = F\,P \tag{3.1}$$

and

$$\begin{bmatrix} P_1 \\ P_2 \\ P_3 \\ P_4 \\ P_5 \\ P_6 \end{bmatrix} = \begin{bmatrix} K_{11} & K_{12} & K_{13} & K_{14} & K_{15} & K_{16} \\ K_{21} & K_{22} & K_{23} & K_{24} & K_{25} & K_{26} \\ K_{31} & K_{32} & K_{33} & K_{34} & K_{35} & K_{36} \\ K_{41} & K_{42} & K_{43} & K_{44} & K_{45} & K_{46} \\ K_{51} & K_{52} & K_{53} & K_{54} & K_{55} & K_{56} \\ K_{61} & K_{62} & K_{63} & K_{64} & K_{65} & K_{66} \end{bmatrix} \begin{bmatrix} \Delta_1 \\ \Delta_2 \\ \Delta_3 \\ \Delta_4 \\ \Delta_5 \\ \Delta_6 \end{bmatrix}$$

$$P = K\,\Delta \tag{3.2}$$

where

P	is the loading vector
Δ	is the displacement vector
F	is the structure flexibility matrix
K	is the structure stiffness matrix

Equations (3.1) and (3.2) form the basis of the *flexibility* and *stiffness* methods of structural analysis respectively.

At first sight the flexibility method looks the more attractive because, once the flexibility matrix F has been found, the structural displacements Δ can be obtained by simple matrix multiplication. By contrast, to find the unknown displacements using the stiffness method involves solving a system of simultaneous equations. However, the stiffness matrix turns out to be much easier to develop than the flexibility matrix. Further, the stiffness method is better suited to computer implementation than the flexibility method. For these (and historical) reasons the stiffness method now completely dominates structural analysis computer programs. As the displacements are the primary unknowns, the stiffness method is sometimes referred to as the *displacement method*.

The flexibility method will often be more economical than the stiffness method for hand calculation. It is the opinion of the author, however, that matrix methods should not be used for hand calculation because traditional methods of structural analysis will almost always be quicker and less error prone. Finally, as the flexibility method is less well suited to computer implementation than the stiffness method it will not be considered further and the remainder of this book will deal exclusively with the stiffness method.

3.4 STIFFNESS COEFFICIENTS

The analysis of structures using the stiffness equation P = KΔ has two distinct phases:

1. Generation of the stiffness matrix K. The elements of K are, in effect, the coefficients of a set of simultaneous equations.

2. Solution of the simultaneous equations. This yields the unknown structural displacements Δ caused by the known applied loads P.

INTRODUCTION TO THE STIFFNESS METHOD

The remainder of this section takes a closer look at the structure stiffness matrix **K**. The solution of simultaneous equations will be considered in Chapter 4.

Consider a linearly elastic body acted upon by a system of forces **P**, which cause displacements **Δ** in the direction of the forces as shown in fig 3.6.

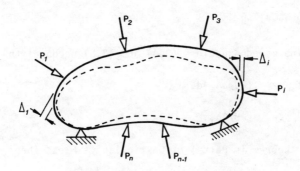

FIGURE 3.6 DEFORMATION OF AN ELASTIC BODY

The total work done, W, during the application of the forces is

$$W = 1/2 \, (P_1\Delta_1 + P_2\Delta_2 + \ldots + P_n\Delta_n) \quad (3.3)$$

As the body is elastic, energy is conserved, and the work done by the external forces is equal to the strain energy, U, gained by the body.

i.e.   W = U

Now if one displacement, say Δ_1, is varied by an infinitesimal amount $d\Delta_1$ while all other displacements are held constant, then the change in the strain energy of the body is

$$dU = (\partial U/\partial \Delta_1)d\Delta_1$$

and the corresponding change in P_1 is

$$dP_1 = (\partial P_1/\partial \Delta_1)d\Delta_1 \quad (3.4)$$

Hence the change in the work done is equal to

$$dW = (P_1 + (P_1 + dP_1))d\Delta_1/2$$

$$= P_1 d\Delta_1 + (\partial P_1/\partial \Delta_1)(d\Delta_1)^2/2$$

In the limit

$$dW = P_1 d\Delta_1 \qquad (3.5)$$

During this variation of displacement energy is conserved and therefore the change in work done is equal to the change in the strain energy, hence

$$dW = dU$$

Substituting using equations (3.4) and (3.5) produces the following:

$$P_1 d\Delta_1 = (\partial U/\partial \Delta_1) d\Delta_1$$

hence

$$P_1 = \partial U/\partial \Delta_1 \qquad (3.6)$$

The above theorem is Part I of Castigliano's First Theorem.

Differentiation of equation (3.3) yields (noting that $U = W$)

$$\partial U/\partial \Delta_1 = 1/2 \{P_1 + (\partial P_1/\partial \Delta_1)\Delta_1 + (\partial P_2/\partial \Delta_1)\Delta_2 + \ldots + (\partial P_n/\partial \Delta_1)\Delta_n\}$$

Using Castigliano's First Theorem to substitute for $\partial U/\partial \Delta_1$ in the above equation gives:

$$P_1 = (\partial P_1/\partial \Delta_1)\Delta_1 + (\partial P_2/\partial \Delta_1)\Delta_2 + \ldots + (\partial P_n/\partial \Delta_1)\Delta_n$$

Varying all other displacements in turn yields:

$$P_2 = (\partial P_1/\partial \Delta_2)\Delta_1 + (\partial P_2/\partial \Delta_2)\Delta_2 + \ldots + (\partial P_n/\partial \Delta_2)\Delta_n$$
$$P_3 = (\partial P_1/\partial \Delta_3)\Delta_1 + (\partial P_2/\partial \Delta_3)\Delta_2 + \ldots + (\partial P_n/\partial \Delta_3)\Delta_n$$
$$\vdots$$
$$P_n = (\partial P_1/\partial \Delta_n)\Delta_1 + (\partial P_2/\partial \Delta_n)\Delta_2 + \ldots + (\partial P_n/\partial \Delta_n)\Delta_n$$

Writing these equations in matrix form:

$$\begin{bmatrix} P_1 \\ P_2 \\ \vdots \\ P_n \end{bmatrix} = \begin{bmatrix} \partial P_1/\partial \Delta_1 & \partial P_2/\partial \Delta_1 & \ldots & \partial P_n/\partial \Delta_1 \\ \partial P_1/\partial \Delta_2 & \partial P_2/\partial \Delta_2 & \ldots & \partial P_n/\partial \Delta_2 \\ \vdots & \vdots & & \vdots \\ \partial P_1/\partial \Delta_n & \partial P_2/\partial \Delta_n & \ldots & \partial P_n/\partial \Delta_n \end{bmatrix} \begin{bmatrix} \Delta_1 \\ \Delta_2 \\ \vdots \\ \Delta_n \end{bmatrix} \qquad (3.7)$$

INTRODUCTION TO THE STIFFNESS METHOD

This is, of course, the stiffness equation

$$P = K \Delta$$

Note that the elastic body is not subject to applied moments. Introducing applied moments causes no difficulty, their omission was simply for the sake of clarity.

Inspection of the stiffness matrix shows that it consists of derivatives that represent the rate of change of force with displacement, i.e. they are *stiffnesses*. If the structure is linearly elastic then these terms are constant. The physical meaning of $\partial P_i / \partial \Delta_j$ is the rate of change of the force at "i" with variation of displacement at "j" ($\partial P_i / \partial \Delta_j$ is a partial derivative which implies that all other displacements are held constant during the variation of displacement at "j"). If Δj is varied by unity, then $\partial P_i / \partial \Delta_j$ becomes *the force at "i" associated with unit change in the displacement at "j" (all other displacements held constant)*. Hence, column "j" of the stiffness matrix is the force system required to maintain unit displacement at "j". This means that the complete stiffness matrix can be generated by calculating the force system required to maintain unit displacement at each freedom in turn.

Physical insight into the stiffness method can be gained by considering unit displacement at each freedom in turn, and then scaling the force systems for the unit displacements by the actual displacements to allow the true structural behaviour to be obtained by superposition. Consider the rigidly jointed frame shown in fig 3.7.

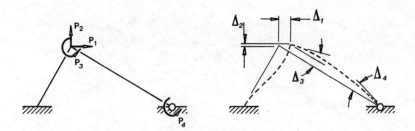

FIGURE 3.7   FORCES AND DISPLACEMENTS

The frame has four freedoms and is acted upon by the loads P_1, P_2, P_3 and P_4 in the directions of the freedoms. These loads cause displacements Δ_1, Δ_2, Δ_3, and Δ_4 in the directions of the loads.

Figure 3.8 shows the force systems required to maintain unit displacement in the direction of each freedom in turn (note that the reactive forces have been omitted and that K_{ij} is the force developed in the direction of freedom "i" when there is unit displacement in the direction of freedom "j").

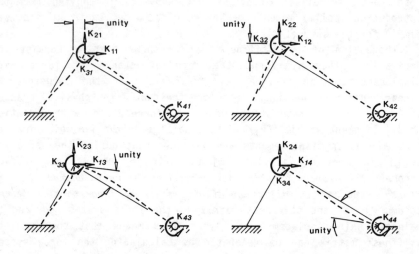

FIGURE 3.8 EVALUATION OF THE STIFFNESS COEFFICIENTS

The force system associated with displacement Δ_i will be equal to Δ_i times the force system associated with unit displacement in the direction of freedom "i". If the individual force systems associated with each displacement are summed then, by the principle of superposition, the result must be equal to the system of applied loads that is shown in fig 3.8.

i.e.
$$\Delta_1 K_{11} + \Delta_2 K_{12} + \Delta_3 K_{13} + \Delta_4 K_{14} = P_1$$
$$\Delta_1 K_{21} + \Delta_2 K_{22} + \Delta_3 K_{23} + \Delta_4 K_{24} = P_2$$
$$\Delta_1 K_{31} + \Delta_2 K_{32} + \Delta_3 K_{33} + \Delta_4 K_{34} = P_3$$
$$\Delta_1 K_{41} + \Delta_2 K_{42} + \Delta_3 K_{43} + \Delta_4 K_{44} = P_4$$

Or, in matrix notation:

i.e.
$$\begin{bmatrix} K_{11} & K_{12} & K_{13} & K_{14} \\ K_{21} & K_{22} & K_{23} & K_{24} \\ K_{31} & K_{32} & K_{33} & K_{34} \\ K_{41} & K_{42} & K_{43} & K_{44} \end{bmatrix} \begin{bmatrix} \Delta_1 \\ \Delta_2 \\ \Delta_3 \\ \Delta_4 \end{bmatrix} = \begin{bmatrix} P_1 \\ P_2 \\ P_3 \\ P_4 \end{bmatrix}$$

The above equation is, of course, the structure stiffness equation.

3.5 DIRECT APPLICATION OF THE STIFFNESS METHOD

The material in the previous three sections has shown that the stiffness method can be applied directly as a four stage procedure:

1. Identify the nodal freedoms.

2. Find the coefficients of the stiffness matrix by applying unit displacement at each freedom in turn.

3. Develop the loading vector from the applied loads.

4. Solve the simultaneous equations to find the structural displacements.

Once the structural displacements are known the element forces can be easily evaluated.

Example 3.1

Find the displacements and the forces at the top of the left-hand column of the portal shown in the left-hand diagram (assume the members to be axially rigid). EI is constant for all members.

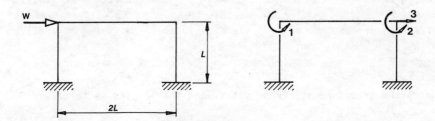

The frame shown is similar to the portal shown in fig 3.5. A computer analysis of such a frame would take the axial strain in the members into account. Consequently the frame would be regarded as having the six freedoms shown in fig 3.5. Hand methods of calculation would, however, regard the members as axially rigid, thus simplifying the problem by reducing the degree of freedom of the structure from six to three as shown in the right-hand diagram. The error introduced by this assumption is small. Axial rigidity is assumed here in order to limit the size of the problem.

The elements of the stiffness matrix are found by evaluation of the forces required to maintain unit displacement at each freedom in turn. The stiffness characteristics of line elements are derived in

COMPUTER ANALYSIS OF STRUCTURAL FRAMEWORKS

section 2.12.

Freedom 1 (apply unit rotation in the direction of freedom 1)

$$K_{11} = 4EI/L + 4EI/2L = 6EI/L$$
$$K_{21} = 2EI/2L = EI/L$$
$$K_{31} = 6EI/L^2$$

Freedom 2 (apply unit rotation in the direction of freedom 2)

$$K_{22} = 4EI/L + 4EI/2L = 6EI/L$$
$$K_{12} = 2EI/2L = EI/L$$
$$K_{32} = 6EI/L^2$$

Freedom 3 (apply unit translation in the direction of freedom 3)

$$K_{33} = 2 \times 12EI/L^3 = 24EI/L^3$$
$$K_{13} = 6EI/L^2$$
$$K_{23} = 6EI/L^2$$

Hence the structure stiffness matrix and the loading vector are

$$K = \begin{bmatrix} 6EI/L & EI/L & 6EI/L^2 \\ EI/L & 6EI/L & 6EI/L^2 \\ 6EI/L^2 & 6EI/L^2 & 24EI/L^3 \end{bmatrix} \qquad P = \begin{bmatrix} 0 \\ 0 \\ W \end{bmatrix}$$

and the structure stiffness equation $K \Delta = P$ is

$$EI/L \begin{bmatrix} 6 & 1 & 6/L \\ 1 & 6 & 6/L \\ 6/L & 6/L & 24/L^2 \end{bmatrix} \begin{bmatrix} \Delta_1 \\ \Delta_2 \\ \Delta_3 \end{bmatrix} = \begin{bmatrix} 0 \\ 0 \\ W \end{bmatrix}$$

and

$$K^{-1} = L/480EI \begin{bmatrix} 108 & 12 & -30L \\ 12 & 108 & -30L \\ -30L & -30L & 35L^2 \end{bmatrix}$$

INTRODUCTION TO THE STIFFNESS METHOD

$$\Delta = K^{-1} P \quad \text{which yields} \quad \begin{bmatrix} \Delta_1 \\ \Delta_2 \\ \Delta_3 \end{bmatrix} = \begin{bmatrix} -WL^2/16EI \\ -WL^2/16EI \\ 7WL^3/96EI \end{bmatrix}$$

The forces at the top of the left-hand column are found by superposition of the effects of the end displacements.

Forces due to Δ_1

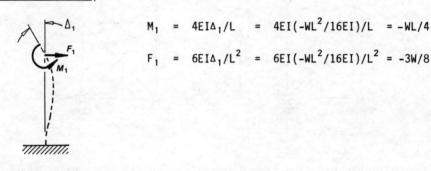

$M_1 = 4EI\Delta_1/L = 4EI(-WL^2/16EI)/L = -WL/4$

$F_1 = 6EI\Delta_1/L^2 = 6EI(-WL^2/16EI)/L^2 = -3W/8$

Forces due to Δ_3

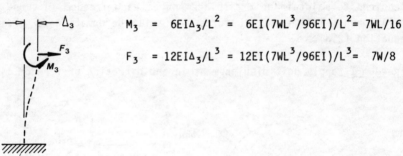

$M_3 = 6EI\Delta_3/L^2 = 6EI(7WL^3/96EI)/L^2 = 7WL/16$

$F_3 = 12EI\Delta_3/L^3 = 12EI(7WL^3/96EI)/L^3 = 7W/8$

Therefore the final forces at the top of the left hand column are:

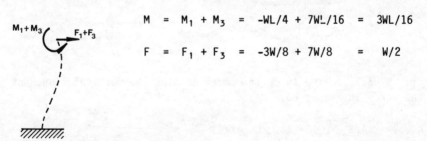

$M = M_1 + M_3 = -WL/4 + 7WL/16 = 3WL/16$

$F = F_1 + F_3 = -3W/8 + 7W/8 = W/2$

COMPUTER ANALYSIS OF STRUCTURAL FRAMEWORKS

Example 3.2

Find the nodal displacements for the pin jointed frame shown below. Use these displacements to find the force in members AC, BC, and CD. Demonstrate the equilibrium of joint C and find the nodal displacements if member AD is removed. $EA = 4 \times 10^6$ N for all members.

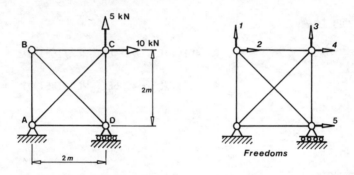

The structure has five freedoms as shown in the right-hand diagram. The structure stiffness matrix is generated from the force systems required to maintain unit displacement at each freedom in turn. The relationship between axial force and axial deformation is given by equation (2.3).

<u>Freedom 1</u> (apply unit displacement in the direction of freedom 1)

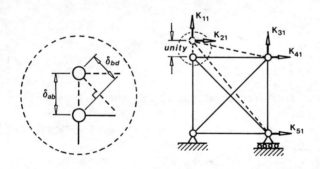

K_{11} is equal to the force in member AB plus the vertical component of the force in member BD. Hence

$$K_{11} = EA(\delta_{ab}/L_{ab} + (\delta_{bd}/L_{bd})/\sqrt{2})$$

$$= 4 \times 10^6[1/2 + ((1/\sqrt{2})/2\sqrt{2})/\sqrt{2}] \qquad = 2.707 \times 10^6 \text{ N/m}$$

INTRODUCTION TO THE STIFFNESS METHOD

$K_{21} = -EA(\delta_{bd}/L_{bd})/\sqrt{2} = -4 \times 10^6 (1/4)/\sqrt{2} = -0.707 \times 10^6$ N/m

$K_{51} = -K_{21} = 0.707 \times 10^6$ N/m

$K_{31} = K_{41} = 0$

Applying unit displacement in a similar manner at the other four freedoms in turn yields the remaining stiffness coefficients, and the final stiffness equation is:

$$10^6 \begin{bmatrix} 2.707 & -0.707 & 0 & 0 & 0.707 \\ -0.707 & 2.707 & 0 & -2.0 & -0.707 \\ 0 & 0 & 2.707 & 0.707 & 0 \\ 0 & -2.0 & 0.707 & 2.707 & 0 \\ 0.707 & -0.707 & 0 & 0 & 2.707 \end{bmatrix} \begin{bmatrix} \Delta_1 \\ \Delta_2 \\ \Delta_3 \\ \Delta_4 \\ \Delta_5 \end{bmatrix} = \begin{bmatrix} 0 \\ 0 \\ 5000 \\ 10000 \\ 0 \end{bmatrix}$$

Solving these equations with any of the equation solving programs given in Chapter 4 yields

$$\begin{bmatrix} \Delta_1 \\ \Delta_2 \\ \Delta_3 \\ \Delta_4 \\ \Delta_5 \end{bmatrix} = \begin{bmatrix} 0.17 \times 10^{-2} \text{ m} \\ 0.83 \times 10^{-2} \text{ m} \\ -0.77 \times 10^{-3} \text{ m} \\ 0.10 \times 10^{-1} \text{ m} \\ 0.17 \times 10^{-2} \text{ m} \end{bmatrix}$$

Member forces

To find the forces in members AC, BC, and CD first calculate extensions

$\delta_{ac} = \Delta_4/\sqrt{2} + \Delta_3/\sqrt{2} = 0.01/\sqrt{2} - 0.00077/\sqrt{2} = 0.00653$ m

$\delta_{bc} = \Delta_4 - \Delta_2 = 0.01 - 0.00830 = 0.0017$ m

$\delta_{cd} = \Delta_3 = -0.00077$ m

and

$F = EA\delta/L$

hence

$F_{ac} = 0.00653 \times 4 \times 10^6 / 2\sqrt{2} = 9230$ N

$F_{bc} = 0.0017 \times 4 \times 10^6 / 2 = 3400$ N

$F_{cd} = -0.00077 \times 4 \times 10^6 / 2 = -1540$ N

Equilibrium of joint C

The freebody diagram for joint C is as shown below, and its equilibrium is checked by summing horizontal and vertical forces.

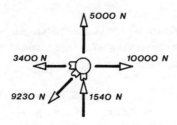

$$\sum H = 3400 - 10000 + 9230/\sqrt{2} \approx 0$$

$$\sum V = 5000 + 1540 - 9230/\sqrt{2} \approx 0$$

Removal of member AD

If member AD is removed then K_{55} becomes 0.707×10^6, and the new the solution vector is

$$\begin{bmatrix} \Delta_1 \\ \Delta_2 \\ \Delta_3 \\ \Delta_4 \\ \Delta_5 \end{bmatrix} = \begin{bmatrix} 0.0 & m \\ 1.66 \times 10^{-2} & m \\ -0.25 \times 10^{-2} & m \\ 1.66 \times 10^{-2} & m \\ 1.66 \times 10^{-2} & m \end{bmatrix}$$

As expected the displacements are much larger, and their relative magnitudes indicate that $F_{ab} = F_{bc} = F_{bd} = 0$. The reader should confirm that these forces must be zero.

Summary

These two examples have shown the stiffness method to be simple in concept and flexible in use. However, because each additional freedom produces another equation which must be solved simultaneously, hand application is limited to all but the simplest of structures. The criteria used to assess hand methods of analysis are very different from those used to assess computer-based methods. A hand technique stands or falls by the amount of arithmetic it generates. By contrast, the success of a computer-based method depends primarily upon its generality and how amenable it is to computer implementation. The generality and amenability to computer implementation offered by the stiffness method has led to its complete dominance of commercial structural analysis software. The remainder of this text is concerned with the techniques required to computerise the stiffness method for the analysis of structural frameworks.

3.6 AXES SYSTEMS

To computerise the stiffness method requires formalisation of the solution procedure. In practice the use of a number of different axes systems in a single analysis is found to facilitate automation of the method. Three different axes systems are commonly used in the analysis of structural frameworks. These are

1. Coordinate axes.
2. Nodal axes.
4. Member axes.

Coordinate Axes

Coordinate axes are used to define the location of the nodes, and hence the structural geometry. The origin and orientation of the coordinate axes are totally arbitrary. Figure 3.9 shows a pitched roof portal and a coordinate axes system which might be used to describe the nodal coordinates.

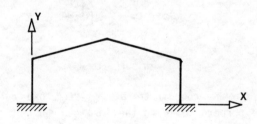

FIGURE 3.9   COORDINATE AXES

Nodal Axes

The forces and displacements at each node are described in terms of an axis system located at the node. If the nodal axes lie parallel to the coordinate axes then the nodal axes are referred to as *global* axes. If all nodal axes systems are global then the solution procedure is simplified. Figure 3.10 shows the global axes for the pitched roof portal shown in fig 3.9 (note that the nodal x-axes are horizontal).

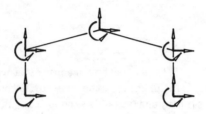

FIGURE 3.10   GLOBAL AXES

The boundary conditions for most structures can be described in terms of the global axes system. Where the structure has inclined supports, however, a *local* axes system may have to be employed. Figure 3.11(a) shows the coordinate axes system adopted for a rectangular portal with an inclined support, and fig 3.11(b) shows how a local axes system has been used to allow the application of the boundary conditions.

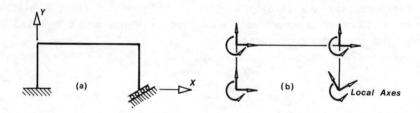

FIGURE 3.11   LOCAL AXES

Of course, if the coordinate axes system had been chosen to lie parallel and perpendicular to the inclined support then the use of a local axes system would, in this case, have been avoided.

Member Axes

Nodal axes systems (global and local) are used to describe the displacements and forces at the nodes. These axes are used when generating the stiffness equation $K \Delta = P$ because, when considering nodal equilibrium, forces and displacements at each node must be expressed in a single axes system. The solution of the stiffness

INTRODUCTION TO THE STIFFNESS METHOD

equation yields the unknown structural displacements, and from these displacements the member end forces are found.

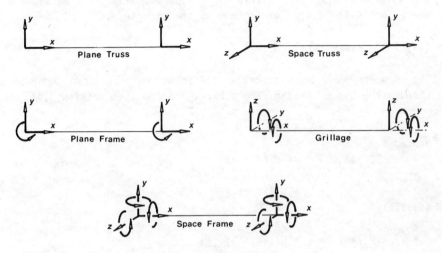

FIGURE 3.12 MEMBER AXES SYSTEMS

In this text the member axes systems shown in fig 3.12 are used to describe the member stress resultants. These axes yield the stress resultants that are most useful to the engineer.

1. Plane truss - axial force.
2. Space truss - axial force.
3. Plane frame - axial force, shear force and bending moment.
4. Grillage   - bending moment, torsion and shear force.
5. Space frame - axial force, bending moment (in two planes), torsion, and shear force (in two planes).

Note that the foregoing assumes that for plane frames, grillages, and space frames the member y-axis coincides with a principal axis of the section. The reader should also note the following:

1. Right hand axes systems are used throughout.
2. Plane structures always lie in the coordinate "x,y" plane.
3. In all cases *the x-axis of member "i,j" runs along the member from end "i" to end "j".*

3.7 PROPERTIES OF THE STRUCTURE STIFFNESS MATRIX

There now follows a description of some of the properties of the structure stiffness matrix.

(i) It is square

The structure stiffness matrix is a square matrix of dimension "n" where "n" is the degree of freedom of the structure.

(ii) It is symmetrical

The symmetry of the structure stiffness equation can be proved through the use of Castigliano's First Theorem (see section 3.4):

$$P_i = \partial U/\partial \Delta_i$$

hence

$$\partial P_i/\partial \Delta_j = \partial(\partial U/\partial \Delta_i)/\partial \Delta_j$$

$$= \partial^2 U/(\partial \Delta_j \partial \Delta_i)$$

similarly

$$\partial P_j/\partial \Delta_i = \partial^2 U/(\partial \Delta_i \partial \Delta_j)$$

As the order of differentiation is immaterial

$$\partial^2 U/(\partial \Delta_i \partial \Delta_j) = \partial^2 U/(\partial \Delta_j \partial \Delta_i)$$

hence

$$\partial P_i/\partial \Delta_j = \partial P_j/\partial \Delta_i$$

i.e.

$$K_{ij} = K_{ji}$$

(iii) It is banded

Consider the multi-bay portal shown in fig 3.13.

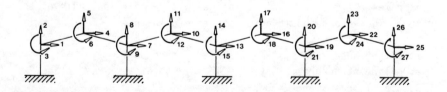

FIGURE 3.13   MULTI-BAY PORTAL

Column "j" of the structure stiffness matrix is, in effect, the force system required to maintain unit displacement in the direction of freedom "j".   Say the elements of column 12 of the structure stiffness matrix are to be found (freedoms are numbered as shown in fig 3.13).

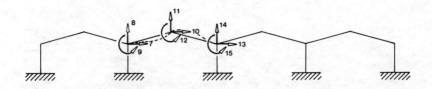

FIGURE 3.14   FORCES FOR UNIT DISPLACEMENT AT FREEDOM 12

Figure 3.14 shows the non-zero forces required to maintain unit rotation at freedom 12. As can be seen, only 9 of the possible 27 stiffness terms in column 12 of the stiffness matrix will be non-zero, and if the freedoms are well numbered the non-zero terms will be clustered around the diagonal (i.e. the matrix will be banded).

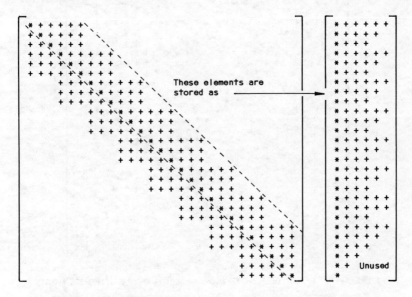

FIGURE 3.15   NON-ZERO ELEMENTS IN THE STRUCTURE STIFFNESS MATRIX

By taking advantage of the band in the structure stiffness matrix savings can be made of both computer storage and time. Indeed the problem size that can be handled by many structural analysis computer programs is limited, not by the degree of freedom, but by the bandwidth of the structure stiffness matrix. Figure 3.15 shows the non-zero elements of the structure stiffness matrix for the multi-bay portal shown in fig 3.13, and also illustrates how computer storage can be reduced by storing only the upper semi-band of the matrix.

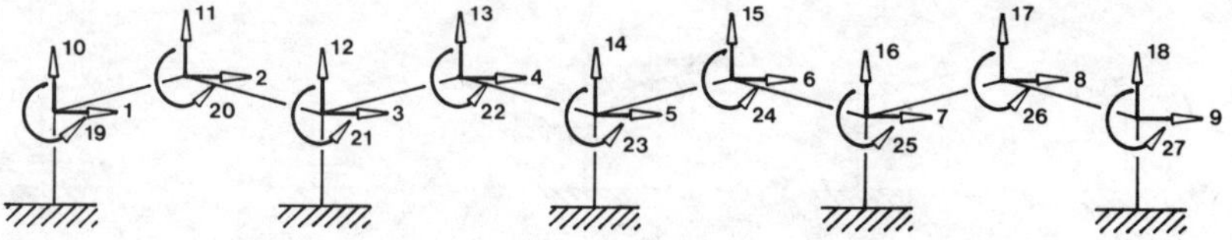

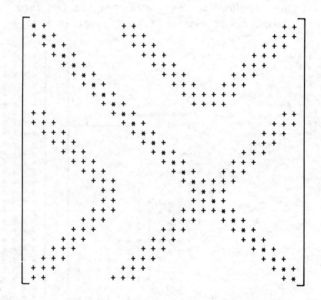

FIGURE 3.16   POORLY NUMBERED FREEDOMS

Had the freedoms been poorly numbered as shown in fig 3.16, then the band in the structure stiffness matrix would be lost, as can be seen in fig 3.17.

FIGURE 3.17   EFFECT OF POOR FREEDOM NUMBERING ON THE STIFFNESS MATRIX

(iv) It is positive definite

The structure stiffness matrix is positive definite as pre- and post-multiplication by an arbitrary non-zero vector always results in a positive quantity since

$$P = K \Delta$$

INTRODUCTION TO THE STIFFNESS METHOD

and strain energy

$$U = 1/2\, P^T \Delta$$

hence

$$U = 1/2\, \Delta^T K^T \Delta$$

but

$$K^T = K$$

therefore

$$U = 1/2\, \Delta^T K \Delta$$

If the structure is in a state of stable equilibrium then the structure gains strain energy on any arbitrary small deformation, proving that the structure stiffness matrix is positive definite. Having a positive definite coefficient matrix is a useful property when solving simultaneous equations.

4 Simultaneous Linear Equations

4.1 INTRODUCTION

Linear simultaneous equations take the form shown below:

$$
\begin{aligned}
a_{11}x_1 + a_{12}x_2 + a_{13}x_3 + \ldots + a_{1n}x_n &= b_1 \\
a_{21}x_1 + a_{22}x_2 + a_{23}x_3 + \ldots + a_{2n}x_n &= b_2 \\
a_{31}x_1 + a_{32}x_2 + a_{33}x_3 + \ldots + a_{3n}x_n &= b_3 \\
&\vdots \\
a_{n1}x_1 + a_{n2}x_2 + a_{n3}x_3 + \ldots + a_{nn}x_n &= b_n
\end{aligned}
$$

where $a_{11}, a_{12}, a_{13}, \ldots a_{nn}$ and $b_1, b_2, b_3, \ldots b_n$ are known constants, and $x_1, x_2, x_3, \ldots x_n$ are required to be found. The above system of equations can be expressed more elegantly in matrix notation as follows:

$$
\begin{bmatrix}
a_{11} & a_{12} & a_{13} & \ldots & a_{1n} \\
a_{21} & a_{22} & a_{23} & \ldots & a_{2n} \\
a_{31} & a_{32} & a_{33} & \ldots & a_{3n} \\
\vdots & & & & \vdots \\
a_{n1} & a_{n2} & a_{n3} & \ldots & a_{nn}
\end{bmatrix}
\begin{bmatrix} x_1 \\ x_2 \\ x_3 \\ \vdots \\ x_n \end{bmatrix}
=
\begin{bmatrix} b_1 \\ b_2 \\ b_3 \\ \vdots \\ b_n \end{bmatrix}
$$

or

$$A X = B$$

where
- A is the coefficient matrix.
- X is the vector of unknowns (the solution vector).
- B is the right-hand side vector.

SIMULTANEOUS LINEAR EQUATIONS

The analysis of linearly elastic structures using the stiffness method produces the system of linear equations $K \Delta = P$ which must be solved simultaneously. The methods used to solve systems of linear equations are classed either as direct methods or iterative methods. In section 3.7 the structure stiffness matrix was seen to have the following properties:

1. It is symmetrical.
2. It is banded.
3. It is positive definite.

These properties make the use of direct methods preferable to iterative methods, and consequently only direct methods will be considered in this text.

4.2 GAUSS-JORDAN ELIMINATION - Program GAUSSJ.EQ

Gauss-Jordan elimination is a direct method that is well suited to digital computation and is found in many stock codes for the solution of simultaneous linear equations. These codes usually employ row interchange to increase numerical accuracy by using the biggest available pivot in the elimination procedure. The stiffness method, however, produces large elements on the diagonal, and this renders row interchange unnecessary, which is fortunate since row interchange destroys banding. In essence Gauss-Jordan elimination solves the system of linear equations by reducing the coefficient matrix to the unit matrix. The method, without row interchange, is illustrated in the following example. Normalisation is used to avoid the pivots becoming uncomfortably large or small (i.e. the pivotal row is divided by the pivot prior to elimination).

Example 4.1

Use Gauss-Jordan elimination to solve this system of equations:

$$\begin{bmatrix} 2 & 1 & 0 \\ 1 & 3 & -2 \\ 0 & -2 & 5 \end{bmatrix} \begin{bmatrix} x_1 \\ x_2 \\ x_3 \end{bmatrix} = \begin{bmatrix} 4 \\ 1 \\ 11 \end{bmatrix}$$

First pivot   -   Divide row 1 by the pivot (the diagonal element)

$$\checkmark \quad \begin{bmatrix} 1 & 0.5 & 0 \\ 1 & 3 & -2 \\ 0 & -2 & 5 \end{bmatrix} \begin{bmatrix} x_1 \\ x_2 \\ x_3 \end{bmatrix} = \begin{bmatrix} 2 \\ 1 \\ 11 \end{bmatrix}$$

Subtract   -   1 x row 1 from row 2
                      0 x row 1 from row 3

$$\begin{bmatrix} 1 & 0.5 & 0 \\ 0 & 2.5 & -2 \\ 0 & -2 & 5 \end{bmatrix} \begin{bmatrix} x_1 \\ x_2 \\ x_3 \end{bmatrix} = \begin{bmatrix} 2 \\ -1 \\ 11 \end{bmatrix}$$

Second pivot   -   Divide row 2 by the pivot

$$\begin{bmatrix} 1 & 0.5 & 0 \\ 0 & 1 & -0.8 \\ 0 & -2 & 5 \end{bmatrix} \begin{bmatrix} x_1 \\ x_2 \\ x_3 \end{bmatrix} = \begin{bmatrix} 2.0 \\ -0.4 \\ 11.0 \end{bmatrix}$$

Subtract   -   0.5 x row 2 from row 1
                      -2 x row 2 from row 3

$$\begin{bmatrix} 1 & 0 & 0.4 \\ 0 & 1 & -0.8 \\ 0 & 0 & 3.4 \end{bmatrix} \begin{bmatrix} x_1 \\ x_2 \\ x_3 \end{bmatrix} = \begin{bmatrix} 2.2 \\ -0.4 \\ 10.2 \end{bmatrix}$$

Third pivot   -   Divide row 3 by the pivot

$$\begin{bmatrix} 1 & 0 & 0.4 \\ 0 & 1 & -0.8 \\ 0 & 0 & 1 \end{bmatrix} \begin{bmatrix} x_1 \\ x_2 \\ x_3 \end{bmatrix} = \begin{bmatrix} 2.2 \\ -0.4 \\ 3.0 \end{bmatrix}$$

Subtract   -   0.4 x row 3 from row 1
                      -0.8 x row 3 from row 2

$$\begin{bmatrix} 1 & 0 & 0 \\ 0 & 1 & 0 \\ 0 & 0 & 1 \end{bmatrix} \begin{bmatrix} x_1 \\ x_2 \\ x_3 \end{bmatrix} = \begin{bmatrix} 1 \\ 2 \\ 3 \end{bmatrix}$$

Which yields

$$\begin{bmatrix} x_1 \\ x_2 \\ x_3 \end{bmatrix} = \begin{bmatrix} 1 \\ 2 \\ 3 \end{bmatrix}$$

This example has shown how each diagonal element is used to eliminate the off-diagonal elements above and below it.

SIMULTANEOUS LINEAR EQUATIONS

Program GAUSSJ.EQ

This program solves a system of linear simultaneous equations by Gauss-Jordan elimination with no row interchange. Solution is for a single right-hand side.

Input Coding is simplified if the coefficient matrix and the right-hand side vector(s) are stored in one array, which is called the augmented matrix. The augmented matrix for the system of equations solved in the previous example is

2	1	0	4
1	3	-2	1
0	-2	5	11

Comments on the Algorithm Note that if a zero pivot is encountered then division by zero occurs and the algorithm breaks down. Singularity and ill-conditioning are discussed in section 4.7.

Output The input data is echoed and after solution the solution vector is output.

Listing

```
1000 REM ***********************************************************************
1010 REM *                                                                     *
1020 REM *     PROGRAM    G A U S S J . E Q    TO SOLVE LINEAR SIMULTANEOUS    *
1030 REM *     EQUATIONS BY GAUSS-JORDAN ELIMINATION WITH NO ROW INTERCHANGE   *
1040 REM *                                                                     *
1050 REM *     J.A.D.BALFOUR                                                   *
1060 REM *                                                                     *
1070 REM ***********************************************************************
1080 REM
1090 REM --- NOTE THAT VARIABLES ARE DEFINED IN APPENDIX A
1100 REM
1110 OPTION BASE 1
1120 DIM A(20,21)
1130 REM
1140 REM --- MAXIMUM NUMBER OF EQUATIONS IS SET BY ABOVE DIMENSION STATEMENT
1150 REM
1160 PRINT : PRINT
1170 PRINT "-------------------------------------------------------------"
1180 PRINT
1190 PRINT "PROGRAM    G A U S S J . E Q    TO SOLVE LINEAR SIMULTANEOUS"
1200 PRINT "EQUATIONS BY GAUSS-JORDAN ELIMINATION WITH NO ROW INTERCHANGE"
1210 PRINT
1220 PRINT "-------------------------------------------------------------"
1230 PRINT : INPUT "NUMBER OF EQUATIONS  =  "; NEQ
1240 PRINT : PRINT "INPUT THE AUGMENTED MATRIX" : PRINT
1250 FOR I = 1 TO NEQ
1260 FOR J = 1 TO NEQ+1
1270 PRINT "ELEMENT ("; I; ","; J; ") = "; : INPUT ""; A(I,J)
1280 NEXT J
1290 NEXT I
1300 REM
1310 REM --- ECHO THE AUGMENTED MATRIX
1320 REM
```

```
1330 PRINT : PRINT
1340 PRINT "+ + + + + + + + + + +"
1350 PRINT "+     AUGMENTED MATRIX   +"
1360 PRINT "+ + + + + + + + + + +"
1370 PRINT
1380 FOR I = 1 TO NEQ
1390 FOR J = 1 TO NEQ+1
1400 PRINT A(I,J),
1410 NEXT J
1420 PRINT : PRINT
1430 NEXT I
1440 REM
1450 REM  ---  CALL SUBROUTINE TO SOLVE THE EQUATIONS
1460 REM
1470 GOSUB 1700
1480 REM
1490 REM  ---  PRINT SOLUTION
1500 REM
1510 PRINT
1520 PRINT "* * * * * * * * * * * *"
1530 PRINT "*    SOLUTION VECTOR    *"
1540 PRINT "* * * * * * * * * * * *"
1550 PRINT
1560 FOR I          = 1 TO NEQ
1570 PRINT A(I,NEQ+1) : PRINT
1580 NEXT I
1590 PRINT
1600 END
1610 REM  * * * * * * * * * * * * * * * * * * * * * * * * * * * * * *
1620 REM  *                                                          *
1630 REM  *    SUBROUTINE TO SOLVE LINEAR SIMULTANEOUS EQUATIONS     *
1640 REM  *    BY GAUSS-JORDAN ELIMINATION WITH NO ROW INTERCHANGE   *
1650 REM  *                                                          *
1660 REM  * * * * * * * * * * * * * * * * * * * * * * * * * * * * * *
1670 REM
1680 REM  ---  LOOP FOR ALL PIVOTS
1690 REM
1700 FOR I    = 1 TO NEQ
1710 TEMP     = A(I,I)
1720 REM
1730 REM  ---  NORMALISE
1740 REM
1750 FOR J    = 1 TO NEQ+1
1760 A(I,J)   = A(I,J) / TEMP
1770 NEXT J
1780 REM
1790 REM  ---  LOOP FOR ALL ROWS USING THE CURRENT PIVOT
1800 REM
1810 FOR J    = 1 TO NEQ
1820 IF J     = I THEN GOTO 1900
1830 TEMP     = A(J,I)
1840 REM
1850 REM  ---  LOOP FOR ALL ELEMENTS IN THE ROW
1860 REM
1870 FOR K    = 1 TO NEQ+1
1880 A(J,K)   = A(J,K) - TEMP*A(I,K)
1890 NEXT K
1900 NEXT J
1910 NEXT I
1920 RETURN
```

SIMULTANEOUS LINEAR EQUATIONS

Sample Run The following sample run shows how example 4.1 would be solved using the program GAUSSJ.EQ. Input from the keyboard is shown in bold typeface.

```
-------------------------------------------------------------
   PROGRAM      G A U S S J . E Q    TO SOLVE LINEAR SIMULTANEOUS
   EQUATIONS BY GAUSS-JORDAN ELIMINATION WITH NO ROW INTERCHANGE
-------------------------------------------------------------

   NUMBER OF EQUATIONS = 3

   INPUT THE AUGMENTED MATRIX

   ELEMENT ( 1 , 1 ) = 2
   ELEMENT ( 1 , 2 ) = 1
   ELEMENT ( 1 , 3 ) = 0
   ELEMENT ( 1 , 4 ) = 4
   ELEMENT ( 2 , 1 ) = 1
   ELEMENT ( 2 , 2 ) = 3
   ELEMENT ( 2 , 3 ) = -2
   ELEMENT ( 2 , 4 ) = 1
   ELEMENT ( 3 , 1 ) = 0
   ELEMENT ( 3 , 2 ) = -2
   ELEMENT ( 3 , 3 ) = 5
   ELEMENT ( 3 , 4 ) = 11


   + + + + + + + + + + +
   +   AUGMENTED MATRIX   +
   + + + + + + + + + + +

        2            1            0            4
        1            3           -2            1
        0           -2            5           11


   * * * * * * * * * * * *
   *    SOLUTION VECTOR    *
   * * * * * * * * * * * *

        1
        2
        3
```

4.3 MULTIPLE RIGHT-HAND SIDES

It is a feature of direct methods that very little extra labour is required to solve for additional right-hand sides.

When analysing structures using the stiffness method the right-hand side vector is the vector of applied loads. Consequently to analyse a structure for a number of loads cases costs very little more than the analysis for a single load case.

If Gauss-Jordan elimination is used then multiple right hand sides are dealt with by operating on all right-hand sides simultaneously. This is best illustrated by an example.

Example 4.2

Solve the following two systems of equations using Gauss-Jordan elimination.

$$\begin{bmatrix} 2 & 1 & 0 \\ 1 & 3 & -2 \\ 0 & -2 & 5 \end{bmatrix} \begin{bmatrix} x_1 \\ x_2 \\ x_3 \end{bmatrix} = \begin{bmatrix} 25 \\ -20 \\ 60 \end{bmatrix}$$

and

$$\begin{bmatrix} 2 & 1 & 0 \\ 1 & 3 & -2 \\ 0 & -2 & 5 \end{bmatrix} \begin{bmatrix} x_1 \\ x_2 \\ x_3 \end{bmatrix} = \begin{bmatrix} -1.0 \\ -8.5 \\ 11.5 \end{bmatrix}$$

The augmented matrix is

$$\begin{bmatrix} 2 & 1 & 0 & 25 & -1.0 \\ 1 & 3 & -2 & -20 & -8.5 \\ 0 & -2 & 5 & 60 & 11.5 \end{bmatrix}$$

First pivot — Divide row 1 by the pivot

$$\begin{bmatrix} 1 & 0.5 & 0 & 12.5 & -0.5 \\ 1 & 3 & -2 & -20 & -8.5 \\ 0 & -2 & 5 & 60 & 11.5 \end{bmatrix}$$

Subtract — 1 x row 1 from row 2
0 x row 1 from row 3

$$\begin{bmatrix} 1 & 0.5 & 0 & 12.5 & -0.5 \\ 0 & 2.5 & -2 & -32.5 & -8.0 \\ 0 & -2 & 5 & 60 & 11.5 \end{bmatrix}$$

Second pivot — Divide row 2 by the pivot

$$\begin{bmatrix} 1 & 0.5 & 0 & 12.5 & -0.5 \\ 0 & 1 & -0.8 & -13.0 & -3.2 \\ 0 & -2 & 5 & 60 & 11.5 \end{bmatrix}$$

Subtract — 0.5 x row 2 from row 1
-2 x row 2 from row 3

$$\begin{bmatrix} 1 & 0 & 0.4 & 19.0 & 1.1 \\ 0 & 1 & -0.8 & -13.0 & -3.2 \\ 0 & 0 & 3.4 & 34.0 & 5.1 \end{bmatrix}$$

SIMULTANEOUS LINEAR EQUATIONS

Third pivot   -   Divide row 3 by the pivot

$$\begin{bmatrix} 1 & 0 & 0.4 & 19.0 & 1.1 \\ 0 & 1 & -0.8 & -13.0 & -3.2 \\ 0 & 0 & 1 & 10.0 & 1.5 \end{bmatrix}$$

Subtract   -   0.4 x row 3 from row 1
              -0.8 x row 3 from row 2

$$\begin{bmatrix} 1 & 0 & 0 & 15.0 & 0.5 \\ 0 & 1 & 0 & -5.0 & -2.0 \\ 0 & 0 & 1 & 10.0 & 1.5 \end{bmatrix}$$

Hence the two solution vectors are

$$\begin{bmatrix} x_1 \\ x_2 \\ x_3 \end{bmatrix} = \begin{bmatrix} 15 \\ -5 \\ 10 \end{bmatrix} \quad \text{and} \quad \begin{bmatrix} x_1 \\ x_2 \\ x_3 \end{bmatrix} = \begin{bmatrix} 0.5 \\ -2.0 \\ 1.5 \end{bmatrix}$$

4.4 GAUSS ELIMINATION - Program GAUSS.EQ

Many commercial structural analysis programs use Gauss elimination to solve the system of equations produced by the stiffness method. The method has two distinct phases:

1. Reduction of the coefficient matrix to triangular form.

2. Back substitution to find the solution vector.

Normalisation is used to avoid the pivots becoming uncomfortably large or small (i.e. the pivotal row is divided by the pivot prior to elimination). These procedures are best illustrated by an example.

Example 4.3

Solve the following system of equations by Gauss elimination:

$$\begin{bmatrix} 2 & 1 & 0 & 0 \\ 1 & 3 & -2 & 0 \\ 0 & -2 & 5 & -1 \\ 0 & 0 & -1 & 3 \end{bmatrix} \begin{bmatrix} x_1 \\ x_2 \\ x_3 \\ x_4 \end{bmatrix} = \begin{bmatrix} 4 \\ 1 \\ 7 \\ 9 \end{bmatrix}$$

First pivot — Normalise row 1 of the augmented matrix, and then subtract
  1 x row 1 from row 2
  0 x row 1 from row 3
  0 x row 1 from row 4

$$\begin{bmatrix} 1 & 0.5 & 0 & 0 & 2 \\ 0 & 2.5 & -2 & 0 & -1 \\ 0 & -2 & 5 & -1 & 7 \\ 0 & 0 & -1 & 3 & 9 \end{bmatrix}$$

Second pivot — Normalise row 2 and then subtract
  -2 x row 2 from row 3
  0 x row 2 from row 4

$$\begin{bmatrix} 1 & 0.5 & 0 & 0 & 2.0 \\ 0 & 1 & -0.8 & 0 & -0.4 \\ 0 & 0 & 3.4 & -1 & 6.2 \\ 0 & 0 & -1 & 3 & 9.0 \end{bmatrix}$$

Third pivot — Normalise row 3 and then subtract
  -1 x row 3 from row 4

$$\begin{bmatrix} 1 & 0.5 & 0 & 0 & 2.0 \\ 0 & 1 & -0.8 & 0 & -0.4 \\ 0 & 0 & 1 & -0.294 & 1.824 \\ 0 & 0 & 0 & 2.706 & 10.824 \end{bmatrix}$$

Fourth pivot — Normalise row 4

$$\begin{bmatrix} 1 & 0.5 & 0 & 0 & 2.0 \\ 0 & 1 & -0.8 & 0 & -0.4 \\ 0 & 0 & 1 & -0.294 & 1.824 \\ 0 & 0 & 0 & 1 & 4.0 \end{bmatrix}$$

The coefficient matrix has now been reduced to triangular form (i.e. all sub-diagonal elements have been eliminated). The solution can now be found by back substitution.

$$\begin{aligned} x_4 &= 4.000 \\ x_3 &= 1.824 - (-0.294)x_4 = 3.000 \\ x_2 &= -0.4 - (-0.8)x_3 - (0)x_4 = 2.000 \\ x_1 &= 2 - (0.5)x_2 - (0)x_3 - (0)x_4 = 1.000 \end{aligned}$$

SIMULTANEOUS LINEAR EQUATIONS

This example illustrates two important characteristics of Gauss elimination:

1. Banding is maintained in the upper triangle
2. The submatrix below the pivot (shown by the broken lines) remains symmetrical.

Advantage can be taken of these facts by storing only the upper semi-band of the coefficient matrix as illustrated in fig 3.15.

Program GAUSS.EQ

This program solves a system of linear simultaneous equations with a symmetrical, banded coefficient matrix by Gauss elimination.

Input Only the upper semi-band of the coefficient matrix is input and the matrix is stored as shown in fig 3.15. The right-hand side vector(s) is stored in a separate array.

Comments on the Algorithm To save space the coefficient matrix is overwritten with the decomposed matrix and the right-hand side vector(s) is overwritten by the solution vector(s) during computation.

Output The upper semi-band of the coefficient matrix and the right-hand side vector(s) are output. After solution the solution vector(s) is output.

Listing

```
1000 REM ****************************************************************
1010 REM *                                                              *
1020 REM *    PROGRAM    G A U S S . E Q    TO SOLVE LINEAR SIMULTANEOUS *
1030 REM *    EQUATIONS BY GAUSS ELIMINATION WITH NO ROW INTERCHANGE     *
1040 REM *                                                              *
1050 REM *    J.A.D.BALFOUR                                             *
1060 REM *                                                              *
1070 REM ****************************************************************
1080 REM
1090 REM  ---  NOTE THAT VARIABLES ARE DEFINED IN APPENDIX A
1100 REM
1110 OPTION BASE 1
1120 DIM A(20,10), B(20,5)
1130 REM
1140 REM  ---  MAXIMUM SEMI-BANDWIDTH, NUMBER OF EQUATIONS, AND NUMBER OF
1150 REM       RIGHT-HAND SIDES SET BY THE ABOVE DIMENSION STATEMENT
1160 REM
1170 PRINT : PRINT
```

```
1180 PRINT "----------------------------------------------------------"
1190 PRINT
1200 PRINT "PROGRAM  G A U S S . E Q  TO SOLVE LINEAR SIMULTANEOUS"
1210 PRINT "EQUATIONS BY GAUSS ELIMINATION WITH NO ROW INTERCHANGE"
1220 PRINT
1230 PRINT "----------------------------------------------------------"
1240 PRINT : PRINT
1250 INPUT "NUMBER OF EQUATIONS        = "; NEQ
1260 INPUT "SEMI-BANDWIDTH             = "; BAND
1270 INPUT "NUMBER OF RIGHT-HAND SIDES = "; NR
1280 PRINT
1290 PRINT "ENTER THE UPPER SEMI-BAND OF THE COEFFICIENT MATRIX"
1300 PRINT
1310 FOR I = 1 TO NEQ
1320 FOR J = 1 TO BAND
1330 K     = I + J - 1
1340 IF K > NEQ THEN GOTO 1370
1350 PRINT "ELEMENT("; I; ","; K; ")  =  "; : INPUT ""; A(I,J)
1360 NEXT J
1370 NEXT I
1380 REM
1390 REM  ---  INPUT THE RIGHT-HAND SIDE VECTOR(S)
1400 REM
1410 FOR J = 1 TO NR
1420 PRINT
1430 PRINT "ENTER RIGHT-HAND SIDE VECTOR NUMBER "; J
1440 PRINT
1450 FOR I = 1 TO NEQ
1460 PRINT "ELEMENT("; I; ")  =  "; : INPUT ""; B(I,J)
1470 NEXT I
1480 NEXT J
1490 REM
1500 REM  ---  ECHO COEFFICIENT MATRIX
1510 REM
1520 PRINT : PRINT
1530 PRINT "+ + + + + + + + + + + + + + + + + + + + + + + + +"
1540 PRINT "+     UPPER SEMI-BAND OF THE COEFFICIENT MATRIX  +"
1550 PRINT "+ + + + + + + + + + + + + + + + + + + + + + + + +"
1560 PRINT
1570 FOR I = 1 TO NEQ
1580 FOR J = 1 TO BAND
1590 PRINT A(I,J),
1600 NEXT J
1610 PRINT : PRINT
1620 NEXT I
1630 REM
1640 REM  ---  ECHO RIGHT-HAND SIDE VECTOR(S)
1650 REM
1660 PRINT
1670 PRINT "+ + + + + + + + + + + + + + + +"
1680 PRINT "+    RIGHT-HAND SIDE VECTOR(S)    +"
1690 PRINT "+ + + + + + + + + + + + + + + +"
1700 PRINT
1710 FOR I = 1 TO NEQ
1720 FOR J = 1 TO NR
1730 PRINT B(I,J),
1740 NEXT J
1750 PRINT : PRINT
1760 NEXT I
1770 REM
1780 REM  ---  CALL SUBROUTINE TO SOLVE THE EQUATIONS
1790 REM
1800 GOSUB 2050
1810 REM
1820 REM  ---  PRINT SOLUTION VECTOR(S)
1830 REM
1840 PRINT
1850 PRINT "* * * * * * * * * * * * *"
1860 PRINT "*    SOLUTION VECTOR(S)    *"
1870 PRINT "* * * * * * * * * * * * *"
1880 PRINT
1890 FOR I = 1 TO NEQ
1900 FOR J = 1 TO NR
```

SIMULTANEOUS LINEAR EQUATIONS

```
1910 PRINT B(I,J),
1920 NEXT J
1930 PRINT : PRINT
1940 NEXT I
1950 END
1960 REM * * * * * * * * * * * * * * * * * * * * * * * * * * * * * * * * *
1970 REM  *                                                                *
1980 REM  *      SUBROUTINE TO SOLVE LINEAR SIMULTANEOUS EQUATIONS BY GAUSS *
1990 REM  *      ELIMINATION WITH NO ROW INTERCHANGE                       *
2000 REM  *                                                                *
2010 REM * * * * * * * * * * * * * * * * * * * * * * * * * * * * * * * * *
2020 REM
2030 REM  --- LOOP FOR ALL PIVOTS
2040 REM
2050 BB       = BAND
2060 FOR I    = 1 TO NEQ
2070 REM
2080 REM  --- CHECK IF IN THE UNUSED TRIANGLE
2090 REM
2100 IF I > NEQ-BAND+1 THEN BB = NEQ-I+1
2110 PIVOT    = A(I,1)
2120 REM
2130 REM  --- NORMALISE
2140 REM
2150 FOR J    = 1 TO BB
2160 A(I,J) = A(I,J) / PIVOT
2170 NEXT J
2180 FOR J    = 1 TO NR
2190 B(I,J) = B(I,J) / PIVOT
2200 NEXT J
2210 REM
2220 REM  --- CHECK IF LAST ROW
2230 REM
2240 IF BB = 1 THEN GOTO 2450
2250 REM
2260 REM  --- ELIMINATE FOR ALL ROWS ABOVE PIVOT AND WITHIN BAND
2270 REM
2280 FOR K    = 2 TO BB
2290 REM
2300 REM  --- CALCULATE ROW NUMBER THEN EVALUATE MULTIPLIER
2310 REM
2320 L        = I + K - 1
2330 MULT     = A(I,K) * PIVOT
2340 REM
2350 REM  --- LOOP FOR ELEMENTS IN THE ELIMINATION ROW
2360 REM
2370 FOR J    = K TO BB
2380 M        = J - K + 1
2390 A(L,M) = A(L,M) - MULT * A(I,J)
2400 NEXT J
2410 FOR J    = 1 TO NR
2420 B(L,J) = B(L,J) - MULT*B(I,J)
2430 NEXT J
2440 NEXT K
2450 NEXT I
2460 REM
2470 REM  --- BACK SUBSTITUTE
2480 REM
2490 FOR I    = 1 TO NEQ-1
2500 BB       = 1
2510 IF I > BAND-1 THEN BB = BAND-1
2520 FOR K    = 1 TO NR
2530 FOR J    = 1 TO BB
2540 B(NEQ-I,K) = B(NEQ-I,K) - A(NEQ-I,J+1)*B(NEQ-I+J,K)
2550 NEXT J
2560 NEXT K
2570 NEXT I
2580 RETURN
```

Sample Run The sample run that follows shows the program GAUSS.EQ being used to solve example 4.3. Input from the keyboard is shown in bold typeface.

```
---------------------------------------------------------
PROGRAM  G A U S S . E Q  TO SOLVE LINEAR SIMULTANEOUS
EQUATIONS BY GAUSS ELIMINATION WITH NO ROW INTERCHANGE
---------------------------------------------------------

NUMBER OF EQUATIONS       = 4
SEMI-BANDWIDTH            = 2
NUMBER OF RIGHT-HAND SIDES = 1

ENTER THE UPPER SEMI-BAND OF THE COEFFICIENT MATRIX

ELEMENT( 1 , 1 )  =  2
ELEMENT( 1 , 2 )  =  1
ELEMENT( 2 , 2 )  =  3
ELEMENT( 2 , 3 )  =  -2
ELEMENT( 3 , 3 )  =  5
ELEMENT( 3 , 4 )  =  -1
ELEMENT( 4 , 4 )  =  3

ENTER RIGHT-HAND SIDE VECTOR NUMBER  1

ELEMENT( 1 )  =  4
ELEMENT( 2 )  =  1
ELEMENT( 3 )  =  7
ELEMENT( 4 )  =  9


+ + + + + + + + + + + + + + + + + + + + + + + + +
+     UPPER SEMI-BAND OF THE COEFFICIENT MATRIX  +
+ + + + + + + + + + + + + + + + + + + + + + + + +

   2              1

   3             -2

   5             -1

   3              0

+ + + + + + + + + + + + + + + + +
+     RIGHT-HAND SIDE VECTOR(S)  +
+ + + + + + + + + + + + + + + + +

   4

   1

   7

   9


* * * * * * * * * * * * *
*    SOLUTION VECTOR(S)  *
* * * * * * * * * * * * *

   1

   2

   3

   4
```

SIMULTANEOUS LINEAR EQUATIONS

4.5 TRIANGULAR DECOMPOSITION - Program UDU.EQ

Crout reduction decomposes a non-singular coefficient matrix to the product of two triangular matrices, as shown in fig 4.1.

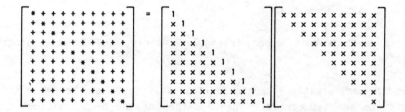

FIGURE 4.1   CROUT REDUCTION

In matrix notation

$$A = L U$$

where **L** is a unit lower triangular matrix and **U** is an upper triangular matrix. The solution algorithm is as follows:

1. Use triangular decomposition to express the coefficient matrix as a product of two triangular matrices. Hence the system of equations **A X = B** becomes

    $$L U X = B \qquad (4.1)$$

2. Let

    $$U X = Y \qquad (4.2)$$

    Hence equation (4.1) can be rewritten as follows:

    $$L Y = B \qquad (4.3)$$

3. Following triangular decomposition, **Y** is found by forward substitution (equation (4.3)). **X** is then found by backward substitution (equation (4.2)).

If the coefficient matrix **A** is symmetrical and positive definite then it can be decomposed into a product of an upper triangular matrix and its transpose, i.e.

$$A = \tilde{U}^T \tilde{U} \qquad (4.4)$$

where $\tilde{U}$ is an upper triangular matrix.

This technique is known as Cholesky decomposition. An alternative decomposition is:

$$A = U^T D U \qquad (4.5)$$

where U is a unit upper triangular matrix, and D is a diagonal matrix.

For linear elastic structures the stiffness matrix is positive definite and symmetrical. Hence either of the decompositions shown in equations (4.4) and (4.5) can be employed. $U^T D U$ decomposition is slightly more efficient than $\tilde{U}^T \tilde{U}$ decomposition. This chapter will consider only $U^T D U$ decomposition.

Let
$$A X = U^T D U X = B$$

then
$$U X = Y \qquad (4.6)$$

$$U^T D Y = B \qquad (4.7)$$

Once A has been decomposed to $U^T D U$, Y is found by forward substitution (equation (4.7)). Finally the solution vector X is found by backward substitution (equation (4.6)).

The $U^T D U$ decomposition algorithm is as follows:

Decomposition

$d_{ij} = 0$ if $i \neq j$
$\phantom{d_{ij}} = a_{ii}$ if $i = 1$ and $j = 1$
$\phantom{d_{ij}} = a_{ii} - \sum u_{ki}^2 d_k$ if $i > 1$ and $j = i$
     summation is for $k = 1$ to $(i-1)$

$u_{ij} = 0$ if $i > j$
$\phantom{u_{ij}} = 1$ if $i = j$
$\phantom{u_{ij}} = a_{ij} / a_{ii}$ if $i = 1$ and $j > i$
$\phantom{u_{ij}} = (a_{ij} - \sum u_{ki} u_{kj} d_k) / d_i$ if $i > 1$ and $j > i$
     summation is for $k = 1$ to $(i-1)$

SIMULTANEOUS LINEAR EQUATIONS

Forward Substitution

$$y_i = b_i / d_i \qquad \text{if } i = 1$$
$$= (b_i - \sum u_{ki} d_k y_k) / d_i \qquad \text{if } i > 1$$
summation is for $k = 1$ to $(i-1)$

Backward Substitution

$$x_i = y_i \qquad \text{if } i = n$$
$$= y_i - \sum u_{i,k} x_k \qquad \text{if } i < n$$
summation is for $k = (i+1)$ to n

Example 4.4

Decompose the matrix shown below using $U^T D\, U$ decomposition.

$$\begin{bmatrix} 2 & 1 & 1 & 0 \\ 1 & 6 & 2 & 1 \\ 1 & 2 & 7 & 2 \\ 0 & 1 & 2 & 4 \end{bmatrix}$$

$$d_1 = 2$$
$$u_{12} = 1/2 = 0.5$$
$$u_{13} = 1/2 = 0.5$$
$$u_{14} = 0/2 = 0.0$$

$$d_2 = 6 - 0.5^2 \times 2 = 5.50$$
$$u_{23} = (2 - 0.5 \times 0.5 \times 2) / 5.5 = 0.273$$
$$u_{24} = (1 - 0.5 \times 0.0 \times 2) / 5.5 = 0.181$$

$$d_3 = 7 - 0.5^2 \times 2 - 0.273^2 \times 5.5$$
$$= 6.090$$
$$u_{34} = (2 - 0.5 \times 0 \times 2 - 0.273 \times 0.181 \times 5.5) / 6.090$$
$$= 0.284$$

$$d_4 = 4 - 0.0^2 \times 2 - 0.181^2 \times 5.5 - 0.284^2 \times 6.090$$
$$= 3.329$$

Note that the composition of a "n x n" coefficient matrix involves "n" major steps. Step "i" entails finding d_{ii} and row "i" of **U**.

Check that $U^T D U = A$

$$\begin{bmatrix} 1 & 0 & 0 & 0 \\ 0.5 & 1 & 0 & 0 \\ 0.5 & 0.273 & 1 & 0 \\ 0 & 0.181 & 0.284 & 1 \end{bmatrix} \begin{bmatrix} 2 & 0 & 0 & 0 \\ 0 & 5.5 & 0 & 0 \\ 0 & 0 & 6.090 & 0 \\ 0 & 0 & 0 & 3.329 \end{bmatrix} \begin{bmatrix} 1 & 0.5 & 0.5 & 0 \\ 0 & 1 & 0.273 & 0.181 \\ 0 & 0 & 1 & 0.284 \\ 0 & 0 & 0 & 1 \end{bmatrix} =$$

$$\begin{bmatrix} 2 & 0 & 0 & 0 \\ 1 & 5.5 & 0 & 0 \\ 1 & 1.502 & 6.090 & 0 \\ 0 & 0.996 & 1.730 & 3.329 \end{bmatrix} \begin{bmatrix} 1 & 0.5 & 0.5 & 0 \\ 0 & 1 & 0.273 & 0.181 \\ 0 & 0 & 1 & 0.284 \\ 0 & 0 & 0 & 1 \end{bmatrix} = \begin{bmatrix} 2.000 & 1.000 & 1.000 & 0.000 \\ 1.000 & 6.000 & 2.002 & 0.996 \\ 1.000 & 2.002 & 7.000 & 2.001 \\ 0.000 & 0.996 & 2.002 & 4.001 \end{bmatrix}$$

Example 4.5

Using the decomposed matrix from example 4.4 solve the following system of equations.

$$\begin{bmatrix} 2 & 1 & 1 & 0 \\ 1 & 6 & 2 & 1 \\ 1 & 2 & 7 & 2 \\ 0 & 1 & 2 & 4 \end{bmatrix} \begin{bmatrix} x_1 \\ x_2 \\ x_3 \\ x_4 \end{bmatrix} = \begin{bmatrix} 1.5 \\ -5.5 \\ 16.0 \\ 7.0 \end{bmatrix}$$

i.e. $A X = B$, which becomes $U^T D U X = B$ on decomposition

$U X = Y$, hence $U^T D Y = B$. Using U^T and D from example 4.4 allows Y to be found by forward substitution as follows:

$$\begin{bmatrix} 2 & 0 & 0 & 0 \\ 1 & 5.5 & 0 & 0 \\ 1 & 1.502 & 6.090 & 0 \\ 0 & 0.996 & 1.730 & 1 \end{bmatrix} \begin{bmatrix} y_1 \\ y_2 \\ y_3 \\ y_4 \end{bmatrix} = \begin{bmatrix} 1.5 \\ -5.5 \\ 16.0 \\ 7.0 \end{bmatrix}$$

$y_1 = 1.5 / 2$ = 0.750
$y_2 = (-5.5 - y_1) / 5.5$ = -1.136
$y_3 = (16.0 - y_1 - 1.502 \, y_2) / 6.090$ = 2.784
$y_4 = (7.0 - 0.996 \, y_2 - 1.730 \, y_3) / 3.329$ = 0.996

Finally the solution is found by backward substitution of $U X = Y$

$$\begin{bmatrix} 1 & 0.5 & 0.5 & 0 \\ 0 & 1 & 0.273 & 0.181 \\ 0 & 0 & 1 & 0.284 \\ 0 & 0 & 0 & 1 \end{bmatrix} \begin{bmatrix} x_1 \\ x_2 \\ x_3 \\ x_4 \end{bmatrix} = \begin{bmatrix} 0.750 \\ -1.136 \\ 2.784 \\ 0.996 \end{bmatrix}$$

SIMULTANEOUS LINEAR EQUATIONS

which yields

$$x_4 = 0.996$$
$$x_3 = 2.784 - 0.284\, x_4 = 2.501$$
$$x_2 = -1.136 - 0.181\, x_4 - 0.273\, x_3 = -1.992$$
$$x_1 = 0.750 - 0.000\, x_4 - 0.500\, x_3 - 0.500\, x_2 = 0.496$$

Program UDU.EQ

This program solves a system of linear simultaneous equations with a symmetrical, positive definite, banded coefficient matrix by $U^T D\, U$ decomposition.

Input Only the upper semi-band of the coefficient matrix is input and the matrix is stored as shown in fig 3.15. The right-hand side vector(s) is stored in a separate array.

Comments on the Algorithm To save space the coefficient matrix is overwritten with the decomposed matrix and the right-hand side vector(s) is overwritten by the solution vector(s) during computation.

Output The upper semi-band of the coefficient matrix and the right-hand side vector(s) are output. Finally the solution vector(s) is output.

```
1000 REM  ****************************************************************
1010 REM  *                                                              *
1020 REM  *     PROGRAM    U D U . E Q     TO SOLVE LINEAR SIMULTANEOUS  *
1030 REM  *     EQUATIONS BY TRIANGULAR DECOMPOSITION                    *
1040 REM  *                                                              *
1050 REM  *     J.A.D.BALFOUR                                            *
1060 REM  *                                                              *
1070 REM  ****************************************************************
1080 REM
1090 REM  ---  NOTE THAT VARIABLES ARE DEFINED IN APPENDIX A
1100 REM
1110 OPTION BASE 1
1120 DIM A(20,10), B(20,5)
1130 REM
1140 REM  ---  MAXIMUM SEMI-BANDWIDTH, NUMBER OF EQUATIONS, AND NUMBER OF
1150 REM       RIGHT-HAND SIDES ARE SET BY THE ABOVE DIMENSION STATEMENT
1160 REM
1170 PRINT : PRINT
1180 PRINT "----------------------------------------------------------------"
1190 PRINT
1200 PRINT "PROGRAM        U D U . E Q      TO SOLVE LINEAR SIMULTANEOUS"
1210 PRINT "EQUATIONS BY TRIANGULAR DECOMPOSITION WITH NO ROW INTERCHANGE"
1220 PRINT
1230 PRINT "----------------------------------------------------------------"
1240 PRINT : PRINT
1250 INPUT "NUMBER OF EQUATIONS      = ", NEQ
1260 INPUT "SEMI-BANDWIDTH           = ", BAND
1270 INPUT "NUMBER OF RIGHT-HAND SIDES = ", NR
1280 PRINT
1290 PRINT "ENTER THE UPPER SEMI-BAND OF THE COEFFICIENT MATRIX "
```

```
1300 PRINT
1310 FOR I = 1 TO NEQ
1320 FOR J = 1 TO BAND
1330 K     = I + J - 1
1340 IF K > NEQ THEN GOTO 1370
1350 PRINT "ELEMENT("; I; ","; K; ") = "; : INPUT "", A(I,J)
1360 NEXT J
1370 NEXT I
1380 REM
1390 REM --- INPUT THE RIGHT-HAND SIDE VECTOR(S)
1400 REM
1410 FOR J = 1 TO NR
1420 PRINT : PRINT "ENTER RIGHT-HAND SIDE VECTOR NUMBER "; J
1430 PRINT
1440 FOR I = 1 TO NEQ
1450 PRINT "ELEMENT("; I; ") = "; : INPUT "", B(I,J)
1460 NEXT I
1470 NEXT J
1480 REM
1490 REM --- ECHO COEFFICIENT MATRIX
1500 REM
1510 PRINT : PRINT
1520 PRINT "+ + + + + + + + + + + + + + + + + + + + + + + +"
1530 PRINT "+    UPPER SEMI-BAND OF THE COEFFICIENT MATRIX    +"
1540 PRINT "+ + + + + + + + + + + + + + + + + + + + + + + +"
1550 PRINT
1560 FOR I = 1 TO NEQ
1570 FOR J = 1 TO BAND
1580 PRINT A(I,J),
1590 NEXT J
1600 PRINT : PRINT
1610 NEXT I
1620 REM
1630 REM --- ECHO RIGHT-HAND SIDE VECTOR(S)
1640 REM
1650 PRINT
1660 PRINT "+ + + + + + + + + + + + + + + +"
1670 PRINT "+    RIGHT-HAND SIDE VECTOR(S)    +"
1680 PRINT "+ + + + + + + + + + + + + + + +"
1690 PRINT
1700 FOR I = 1 TO NEQ
1710 FOR J = 1 TO NR
1720 PRINT B(I,J),
1730 NEXT J
1740 PRINT : PRINT
1750 NEXT I
1760 REM
1770 REM --- CALL SUBROUTINE TO DECOMPOSE THE COEFFICIENT MATRIX
1780 REM
1790 GOSUB 2080
1800 REM
1810 REM --- CALL SUBROUTINE TO SOLVE THE EQUATIONS
1820 REM
1830 GOSUB 2420
1840 REM
1850 REM --- PRINT SOLUTION VECTOR(S)
1860 REM
1870 PRINT
1880 PRINT "* * * * * * * * * * * * * *"
1890 PRINT "*    SOLUTION VECTOR(S)    *"
1900 PRINT "* * * * * * * * * * * * * *"
1910 PRINT
1920 FOR I = 1 TO NEQ
1930 FOR J = 1 TO NR
1940 PRINT B(I,J),
1950 NEXT J
1960 PRINT : PRINT
1970 NEXT I
1980 END
```

SIMULTANEOUS LINEAR EQUATIONS

```
1990 REM   * * * * * * * * * * * * * * * * * * * * * * * * * * * * * * * * * *
2000 REM   *                                                                  *
2010 REM   *    SUBROUTINE TO DECOMPOSE A SYMMETRICAL, BANDED, POSITIVE       *
2020 REM   *    DEFINITE COEFFICIENT MATRIX USING TRIANGULAR DECOMPOSITION    *
2030 REM   *                                                                  *
2040 REM   * * * * * * * * * * * * * * * * * * * * * * * * * * * * * * * * * *
2050 REM
2060 REM   ---   LOOP FOR ALL ROWS
2070 REM
2080 FOR I  = 1 TO NEQ
2090 IF I   = 1 THEN GOTO 2210
2100 BB     = BAND - 1
2110 IF I   < BAND THEN BB = I-1
2120 REM
2130 REM   ---   EVALUATE THE DIAGONAL ELEMENT
2140 REM
2150 FOR L  = 1 TO BB
2160 A(I,1) = A(I,1) - A(I-L,L+1)*A(I-L,L+1)*A(I-L,1)
2170 NEXT L
2180 REM
2190 REM   ---   EVALUATE THE OFF-DIAGONAL ELEMENTS
2200 REM
2210 FOR J  = 2 TO BAND
2220 IF I   = 1 THEN GOTO 2280
2230 BB     = BAND - J
2240 IF I   < BAND THEN BB = I-1
2250 FOR L  = 1 TO BB
2260 A(I,J) = A(I,J) - A(I-L,L+1)*A(I-L,J+L)*A(I-L,1)
2270 NEXT L
2280 A(I,J) = A(I,J) / A(I,1)
2290 NEXT J
2300 NEXT I
2310 RETURN
2320 REM   * * * * * * * * * * * * * * * * * * * * * * * * * * * * * * * * * *
2330 REM   *                                                                  *
2340 REM   *    SUBROUTINE TO SOLVE A SYSTEM OF LINEAR SIMULTANEOUS EQUATIONS *
2350 REM   *    BY FORWARD AND BACKWARD SUBSTITUTION USING A COEFFICIENT     *
2360 REM   *    MATRIX ALREADY DECOMPOSED BY TRIANGULAR DECOMPOSITION        *
2370 REM   *                                                                  *
2380 REM   * * * * * * * * * * * * * * * * * * * * * * * * * * * * * * * * * *
2390 REM
2400 REM   ---   FORWARD SUBSTITUTION
2410 REM
2420 FOR I  = 1 TO NEQ
2430 BB     = I-1
2440 IF I   > BAND-1 THEN BB = BAND-1
2450 FOR J  = 1 TO NR
2460 SUM    = 0
2470 IF I   = 1 THEN GOTO 2510
2480 FOR L  = 1 TO BB
2490 SUM    = SUM + A(I-L,L+1)*A(I-L,1)*B(I-L,J)
2500 NEXT L
2510 B(I,J) = (B(I,J) - SUM) / A(I,1)
2520 NEXT J
2530 NEXT I
2540 REM
2550 REM   ---   BACKWARD SUBSTITUTION
2560 REM
2570 FOR I = 1 TO NEQ-1
2580 BB     = I
2590 IF I   > BAND-1 THEN BB = BAND-1
2600 FOR J = 1 TO NR
2610 FOR L = 1 TO BB
2620 B(NEQ-I,J) = B(NEQ-I,J) - A(NEQ-I,L+1)*B(NEQ-I+L,J)
2630 NEXT L
2640 NEXT J
2650 NEXT I
2660 RETURN
```

COMPUTER ANALYSIS OF STRUCTURAL FRAMEWORKS

Sample Run There follows a sample run that shows the program UDU.EQ being used to solve example 4.5. Input from the keyboard is shown in bold typeface.

```
-------------------------------------------------------------
PROGRAM          U D U . E Q      TO SOLVE LINEAR SIMULTANEOUS
EQUATIONS BY TRIANGULAR DECOMPOSITION WITH NO ROW INTERCHANGE
-------------------------------------------------------------

NUMBER OF EQUATIONS         =  4
SEMI-BANDWIDTH              =  3
NUMBER OF RIGHT-HAND SIDES  =  1
ENTER THE UPPER SEMI-BAND OF THE COEFFICIENT MATRIX

ELEMENT( 1 , 1 )  =  2
ELEMENT( 1 , 2 )  =  1
ELEMENT( 1 , 3 )  =  1
ELEMENT( 2 , 2 )  =  6
ELEMENT( 2 , 3 )  =  2
ELEMENT( 2 , 4 )  =  1
ELEMENT( 3 , 3 )  =  7
ELEMENT( 3 , 4 )  =  2
ELEMENT( 4 , 4 )  =  4

ENTER RIGHT-HAND SIDE VECTOR NUMBER  1

ELEMENT( 1 )  =  1.5
ELEMENT( 2 )  =  -5.5
ELEMENT( 3 )  =  16
ELEMENT( 4 )  =  7

+ + + + + + + + + + + + + + + + + + + + + + + + +
+    UPPER SEMI-BAND OF THE COEFFICIENT MATRIX   +
+ + + + + + + + + + + + + + + + + + + + + + + + +

  2            1            1

  6            2            1

  7            2            0

  4            0            0

+ + + + + + + + + + + + + + + +
+    RIGHT-HAND SIDE VECTOR(S)   +
+ + + + + + + + + + + + + + + +

  1.5

 -5.5

  16

  7

* * * * * * * * * * * * * *
*     SOLUTION VECTOR(S)    *
* * * * * * * * * * * * * *

  .5

 -2

  2.5

  1
```

82

4.6 MATRIX INVERSION USING GAUSS-JORDAN ELIMINATION

The inverse of the square matrix **A** is defined as A^{-1} where

$$A A^{-1} = I \qquad (4.8)$$

I is the identity matrix. The above multiplication is commutative, i.e.

$$A^{-1}A = I$$

Multiplication of both sides of a system of simultaneous equations by the inverse of the coefficient matrix yields

$$A^{-1}A X = A^{-1}B$$

or

$$X = A^{-1}B$$

Hence a solution to a system of simultaneous linear equations can be obtained by multiplication of the right-hand side vector by the inverse of the coefficient matrix.

The inverse of the coefficient matrix is a convenient concept in matrix algebra. However, matrix inversion has a number of practical disadvantages. The evaluation of the inverse of a matrix is a numerically intensive procedure and this makes the solution of simultaneous equations by matrix inversion uncompetitive. If the coefficient matrix is banded then the banding is destroyed during inversion. This is an obvious disadvantage in the stiffness method which tends to produce narrowly banded matrices. Also, because matrix inversion is more numerically intensive than rival methods of equation solving, the effects of ill-conditioning are more likely to be problematic (see section 4.7). The large number of arithmetic operations associated with matrix inversion is, however, a useful property when investigating the effects of ill-conditioning. A matrix inversion program has, therefore, been included to facilitate the investigation of ill-conditioned coefficient matrices. A matrix inversion program can, of course, be used to solve simultaneous equations although the practice is to be discouraged for the reasons outlined above.

By definition, multiplication of the first column of the inverse of the coefficient matrix by the coefficient matrix itself must produce the first column of the identity matrix (see equation (4.8)).

i.e. $\qquad A X_1 = B_1$

where **A** is the coefficient matrix
       X_1 is the first column of the inverse of **A**
       B_1 is the first column of the identity matrix.

Similarly

$$A X_2 = B_2$$

where
       A   is the coefficient matrix
       X_2 is the second column of the inverse of **A**
       B_2 is the second column of the identity matrix.

This means that solving for a set of right-hand side vectors which, collectively, form the identity matrix, produces a set of solution vectors that collectively comprise the inverse of the coefficient matrix. Any solution method can be used to find an inverse in this way, but Gauss-Jordan elimination is, perhaps, the most straightforward.

Example 4.6

Use Gauss-Jordan elimination to find the solution to the equations in example 4.1 by matrix inversion.

The augmented matrix is as follows.

$$\begin{bmatrix} 2 & 1 & 0 & | & 1 & 0 & 0 \\ 1 & 3 & -2 & | & 0 & 1 & 0 \\ 0 & -2 & 5 & | & 0 & 0 & 1 \end{bmatrix}$$

First pivot - Divide row 1 by the pivot

$$\begin{bmatrix} 1 & 0.5 & 0 & 0.5 & 0 & 0 \\ 1 & 3 & -2 & 0 & 1 & 0 \\ 0 & -2 & 5 & 0 & 0 & 1 \end{bmatrix}$$

Subtract -   1 x row 1 from row 2
              0 x row 1 from row 3

$$\begin{bmatrix} 1 & 0.5 & 0 & 0.5 & 0 & 0 \\ 0 & 2.5 & -2 & -0.5 & 1 & 0 \\ 0 & -2 & 5 & 0 & 0 & 1 \end{bmatrix}$$

SIMULTANEOUS LINEAR EQUATIONS

Second pivot - Divide row 2 by the pivot

$$\begin{bmatrix} 1 & 0.5 & 0 & 0.5 & 0 & 0 \\ 0 & 1 & -0.8 & -0.2 & 0.4 & 0 \\ 0 & -2 & 5 & 0 & 0 & 1 \end{bmatrix}$$

Subtract - 0.5 x row 2 from row 1
-2 x row 2 from row 3

$$\begin{bmatrix} 1 & 0 & 0.4 & 0.6 & -0.2 & 0 \\ 0 & 1 & -0.8 & -0.2 & 0.4 & 0 \\ 0 & 0 & 3.4 & -0.4 & 0.8 & 1 \end{bmatrix}$$

Third pivot - Divide row 3 by the pivot

$$\begin{bmatrix} 1 & 0 & 0.4 & 0.6 & -0.2 & 0 \\ 0 & 1 & -0.8 & -0.2 & 0.4 & 0 \\ 0 & 0 & 1 & -0.118 & 0.235 & 0.294 \end{bmatrix}$$

Subtract - 0.4 x row 3 from row 1
-0.8 x row 3 from row 2

$$\begin{bmatrix} 1 & 0 & 0 & 0.647 & -0.294 & -0.118 \\ 0 & 1 & 0 & -0.294 & 0.588 & 0.235 \\ 0 & 0 & 1 & -0.118 & 0.235 & 0.294 \end{bmatrix}$$

Using the inverse to solve the equations (i.e. $X = A^{-1}B$, where B is taken from example 4.1) yields

$$\begin{bmatrix} x_1 \\ x_2 \\ x_3 \end{bmatrix} = \begin{bmatrix} 0.647 & -0.294 & -0.118 \\ -0.294 & 0.588 & 0.235 \\ -0.118 & 0.235 & 0.294 \end{bmatrix} \begin{bmatrix} 4 \\ 1 \\ 11 \end{bmatrix} = \begin{bmatrix} 0.996 \\ 1.997 \\ 2.997 \end{bmatrix}$$

Note that the solution is not exact as only three significant figures have been used in the calculations.

4.7 ILL-CONDITIONED EQUATIONS - Program INVERT.EQ

If arithmetical operations could be conducted using an infinite number of significant figures, then any system of simultaneous equations could be classified as being either

1. Singular (i.e. having no unique solution), or
2. Non-singular (i.e. having a unique solution).

A singular system of equations is characterised by having a coefficient matrix with a determinant of zero, whereas a non-singular system of equations has a coefficient matrix with a non-zero determinant.

In practice, however, computer arithmetic is conducted using a finite number of significant figures. Typically, single precision arithmetic uses about 7 significant figures and double precision arithmetic uses about 14 significant figures. This means that arithmetical results obtained using a computer have only a limited degree of accuracy. Hence, if the determinant of a coefficient matrix is evaluated using a computer, and the result turns out to be zero, then it is not possible to say that the equations are singular. Due to the limited accuracy of the computation a number that is actually very close to zero may be held as zero. Consequently it is only possible to conclude that if a determinant is calculated to be zero, or very close to zero, then the equations are singular or almost singular.

Example 4.7

Solve the system of equations shown below

(i) Using all of the available significant figures
(ii) Using only five significant figures.

$$\begin{bmatrix} 1.000000 & 1.356589 \\ 1.356589 & 1.840000 \end{bmatrix} \begin{bmatrix} x_1 \\ x_2 \end{bmatrix} = \begin{bmatrix} 0.006516518 \\ 0.003737453 \end{bmatrix}$$

(i) Reduce to triangular form using Gauss elimination

$$\begin{bmatrix} 1.000000 & 1.356589 \\ 0 & -0.0003337150 \end{bmatrix} \begin{bmatrix} x_1 \\ x_2 \end{bmatrix} = \begin{bmatrix} 0.006516518 \\ -0.005102784 \end{bmatrix}$$

hence

$$\begin{bmatrix} x_1 \\ x_2 \end{bmatrix} = \begin{bmatrix} -20.73687 \\ 15.29084 \end{bmatrix}$$

(ii) Reduce to triangular form using Gauss elimination

$$\begin{bmatrix} 1.0000 & 1.3565 \\ 0 & -0.000092250 \end{bmatrix} \begin{bmatrix} x_1 \\ x_2 \end{bmatrix} = \begin{bmatrix} 0.0065165 \\ -0.0051021 \end{bmatrix}$$

hence

$$\begin{bmatrix} x_1 \\ x_2 \end{bmatrix} = \begin{bmatrix} -75.019 \\ 55.308 \end{bmatrix}$$

Example 4.7 shows quite clearly that the number of significant figures used during computation can have a dramatic effect upon the results obtained.

If small changes to the coefficient matrix produce large changes in the solution, then the equations are said to be ill-conditioned. The reader should note that example 4.7 demonstrates the effects of ill-conditioning; it does not investigate ill-conditioning itself, and a clear distinction should be made between the presence and the effects of ill-conditioning.

In general, the stiffness method produces well-conditioned systems of equations to which a sufficiently accurate solution can be found using single precision computer arithmetic. The governing equations only become badly behaved under "special" circumstances (e.g. when a structure is close to elastic instability). The reader should note that ill-conditioned equations have physical significance, and if ill-conditioned equations appear unexpectedly, then the cause of the ill-conditioning needs to be found and understood.

The Presence of Ill-conditioning

There is no absolute definition of an ill-conditioned matrix. It can be argued that a matrix is only ill-conditioned if a satisfactorily accurate solution cannot be obtained using the number of significant figures offered by the computer being used (i.e. a matrix is only ill-conditioned if the effects of ill-conditioning are unacceptable). Such a definition is not very useful. It is better, although equally vague, to say that ill-conditioning is present if the solution vector is likely to be very sensitive to the number of significant figures used during computation. There are a number of tests that can be applied to a coefficient matrix to investigate its condition. Of the available tests, inspection of the inverse of the coefficient matrix is perhaps the most useful. In general it can be said that the bigger the elements of the inverse are in comparison with the elements of the original matrix, the poorer is the condition of the matrix. If the elements of the inverse are many orders of magnitude greater than the elements of the coefficient matrix itself, then ill-conditioning may be a problem and should be investigated.

Example 4.8

Use inspection of the inverse to investigate the condition of the coefficient matrix used in example 4.7.

The coefficient matrix is as follows:

$$A = \begin{bmatrix} 1.0000000 & 1.356589 \\ 1.356589 & 1.840000 \end{bmatrix}$$

and its inverse is

$$A^{-1} = -(1/0.000333715) \begin{bmatrix} 1.840000 & -1.356589 \\ -1.356589 & 1.000000 \end{bmatrix}$$

$$= \begin{bmatrix} -5513.686 & 4065.112 \\ 4065.112 & -2996.568 \end{bmatrix}$$

The elements of the inverse are approximately 3000 times bigger than the elements of the coefficient matrix, indicating that ill-conditioning is a potential problem.

Example 4.8 shows that the elements of the inverse of an ill-conditioned matrix are large because its determinant is small. As a system of equations approaches singularity the determinant of the coefficient matrix approaches, zero causing the elements of the inverse to be magnified.

Example 4.9

The symmetric matrix shown below is known as the Hilbert matrix.

$$\begin{bmatrix} 1 & 1/2 & 1/3 & 1/4 & \cdots & 1/n \\ 1/2 & 1/3 & 1/4 & 1/5 & \cdots & 1/(n+1) \\ 1/3 & 1/4 & 1/5 & 1/6 & \cdots & 1/(n+2) \\ \cdot & & & & & \cdot \\ \cdot & & & & & \cdot \\ 1/n & 1/(n+1) & 1/(n+2) & 1/(n+3) & \cdots & 1/(2n-1) \end{bmatrix}$$

Investigate the condition of (3x3) and (5x5) Hilbert matrices using the magnitude of the elements of the inverse as a measure of condition.

$$H_3 = \begin{bmatrix} 1.000000 & 0.500000 & 0.333333 \\ 0.500000 & 0.333333 & 0.250000 \\ 0.333333 & 0.250000 & 0.200000 \end{bmatrix}$$

and

$$H_3^{-1} = \begin{bmatrix} 9.00065 & -36.0034 & 30.0032 \\ -36.0034 & 192.0036 & -180.016 \\ 30.0032 & -180.016 & 180.015 \end{bmatrix}$$

SIMULTANEOUS LINEAR EQUATIONS

$$H_5 = \begin{bmatrix} 1.00000 & 0.500000 & 0.333333 & 0.250000 & 0.200000 \\ 0.500000 & 0.333333 & 0.250000 & 0.200000 & 0.166667 \\ 0.333333 & 0.250000 & 0.200000 & 0.166667 & 0.142857 \\ 0.250000 & 0.200000 & 0.166667 & 0.142857 & 0.125000 \\ 0.200000 & 0.166667 & 0.142857 & 0.125000 & 0.111111 \end{bmatrix}$$

and

$$H_5^{-1} = \begin{bmatrix} 26.9278 & -335.949 & 1204.81 & -1633.57 & 744.183 \\ -335.949 & 5469.83 & -21783.1 & 31228.4 & -14725.2 \\ 1204.80 & -21783.2 & 91785.8 & -136306. & 65840.8 \\ -1633.57 & 31228.6 & -136307. & 207403. & -101980. \\ 744.186 & -14725.4 & 65841.3 & -101980. & 50832.2 \end{bmatrix}$$

Both inverses exhibit elements that are much bigger than the elements of the original matrices, indicating that the original matrices are poorly conditioned. The Hilbert matrix is, in fact, well known to be extremely poorly conditioned.

The Effects of Ill-conditioning

The inverse of the coefficient matrix can be used to investigate the severity of the effects of ill-conditioning. There are two commonly employed tests:

1. Compare $A A^{-1}$ with the identity matrix I

2. Compare $[A^{-1}]^{-1}$ with the original coefficient matrix A.

If the coefficient matrix is well-conditioned then the above comparisons will be almost exact. If, on the other hand, the coefficient matrix is poorly conditioned then the effect of ill-conditioning will be observed as a significant difference between the two matrices. Of the two tests, inversion of the inverse is the more severe as it entails a greater number of arithmetical operations. For instance, the inverse of the inverse of a 5 x 5 Hilbert matrix (shown below) is noticeably different from the original matrix (given in example 4.9).

$$\begin{bmatrix} 0.999968 & 0.499755 & 0.333139 & 0.249839 & 0.199863 \\ 0.499760 & 0.333158 & 0.249863 & 0.199887 & 0.166571 \\ 0.333142 & 0.249863 & 0.199893 & 0.166579 & 0.142783 \\ 0.249841 & 0.199886 & 0.166579 & 0.142785 & 0.124939 \\ 0.199864 & 0.166570 & 0.142782 & 0.124939 & 0.111059 \end{bmatrix}$$

It should also be noted that although any technique for solving simultaneous equations can be used to invert the coefficient matrix, it is better to invert using the same computational procedure as will be used to solve the equations (e.g. if Gauss elimination is to be used to solve the equations then the inverse should also be obtained using Gauss elimination).

As the effects of ill-conditioning arise purely from the truncation of real numbers during computation, it follows that the best defence against the effects of ill-conditioning is to use many significant figures. Most computer languages have the facility to conduct arithmetic to a number of levels of precision. For most calculations single precision arithmetic, which uses about 7 significant figures, is sufficiently accurate. Calculations requiring a higher degree of accuracy are usually conducted using double precision arithmetic, which offers, as the name suggests, approximately twice as many significant figures as single precision arithmetic. Some machines and languages have a tertiary level of accuracy known as extended precision, allowing calculations to be performed using about twice as many significant figures as double precision arithmetic (perhaps 28 significant figures). While high precision arithmetic offers better accuracy than single precision arithmetic, it takes longer and uses more computer memory. In general, high precision arithmetic should only be used if satisfactory results cannot be obtained using single precision arithmetic. If ill-conditioning is suspected, then comparison of the solutions obtained using two different levels of precision gives some insight into the arithmetical stability of the problem in hand.

Program   INVERT.EQ

This program inverts a coefficient matrix by Gauss-Jordan elimination without row interchange. The matrix passed to the inversion routine is overwritten by the inverse. The program offers the following three options for the study of ill-conditioning:

1. Matrix inversion

2. Inversion of the inverse of the coefficient matrix

3. Multiplication of the inverse of the coefficient matrix by the coefficient matrix itself.

Input   The coefficient matrix is input row by row.

SIMULTANEOUS LINEAR EQUATIONS

Comments on the Algorithm The inversion routine overwrites the coefficient matrix with its inverse. To allow comparison of the product of coefficient matrix and its inverse with the identity matrix, a copy of the coefficient matrix is taken before inversion. As the inversion routine does not employ row interchange a zero pivot will cause the algorithm to break down.

Output The coefficient matrix is echoed and the user stipulates whether the inverse, the inverse of the inverse, or the product of the inverse and the coefficient matrix, is to be output.

Listing

```
1000 REM ************************************************************************
1010 REM *                                                                      *
1020 REM *      PROGRAM    I N V E R T . E Q    TO INVERT A MATRIX BY           *
1030 REM *      GAUSS-JORDAN ELIMINATION WITH NO ROW INTERCHANGE                *
1040 REM *                                                                      *
1050 REM *      J.A.D.BALFOUR                                                   *
1060 REM *                                                                      *
1070 REM ************************************************************************
1080 REM
1090 REM --- NOTE THAT VARIABLES ARE DEFINED IN APPENDIX A
1100 REM
1110 OPTION BASE 1
1120 DIM A(20,20), C(20,20), D(20,20)
1130 REM
1140 REM --- THE ABOVE DIMENSION STATEMENT SETS THE MAX. PROBLEM SIZE
1150 REM
1160 PRINT : PRINT
1170 PRINT "---------------------------------------------------"
1180 PRINT
1190 PRINT "PROGRAM    I N V E R T . E Q    TO INVERT A MATRIX"
1200 PRINT "BY GAUSS-JORDAN ELIMINATION WITH NO ROW INTERCHANGE"
1210 PRINT
1220 PRINT "---------------------------------------------------"
1230 PRINT
1240 INPUT "SIZE OF THE COEFFICIENT MATRIX  = ", NEQ
1250 PRINT
1260 PRINT "INPUT THE COEFFICIENT MATRIX"
1270 PRINT
1280 FOR I = 1 TO NEQ
1290 FOR J = 1 TO NEQ
1300 PRINT "ELEMENT ("; I; ","; J; ")  = "; : INPUT "", A(I,J)
1310 NEXT J
1320 NEXT I
1330 REM
1340 REM --- ECHO AND COPY THE COEFFICIENT MATRIX
1350 REM
1360 PRINT
1370 PRINT "+ + + + + + + + + + + + +"
1380 PRINT "+    COEFFICIENT MATRIX +"
1390 PRINT "+ + + I + + + + + + + + +"
1400 PRINT
1410 FOR I = 1 TO NEQ
1420 FOR J = 1 TO NEQ
1430 PRINT A(I,J),
1440 C(I,J) = A(I,J)
1450 NEXT J
1460 PRINT : PRINT
1470 NEXT I
1480 PRINT
1490 PRINT "DO YOU WISH TO"
1500 PRINT "(1) INVERT THE MATRIX ONCE"
```

```
1510 PRINT "(2) INVERT THE MATRIX TWICE"
1520 PRINT "(3) INVERT THE MATRIX ONCE AND MULTIPLY BY THE ORIGINAL MATRIX"
1530 PRINT "(4) STOP"
1540 INPUT ANS
1550 REM
1560 REM  ---  INVERT THE COEFFICIENT MATRIX
1570 REM
1580 GOSUB 1990
1590 PRINT
1600 ON ANS GOTO 1610, 1650, 1700, 1980
1610 PRINT "* * * * * * * * * * * * * * * * * * * *"
1620 PRINT "*    INVERSE OF THE COEFFICIENT MATRIX    *"
1630 PRINT "* * * * * * * * * * * * * * * * * * * *"
1640 GOTO  1890
1650 GOSUB 1990
1660 PRINT "* * * * * * * * * * * * * * * * * * * * * * * * * * *"
1670 PRINT "*    INVERSE OF THE INVERSE OF THE COEFFICIENT MATRIX    *"
1680 PRINT "* * * * * * * * * * * * * * * * * * * * * * * * * * *"
1690 GOTO  1890
1700 PRINT "* * * * * * * * * * * * * * * * * * * * * * * * * * *"
1710 PRINT "*    PRODUCT OF THE INVERSE AND THE COEFFICIENT MATRIX    *"
1720 PRINT "* * * * * * * * * * * * * * * * * * * * * * * * * * *"
1730 FOR I = 1 TO NEQ
1740 FOR J = 1 TO NEQ
1750 D(I,J) = A(I,J)
1760 NEXT J
1770 NEXT I
1780 FOR I = 1 TO NEQ
1790 FOR J = 1 TO NEQ
1800 A(I,J) = 0
1810 FOR K = 1 TO NEQ
1820 A(I,J) = A(I,J) + C(I,K)*D(K,J)
1830 NEXT K
1840 NEXT J
1850 NEXT I
1860 REM
1870 REM  ---  PRINT RESULTS
1880 REM
1890 PRINT
1900 FOR I = 1 TO NEQ
1910 FOR J = 1 TO NEQ
1920 PRINT A(I,J),
1930 A(I,J) = C(I,J)
1940 NEXT J
1950 PRINT : PRINT
1960 NEXT I
1970 GOTO  1480
1980 END
1990 REM  * * * * * * * * * * * * * * * * * * * * * * * * * * * * *
2000 REM  *                                                        *
2010 REM  *    SUBROUTINE TO INVERT A MATRIX BY GAUSS-JORDAN ELIMINATION    *
2020 REM  *    WITH NO ROW INTERCHANGE                             *
2030 REM  *                                                        *
2040 REM  * * * * * * * * * * * * * * * * * * * * * * * * * * * * *
2050 REM
2060 REM  ---  LOOP FOR ALL PIVOTS
2070 REM
2080 FOR I = 1 TO NEQ
2090 TEMP   = A(I,I)
2100 REM
2110 REM  ---  DIVIDE THE PIVOTAL ROW BY THE PIVOT
2120 REM
2130 FOR J = 1 TO NEQ
2140 A(I,J) = A(I,J) / TEMP
2150 NEXT J
2160 A(I,I) = 1 / TEMP
2170 REM
2180 REM  ---  LOOP FOR ALL ROWS
2190 REM
2200 FOR J = 1 TO NEQ
2210 IF J   = I THEN GOTO 2310
2220 FACT   = A(J,I)
2230 REM
```

SIMULTANEOUS LINEAR EQUATIONS

```
2240 REM  ---  LOOP FOR ALL ELEMENTS IN THE ROW
2250 REM
2260 FOR K   = 1 TO NEQ
2270 TEMP    = A(J,K)
2280 IF K    = I THEN TEMP = 0
2290 A(J,K)  = TEMP - FACT*A(I,K)
2300 NEXT K
2310 NEXT J
2320 NEXT I
2330 RETURN
```

Sample Run The following sample run shows the program INVERT.EQ being used to investigate the presence of ill-conditioning in a Hilbert matrix of order 5. Keyboard input is shown in bold typeface.

```
-------------------------------------------------
PROGRAM       I N V E R T . E Q    TO INVERT A MATRIX
BY GAUSS-JORDAN ELIMINATION WITH NO ROW INTERCHANGE
-------------------------------------------------

SIZE OF THE COEFFICIENT MATRIX = 5

INPUT THE COEFFICIENT MATRIX

ELEMENT ( 1 , 1 )  =  1
ELEMENT ( 1 , 2 )  =  .5
ELEMENT ( 1 , 3 )  =  .333333
ELEMENT ( 1 , 4 )  =  .25
ELEMENT ( 1 , 5 )  =  .2
ELEMENT ( 2 , 1 )  =  .5
ELEMENT ( 2 , 2 )  =  .333333
ELEMENT ( 2 , 3 )  =  .25
ELEMENT ( 2 , 4 )  =  .2
ELEMENT ( 2 , 5 )  =  .166667
ELEMENT ( 3 , 1 )  =  .333333
ELEMENT ( 3 , 2 )  =  .25
ELEMENT ( 3 , 3 )  =  .2
ELEMENT ( 3 , 4 )  =  .166667
ELEMENT ( 3 , 5 )  =  .142857
ELEMENT ( 4 , 1 )  =  .25
ELEMENT ( 4 , 2 )  =  .2
ELEMENT ( 4 , 3 )  =  .166667
ELEMENT ( 4 , 4 )  =  .142857
ELEMENT ( 4 , 5 )  =  .125
ELEMENT ( 5 , 1 )  =  .2
ELEMENT ( 5 , 2 )  =  .166667
ELEMENT ( 5 , 3 )  =  .142857
ELEMENT ( 5 , 4 )  =  .125
ELEMENT ( 5 , 5 )  =  .111111

+ + + + + + + + + + + +
+   COEFFICIENT MATRIX   +
+ + + + + + + + + + + +

 1           .5          .333333     .25         .2

 .5          .333333     .25         .2          .166667

 .333333     .25         .2          .166667     .142857

 .25         .2          .166667     .142857     .125

 .2          .166667     .142857     .125        .111111
```

COMPUTER ANALYSIS OF STRUCTURAL FRAMEWORKS

```
DO YOU WISH TO
(1) INVERT THE MATRIX ONCE
(2) INVERT THE MATRIX TWICE
(3) INVERT THE MATRIX ONCE AND MULTIPLY BY THE ORIGINAL MATRIX
(4) STOP
? 1
```

```
* * * * * * * * * * * * * * * * * * * * *
*    INVERSE OF THE COEFFICIENT MATRIX   *
* * * * * * * * * * * * * * * * * * * * *
```

26.9278	-335.949	1204.81	-1633.57	744.183
-335.947	5469.83	-21783.1	31228.4	-14725.2
1204.8	-21783.2	91785.8	-136306	65840.8
-1633.57	31228.6	-136307	207403	-101980
744.186	-14725.4	65841.3	-101980	50832.2

```
DO YOU WISH TO
(1) INVERT THE MATRIX ONCE
(2) INVERT THE MATRIX TWICE
(3) INVERT THE MATRIX ONCE AND MULTIPLY BY THE ORIGINAL MATRIX
(4) STOP
? 2
```

```
* * * * * * * * * * * * * * * * * * * * * * * * * * * * *
*    INVERSE OF THE INVERSE OF THE COEFFICIENT MATRIX    *
* * * * * * * * * * * * * * * * * * * * * * * * * * * * *
```

.999668	.499755	.333139	.249839	.199863
.49976	.333158	.249863	.199887	.166571
.333142	.249863	.199893	.166579	.142783
.249841	.199886	.166579	.142785	.124939
.199864	.16657	.142782	.124939	.111059

```
DO YOU WISH TO
(1) INVERT THE MATRIX ONCE
(2) INVERT THE MATRIX TWICE
(3) INVERT THE MATRIX ONCE AND MULTIPLY BY THE ORIGINAL MATRIX
(4) STOP
? 3
```

```
* * * * * * * * * * * * * * * * * * * * * * * * * * * * *
*    PRODUCT OF THE INVERSE AND THE COEFFICIENT MATRIX   *
* * * * * * * * * * * * * * * * * * * * * * * * * * * * *
```

.999985	0	1.95313E-03	0	0
-3.05176E-05	1.00049	-9.76563E-04	3.90625E-03	0
-2.28882E-05	0	.999023	9.76563E-04	9.76563E-04
-2.28882E-05	1.2207E-04	0	1.00195	0
0	-1.2207E-04	1.46484E-03	-9.76563E-04	1.00098

```
DO YOU WISH TO
(1) INVERT THE MATRIX ONCE
(2) INVERT THE MATRIX TWICE
(3) INVERT THE MATRIX ONCE AND MULTIPLY BY THE ORIGINAL MATRIX
(4) STOP
? 4
```

5 Plane Trusses

5.1 INTRODUCTION

In practice most trusses are rigidly jointed. If the members are connected together by welding or by two or more bolts, then the joints are effectively rigid. However, even when rigidly jointed, triangulated frameworks transmit load primarily by axial force in the members. Hence bending stresses are usually of minor importance compared with axial stresses, thus allowing the assumption of pinned joints to be safely made. Before the use of computers became widespread in design offices, trusses were normally assumed to be pin jointed to simplify the structural analysis. Today computers allow advantage to be taken of the additional stiffness that comes from rigid connections, and trusses are usually analysed as rigidly jointed frameworks. The pin jointed truss is therefore no longer of service in the design office. It does, however, still have a role to play in engineering education.

The techniques employed when using the stiffness method to analyse pin jointed trusses are identical to those used in the analysis of rigidly jointed frameworks. For educational purposes pin jointed trusses have the advantage of having a structural action that is inherently simpler than that of rigidly jointed frames, and this makes the understanding of the fundamentals of the stiffness method easier. The reader should be aware that in this text pin jointed trusses are used purely as an introduction to the stiffness method. In practice trusses that are rigidly jointed are almost always better analysed as rigidly jointed frameworks.

5.2 THE ELEMENT STIFFNESS MATRIX IN THE MEMBER AXES SYSTEM

The forces at the ends of a truss member are related to the displacements at the ends by the element stiffness matrix.

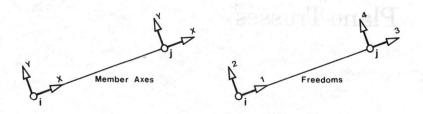

FIGURE 5.1   MEMBER AXES AND FREEDOMS

Figure 5.1 shows a typical plane truss element "i,j". By treating the element as a simple structure with four freedoms, the element stiffness matrix can be found. To obtain the element stiffness matrix the force system required to maintain unit displacement at each freedom in turn must be evaluated. For a pin jointed member only axial forces are involved and the stiffness terms can be found by inspection. Note also that the axial stiffness of a line element is considered in section 2.12.

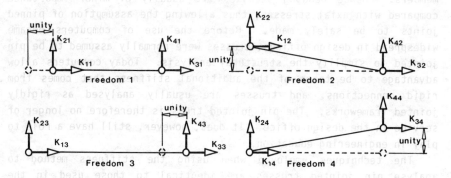

FIGURE 5.2   EVALUATION OF STIFFNESS TERMS

$$k_{11} = k_{33} = EA/L \quad \text{and} \quad k_{13} = k_{31} = -EA/L$$

hence

$$\begin{bmatrix} f_{ijx} \\ f_{ijy} \\ f_{jix} \\ f_{jiy} \end{bmatrix} = \begin{bmatrix} EA/L & 0 & -EA/L & 0 \\ 0 & 0 & 0 & 0 \\ -EA/L & 0 & EA/L & 0 \\ 0 & 0 & 0 & 0 \end{bmatrix} \begin{bmatrix} \delta_{ijx} \\ \delta_{ijy} \\ \delta_{jix} \\ \delta_{jiy} \end{bmatrix} \quad (5.1)$$

The subscripts for force and displacement can be decoded as follows. The first two characters identify the element by the nodes at its ends, with the first character indicating the end under consideration. The third subscript defines the direction of the force or displacement (in the member axes system). If equation (5.1) is split into submatrices as indicated by the broken lines then

$$\begin{bmatrix} f_{ij} \\ \hline f_{ji} \end{bmatrix} = \begin{bmatrix} k_{ii}^j & | & k_{ij} \\ \hline k_{ji} & | & k_{jj}^i \end{bmatrix} \begin{bmatrix} \delta_{ij} \\ \hline \delta_{ji} \end{bmatrix}$$

hence
$$f_{ij} = k_{ii}^j \delta_{ij} + k_{ij} \delta_{ji}$$
and
$$f_{ji} = k_{ji} \delta_{ij} + k_{jj}^i \delta_{ji} \qquad (5.2)$$

The product $k_{ii}^j \delta_{ij}$ gives the forces at end "i" of member "i,j" due to displacements at end "i", and $k_{ij} \delta_{ji}$ gives the forces at end "i" due to the displacements at end "j". Of course the resultant force vector f_{ij} is the matrix sum of these products.

5.3 TRANSFORMATION OF FORCE AND DISPLACEMENT

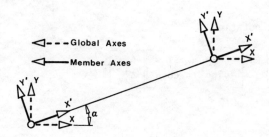

FIGURE 5.3   PLANE TRUSS MEMBER LYING AT AN ANGLE TO THE GLOBAL X-AXIS

Equation (5.1) shows how the member end forces are related to the member end displacements by the element stiffness matrix. In general the member axes associated with equation (5.1) will not coincide with the axes used to describe nodal forces and displacements (i.e. the nodal axes). Solution of the stiffness equation yields the nodal displacements in terms of the global axes, and these displacements must be transformed into the member axes system shown in fig 5.1 before equation (5.1) can be used to find the member end forces. Figure 5.3 shows a typical plane truss element. Note that α is the clockwise rotation of the member axes that will make them coincide with the global axes.

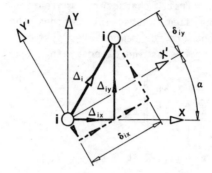

FIGURE 5.4   DISPLACEMENT OF NODE "i" DURING LOADING

Assume that, during loading, the node "i" displaces Δ_{ix} in the global x-direction, and Δ_{iy} in the global y-direction, as shown in fig 5.4. If these global displacements are split into components lying parallel and perpendicular to the member x-axis, then the displacements in the primed axes system (i.e. the member axes system) are:

$$\delta_{ijx} = \cos\alpha\, \Delta_{ix} + \sin\alpha\, \Delta_{iy}$$
$$\delta_{ijy} = -\sin\alpha\, \Delta_{ix} + \cos\alpha\, \Delta_{iy}$$

Or, in matrix notation

$$\begin{bmatrix} \delta_{ijx} \\ \delta_{ijy} \end{bmatrix} = \begin{bmatrix} \cos\alpha & \sin\alpha \\ -\sin\alpha & \cos\alpha \end{bmatrix} \begin{bmatrix} \Delta_{ix} \\ \Delta_{iy} \end{bmatrix}$$

i.e.
$$\delta_{ij} = T_{ij}\, \Delta_i \qquad (5.3)$$

and similarly at end "j" of the member

$$\begin{bmatrix} \delta_{jix} \\ \delta_{jiy} \end{bmatrix} = \begin{bmatrix} \cos\alpha & \sin\alpha \\ -\sin\alpha & \cos\alpha \end{bmatrix} \begin{bmatrix} \Delta_{jx} \\ \Delta_{jy} \end{bmatrix}$$

i.e.
$$\delta_{ji} = T_{ij}\, \Delta_j \qquad (5.4)$$

where
- T_{ij} is the transformation matrix for member "i,j"
- δ_{ij} is a displacement vector at node "i" in the member axes system for member "i,j"
- δ_{ji} is a displacement vector at node "j" in the member axes system for member "i,j"

PLANE TRUSSES

Δ_i is a displacement vector at node "i" in the global axes system

Δ_j is a displacement vector at node "j" in the global axes system.

The transformation matrix T_{ij} can also be used to transform forces from the global axes system to the member axes system.

$$\begin{bmatrix} f_{ijx} \\ f_{ijy} \end{bmatrix} = \begin{bmatrix} \cos\alpha & \sin\alpha \\ -\sin\alpha & \cos\alpha \end{bmatrix} \begin{bmatrix} F_{ijx} \\ F_{ijy} \end{bmatrix}$$

i.e.

$$f_{ij} = T_{ij} F_{ij} \qquad (5.5)$$

and similarly at end "j" of the member

$$\begin{bmatrix} f_{jix} \\ f_{jiy} \end{bmatrix} = \begin{bmatrix} \cos\alpha & \sin\alpha \\ -\sin\alpha & \cos\alpha \end{bmatrix} \begin{bmatrix} F_{jix} \\ F_{jiy} \end{bmatrix}$$

i.e.

$$f_{ji} = T_{ij} F_{ji} \qquad (5.6)$$

where

T_{ij} is the transformation matrix for member "i.j"
f_{ij} is a force vector at node "i" in the member axes system
f_{ji} is a force vector at node "j" in the member axes system
F_{ij} is a force vector at node "i" in the global axes system
F_{ji} is a force vector at node "j" in the global axes system

The reverse transformation (i.e. from member axes to global axes) is effected by the inverse of T_{ij}. Say that the displacements at node "i" are required in the global axes system, but are known in the member axes system for member "i,j". From equation (5.3)

$$\delta_{ij} = T_{ij} \Delta_i$$

Multiplying both sides by the inverse of the transformation matrix:

$$T_{ij}^{-1} \delta_{ij} = T_{ij}^{-1} T_{ij} \Delta_i$$

hence
$$\Delta_i = T_{ij}^{-1} \delta_{ij}$$

Summary

The transformation matrix T_{ij} transforms force and displacement vectors at nodes "i" and "j" from *the global axes system to the member axes system* (for member "i,j") where,

$$T_{ij} = \begin{bmatrix} \cos\alpha & \sin\alpha \\ -\sin\alpha & \cos\alpha \end{bmatrix}$$

The inverse of T_{ij} transforms vectors of force and displacement at nodes "i" and "j" from *the member axes system to the global axes system* where,

$$T_{ij}^{-1} = \begin{bmatrix} \cos\alpha & -\sin\alpha \\ \sin\alpha & \cos\alpha \end{bmatrix}$$

5.4 THE ELEMENT STIFFNESS MATRIX IN THE GLOBAL AXES SYSTEM

In section 5.2 the relationship between the forces at end "i" of member "i,j" and the displacements at its ends was shown to be

$$f_{ij} = k_{ii}^j \delta_{ij} + k_{ij} \delta_{ji}$$

Section 5.3 showed that the inverse of the transformation matrix T_{ij} could transform member end forces from the member axes system used in the above equation to the global axes system. Hence the member end forces can be expressed in terms of the global axes system as follows:

$$\begin{aligned} F_{ij} &= T_{ij}^{-1} f_{ij} \\ &= T_{ij}^{-1} k_{ii}^j \delta_{ij} + T_{ij}^{-1} k_{ij} \delta_{ji} \end{aligned} \quad (5.7)$$

When generating the structure stiffness equation the forces and displacements at each node must be expressed in a common axes system. Consequently the member end displacements, which have hitherto been expressed in the member axes system, must be expressed in the nodal axes system using the transformations given by equations (5.3) and (5.4):

$$\delta_{ij} = T_{ij} \Delta_i \quad \text{and} \quad \delta_{ji} = T_{ij} \Delta_j$$

PLANE TRUSSES

Substituting the preceding equation into equation (5.7):

$$F_{ij} = T_{ij}^{-1} k_{ii}^{j} T_{ij} \Delta_i + T_{ij}^{-1} k_{ij} T_{ij} \Delta_j$$

or

$$F_{ij} = K_{ii}^{j} \Delta_i + K_{ij} \Delta_j \qquad (5.8)$$

where

$$K_{ii}^{j} = T_{ij}^{-1} k_{ii}^{j} T_{ij}$$

and

$$K_{ij} = T_{ij}^{-1} k_{ij} T_{ij}$$

The matrix product $K_{ii}^{j} \Delta_i$ gives the forces (in the global axes system) at end "i" of member "i,j" associated with displacements at end "i" (also in the global axes system). Similarly the matrix product $K_{ij} \Delta_j$ gives the member end forces at end "i" of member "i,j" associated with displacements at end "j". The matrices K_{ii}^{j} and K_{ij} are known as global element stiffness submatrices. It is computationally advantageous to expand these matrices as follows:

$$K_{ii}^{j} = T_{ij}^{-1} k_{ii}^{j} T_{ij}$$

$$= \begin{bmatrix} \cos\alpha & -\sin\alpha \\ \sin\alpha & \cos\alpha \end{bmatrix} \begin{bmatrix} EA/L & 0 \\ 0 & 0 \end{bmatrix} \begin{bmatrix} \cos\alpha & \sin\alpha \\ -\sin\alpha & \cos\alpha \end{bmatrix}$$

$$= \begin{bmatrix} EA/L \cos^2\alpha & EA/L \cos\alpha \sin\alpha \\ EA/L \cos\alpha \sin\alpha & EA/L \sin^2\alpha \end{bmatrix} \qquad (5.9)$$

and

$$K_{ij} = T_{ij}^{-1} k_{ij} T_{ij}$$

$$= \begin{bmatrix} -EA/L \cos^2\alpha & -EA/L \cos\alpha \sin\alpha \\ -EA/L \cos\alpha \sin\alpha & -EA/L \sin^2\alpha \end{bmatrix} \qquad (5.10)$$

Similar expressions can be developed for end "j":

$$F_{ji} = T_{ij}^{-1} f_{ji}$$

$$= T_{ij}^{-1} k_{ji} T_{ij} \Delta_i + T_{ij}^{-1} k_{jj}^{i} T_{ij} \Delta_j$$

or

$$F_{ji} = K_{ji} \Delta_i + K_{jj}^{i} \Delta_j$$

where

$$K_{jj}^{i} = T_{ij}^{-1} k_{jj}^{i} T_{ij}$$

$$= \begin{bmatrix} EA/L \cos^2\alpha & EA/L \cos\alpha \sin\alpha \\ EA/L \cos\alpha \sin\alpha & EA/L \sin^2\alpha \end{bmatrix} \qquad (5.11)$$

and

$$K_{ji} = T_{ij}^{-1} k_{ji} T_{ij}$$

$$= \begin{bmatrix} -EA/L \cos^2\alpha & -EA/L \cos\alpha \sin\alpha \\ -EA/L \cos\alpha \sin\alpha & -EA/L \sin^2\alpha \end{bmatrix} \quad (5.12)$$

Equations (5.8) to (5.11) show that only the K_{ii}^j submatrix need be evaluated as the other submatrices can be obtained by multiplying the corresponding elements of K_{ii}^j as follows:

Find K_{ij} and K_{ji} by multiplying corresponding elements of K_{ii}^j by $\begin{bmatrix} -1 & -1 \\ -1 & -1 \end{bmatrix}$

Find K_{jj}^i by multiplying corresponding elements of K_{ii}^j by $\begin{bmatrix} 1 & 1 \\ 1 & 1 \end{bmatrix}$

(5.13)

Note that this is not a matrix multiplication, but simply the multiplication of corresponding elements.

5.5 NODAL EQUILIBRIUM

Figure 5.5(a) shows a simple triangulated framework, and fig 5.5(b) shows how it might be idealised.

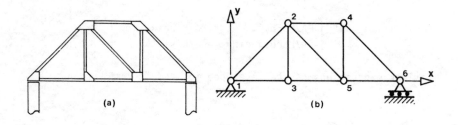

FIGURE 5.5   A SIMPLE TRIANGULATED FRAMEWORK

If node 2 is removed from the truss its freebody diagram is as shown in fig 5.6(a), and if the node is in equilibrium then there can be no net force acting upon it. Resolving each member force into components as shown in fig 5.6(b), and summing horizontal and vertical forces acting on the node produces the following equations:

PLANE TRUSSES

and
$$P_{2x} + F'_{21x} + F'_{23x} + F'_{24x} + F'_{25x} = 0$$
$$P_{2y} + F'_{21y} + F'_{23y} + F'_{24y} + F'_{25y} = 0$$

or,
$$P_2 + F'_{21} + F'_{23} + F'_{24} + F'_{25} = 0$$

where
$$P_2 = \begin{bmatrix} P_{2x} \\ P_{2y} \end{bmatrix} \quad F'_{21} = \begin{bmatrix} F_{21x} \\ F_{21y} \end{bmatrix} \quad \cdots \quad F'_{25} = \begin{bmatrix} F_{25x} \\ F_{25y} \end{bmatrix}$$

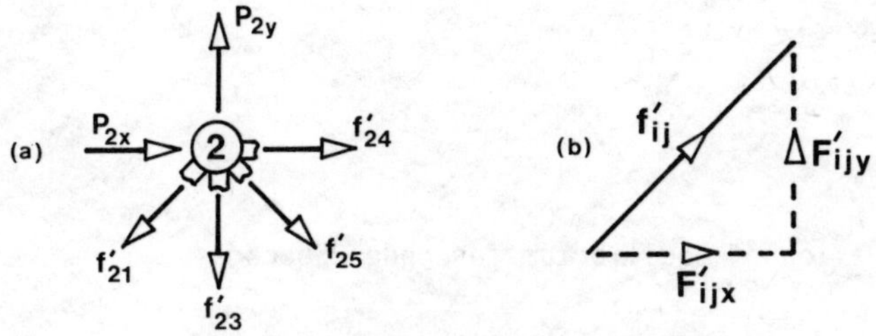

FIGURE 5.6 FREEBODY DIAGRAM FOR NODE 2

Hence, the equilibrium equation for node "i", which has members "i,a", "i,b", "i,c", ... "i,n" framing into it, is:

$$P_i + F'_{ia} + F'_{ib} + F'_{ic} \cdots + F'_{in} = 0 \qquad (5.14)$$

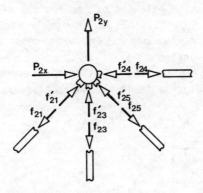

FIGURE 5.7 NODAL FORCES AND MEMBER END FORCES

One of the fundamental principles of static equilibrium is that for every action there must be an equal and opposite reaction. Consequently the force exerted by a member on a node is equal and opposite to the force the node exerts upon that member. Figure 5.7

illustrates this effect.

Hence
$$f_{21} = -f'_{21}$$
$$f_{23} = -f'_{23}$$
$$f_{24} = -f'_{24}$$
$$f_{25} = -f'_{25}$$

Expressing the forces as components in the global axes system

$$F_{21x} = -F'_{21x}$$
$$F_{23y} = -F'_{23y}$$
$$\vdots$$
$$F_{25y} = -F'_{25y}$$

and the equation of equilibrium for node 2 becomes

$$P_2 = F_{21} + F_{23} + F_{24} + F_{25}$$

For node "i" with members "i,a", "i,b", "i,c", ... "i,n" framing into it, the above equation becomes:

$$P_i = F_{ia} + F_{ib} + F_{ic} \ldots + F_{in} \qquad (5.15)$$

Equation (5.15) simply states that the external applied load vector P_i at node "i" must be balanced by the vectors of internal member end forces F_{ia}, F_{ib}, F_{ic}, ... F_{in}. Obviously for the summation to be meaningful the force components must be described in a consistent axes system (see section 3.6).

Equation (5.8) gives the relationship between member end forces and the nodal displacements in the gobal axes system:

$$F_{ij} = K^j_{ii} \Delta_i + K_{ij} \Delta_j$$

where
$$K^j_{ii} = T^{-1}_{ij} k^j_{ii} T_{ij}$$
and
$$K_{ij} = T^{-1}_{ij} k_{ij} T_{ij}$$

Substituting for the member end forces in equation (5.15) gives

$$P_i = K^a_{ii} \Delta_i + K_{ia} \Delta_a + K^b_{ii} \Delta_i + K_{ib} \Delta_b +$$
$$K^c_{ii} \Delta_i + K_{ic} \Delta_c + \ldots + K^n_{ii} \Delta_i + K_{in} \Delta_n$$

PLANE TRUSSES

or

$$P_i = K_{ii}\Delta_i + K_{ia}\Delta_a + K_{ib}\Delta_b \ldots + K_{in}\Delta_n \quad (5.16)$$

where

$$K_{ii} = \sum K_{ii}^j$$

The summation is for all members connected to node "i".

5.6 THE INITIAL STRUCTURE STIFFNESS MATRIX - Programs ESTIFF.PT and ISTIFF.PT

In section 5.5 the equation of nodal equilibrium was shown to be.

$$P_i = K_{ii}\Delta_i + K_{ia}\Delta_a + K_{ib}\Delta_b \ldots + K_{in}\Delta_n$$

where

$$K_{ii} = \sum K_{ii}^j$$

The nodal displacement vectors Δ_1, Δ_2, ... Δ_n, consist of both known and unknown displacements (i.e. the boundary conditions and the unknown nodal displacements). Similarly the nodal force vectors P_1, P_2, ... P_n will contain known and unknown forces (i.e. the known applied loads and the unknown reactive forces). Hence it is evident that the above equation takes no account of the boundary conditions.

Presentation of the remainder of the theory in this section is simplified if the boundary conditions are not applied until after the structure stiffness matrix has been generated. Prior to the application of the boundary conditions the equations of nodal equilibrium for all nodes comprise a system of simultaneous equations which is collectively known as *the initial structure stiffness equation*.

$$\begin{bmatrix} P_1 \\ P_2 \\ \cdot \\ \cdot \\ \cdot \\ P_n \end{bmatrix} = \begin{bmatrix} K_{11} & K_{12} & K_{13} & \ldots & K_{1n} \\ K_{21} & K_{22} & K_{23} & \ldots & K_{2n} \\ \cdot & & & & \\ \cdot & & & & \\ \cdot & & & & \\ K_{n1} & K_{n2} & K_{n3} & \ldots & K_{nn} \end{bmatrix} \begin{bmatrix} \Delta_1 \\ \Delta_2 \\ \cdot \\ \cdot \\ \cdot \\ \Delta_n \end{bmatrix}$$

The above system of simultaneous equations is expressed in terms of submatrices where the vectors of nodal force and displacement are of length 2 and each stiffness submatrix is a 2x2 square matrix. Hence the dimension of the initial structure stiffness matrix is two times the number of nodes. The procedure to generate the initial structure stiffness matrix is as follows:

COMPUTER ANALYSIS OF STRUCTURAL FRAMEWORKS

1. Set all elements of the structure stiffness matrix equal to zero.

2. For member "i,j" calculate the K_{ii}^j, K_{ij}, K_{ji}, and K_{jj}^i stiffness submatrices using equations (5.9) to (5.12).

3. Add these stiffness submatrices to the following locations in the initial structure stiffness matrix:

K_{ii}^j is added to locations $\begin{bmatrix} (2i-1),(2i-1) & (2i-1),(2i) \\ (2i),(2i-1) & (2i),(2i) \end{bmatrix}$

K_{ij} is added to locations $\begin{bmatrix} (2i-1),(2j-1) & (2i-1),(2j) \\ (2i),(2j-1) & (2i),(2j) \end{bmatrix}$

K_{ji} is added to locations $\begin{bmatrix} (2j-1),(2i-1) & (2j-1),(2i) \\ (2j),(2i-1) & (2j),(2i) \end{bmatrix}$

K_{jj}^i is added to locations $\begin{bmatrix} (2j-1),(2j-1) & (2j-1),(2j) \\ (2j),(2j-1) & (2j),(2j) \end{bmatrix}$

4. Repeat steps 2 and 3 for all other elements.

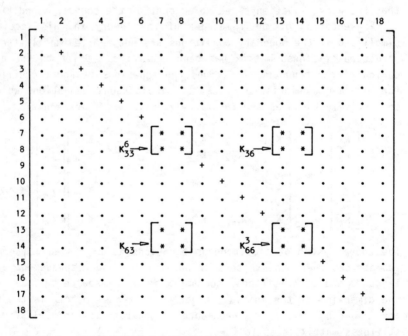

FIGURE 5.8   ADDING ELEMENT 3,6 TO THE INITIAL STRUCTURE STIFFNESS MATRIX

PLANE TRUSSES

Figure 5.8 shows the locations where the global element stiffness submatrices for element 3,6 would be added to the initial structure stiffness matrix. Note that the initial structure stiffness matrix is 18 x 18, indicating that the structure has 9 nodes.

In practice, advantage would be taken of the symmetry of the initial structure stiffness matrix and the similarity of the global element stiffness submatrices.

Example 5.1

Generate the initial structure stiffness equation, in terms of submatrices, for the truss shown below.

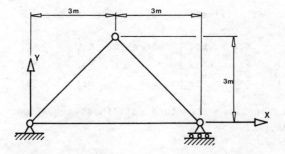

First remove the boundary conditions, leaving the structure completely unrestrained.

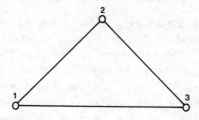

Zero the structure stiffness matrix (note that each zero represents a 2 x 2 submatrix of zeros).

$$\begin{bmatrix} 0 & 0 & 0 \\ 0 & 0 & 0 \\ 0 & 0 & 0 \end{bmatrix}$$

Add the element stiffness submatrices for member 1,2

$$\begin{bmatrix} K_{11} & K_{12} & 0 \\ K_{21} & K_{22} & 0 \\ 0 & 0 & 0 \end{bmatrix} \quad \text{where} \quad \begin{aligned} K_{11} &= K_{11}^2 \\ K_{22} &= K_{22}^1 \end{aligned}$$

(note that the superscripts, taken in conjunction with subscripts, identify the element under consideration).

Add the element stiffness submatrices for member 1,3

$$\begin{bmatrix} K_{11} & K_{12} & K_{13} \\ K_{21} & K_{22} & 0 \\ K_{31} & 0 & K_{33} \end{bmatrix} \quad \text{where} \quad \begin{aligned} K_{11} &= K_{11}^2 + K_{11}^3 \\ K_{22} &= K_{22}^1 \\ K_{33} &= K_{33}^1 \end{aligned}$$

Add the element stiffness submatrices for member 2,3

$$\begin{bmatrix} K_{11} & K_{12} & K_{13} \\ K_{21} & K_{22} & K_{23} \\ K_{31} & K_{32} & K_{33} \end{bmatrix} \quad \text{where} \quad \begin{aligned} K_{11} &= K_{11}^2 + K_{11}^3 \\ K_{22} &= K_{22}^1 + K_{22}^3 \\ K_{33} &= K_{33}^1 + K_{33}^2 \end{aligned}$$

Hence the initial structure stiffness equation, in terms of submatrices, is:

$$\begin{bmatrix} K_{11} & K_{12} & K_{13} \\ K_{21} & K_{22} & K_{23} \\ K_{31} & K_{32} & K_{33} \end{bmatrix} \begin{bmatrix} \Delta_1 \\ \Delta_2 \\ \Delta_3 \end{bmatrix} = \begin{bmatrix} P_1 \\ P_2 \\ P_3 \end{bmatrix}$$

Note that the initial structure stiffness matrix is singular (i.e. there is no unique solution to the initial stiffness relationship). This is because until the structure is adequately restrained there are an infinite number of possible equilibrium configurations.

Take, for example, the truss analysed in the previous example (example 5.1). Imagine the truss to be floating, completely unrestrained and unloaded, at some location in space. Now apply the loads and the reactions necessary for equilibrium. Under the action of these forces the truss would assume its true displaced shape. However, because the restraints (i.e. points of zero displacement) have not been specified the truss and its loads could be moved to any other point in space without disturbing its equilibrium. Hence, until a structure is adequately restrained it can undergo rigid body translation without disturbing its equilibrium. This condition is characterised by a singular structure stiffness matrix. Note also that the element stiffness matrix is always singular because it is, in effect, the initial structure stiffness matrix for a single element structure.

PLANE TRUSSES

Example 5.2

If the frame considered in example 5.1 is constructed from steel angle sections having a cross-sectional area of 3000 mm^2, numerically evaluate the initial structure stiffness matrix. Use 200 kN/mm^2 as the elastic constant for steel.

As the structure has three nodes the initial structure stiffness matrix will be 6 x 6. The initial structure stiffness matrix is first set to zero, and then the element stiffness submatrices are added for each element in turn.

<u>Element 1,2</u>   A = 3000 mm^2, L = 4243 mm, E = 200 kN/mm^2, α = 45°

$$\mathbf{K}^2_{11} = \mathbf{T}^{-1}_{12} \mathbf{k}^2_{11} \mathbf{T}_{12}$$

$$= \begin{bmatrix} \cos\alpha & -\sin\alpha \\ \sin\alpha & \cos\alpha \end{bmatrix} \begin{bmatrix} EA/L & 0 \\ 0 & 0 \end{bmatrix} \begin{bmatrix} \cos\alpha & \sin\alpha \\ -\sin\alpha & \cos\alpha \end{bmatrix}$$

$$= \begin{bmatrix} EA/L \cos^2\alpha & EA/L \cos\alpha \sin\alpha \\ EA/L \cos\alpha \sin\alpha & EA/L \sin^2\alpha \end{bmatrix}$$

$$= \begin{bmatrix} 70.7 & 70.7 \\ 70.7 & 70.7 \end{bmatrix}$$

Adding $\mathbf{K}^2_{11}$, $\mathbf{K}_{12}$, and $\mathbf{K}^1_{22}$ to the structure stiffness matrix gives

$$\begin{bmatrix} 70.7 & 70.7 & -70.7 & -70.7 & 0 & 0 \\ & 70.7 & -70.7 & -70.7 & 0 & 0 \\ & & 70.7 & 70.7 & 0 & 0 \\ & \text{symmetrical} & & 70.7 & 0 & 0 \\ & & & & 0 & 0 \\ & & & & & 0 \end{bmatrix}$$

Note that $\mathbf{K}_{21}$ does not have to be added because the structure stiffness matrix is known to be symmetrical, and that $\mathbf{K}_{12}$ and $\mathbf{K}^1_{22}$ were found from $\mathbf{K}^2_{11}$ using the relationships developed in section (5.4).

<u>Element 1,3</u>   A = 3000 mm^2, L = 6000 mm, E = 200 kN/mm^2, α = 0°

$$\mathbf{K}^3_{11} = \begin{bmatrix} 100.0 & 0 \\ 0 & 0 \end{bmatrix}$$

Adding $\mathbf{K}^3_{11}$, $\mathbf{K}_{13}$, and $\mathbf{K}^1_{33}$ to the structure stiffness matrix gives

$$\begin{bmatrix} 170.7 & 70.7 & -70.7 & -70.7 & -100.0 & 0 \\ & 70.7 & -70.7 & -70.7 & 0 & 0 \\ & & 70.7 & 70.7 & 0 & 0 \\ & \text{symmetrical} & & 70.7 & 0 & 0 \\ & & & & 100.0 & 0 \\ & & & & & 0 \end{bmatrix}$$

<u>Element 2,3</u> $A = 3000 \text{ mm}^2$, $L = 4243 \text{ mm}$, $E = 200 \text{ kN/mm}^2$, $\alpha = 315°$

$$K_{22}^3 = \begin{bmatrix} 70.7 & -70.7 \\ -70.7 & 70.7 \end{bmatrix}$$

Adding K_{22}^3, K_{23}, and K_{33}^2 completes the generation of the initial structure stiffness matrix.

$$\begin{bmatrix} 170.7 & 70.7 & -70.7 & -70.7 & -100.0 & 0 \\ & 70.7 & -70.7 & -70.7 & 0 & 0 \\ & & 141.4 & 0 & -70.7 & 70.7 \\ & \text{symmetrical} & & 141.4 & 70.7 & -70.7 \\ & & & & 170.7 & -70.7 \\ & & & & & 70.7 \end{bmatrix}$$

The calculation of element stiffness submatrices is a tedious and error prone procedure that is best done with the aid of a computer.

Program ESTIFF.PT

This program calculates the element stiffness matrix for a plane truss element in terms of the member and the global axes systems.

Input   The program first requests the number of elements to be considered. It then loops for each element in turn and requests the following data:

    A   -   the cross-sectional area of the element
    L   -   the length of the element
    E   -   the elastic constant for the element
    α   -   the angle that the element makes with the global
            x-axis, defined as shown:

A, L, and E must be in *consistent* units, and the unit of force used used for E must correspond to the unit of force used for the applied loading. α is in degrees.

PLANE TRUSSES

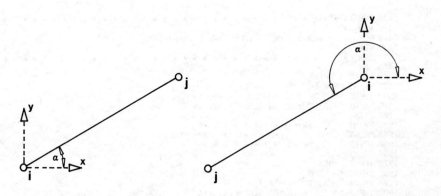

Output The input is echoed and the element stiffness matrix is then output, first in terms of the member axes system, and then in terms of the global axes system, as illustrated in the sample run that follows the program listing.

```
1000 REM  ************************************************************
1010 REM  *                                                          *
1020 REM  *    PROGRAM    E S T I F F . P T    TO CALCULATE THE ELEMENT *
1030 REM  *    STIFFNESS MATRICES IN THE MEMBER AND THE GLOBAL AXES   *
1040 REM  *    SYSTEMS FOR A PLANE TRUSS MEMBER                       *
1050 REM  *                                                          *
1060 REM  *       J.A.D.BALFOUR                                       *
1070 REM  *                                                          *
1080 REM  ************************************************************
1090 REM
1100 REM  ---  NOTE THAT VARIABLES ARE DEFINED IN APPENDIX A
1110 REM
1120 OPTION BASE 1
1130 DIM ESTIFF(4,4)
1140 PRINT : PRINT
1150 PRINT "-----------------------------------------------"
1160 PRINT
1170 PRINT "PROGRAM    E S T I F F . P T    TO CALCULATE THE"
1180 PRINT "ELEMENT STIFFNESS MATRICES IN THE MEMBER AND THE"
1190 PRINT "GLOBAL AXES SYSTEMS FOR A PLANE TRUSS MEMBER"
1200 PRINT
1210 PRINT "-----------------------------------------------"
1220 PRINT : PRINT
1230 INPUT "NUMBER OF ELEMENTS   =  ", NMEMB
1240 REM
1250 REM  ---  LOOP FOR ALL ELEMENTS
1260 REM
1270 FOR K = 1 TO NMEMB
1280 REM
1290 REM  ---  INPUT THE ELEMENT DATA
1300 REM
1310 GOSUB 1420
1320 REM
1330 REM  ---  CALCULATE THE ELEMENT STIFFNESS MATRIX IN THE MEMBER AXES SYSTEM
1340 REM
1350 GOSUB 1630
1360 REM
1370 REM  ---  CALCULATE THE ELEMENT STIFFNESS MATRIX IN THE GLOBAL AXES SYSTEM
1380 REM
1390 GOSUB 1980
1400 NEXT K
1410 END
```

```
1420 REM  * * * * * * * * * * * * * * * * * * * * * * * * * * * * * * * *
1430 REM  *                                                              *
1440 REM  *     SUBROUTINE TO INPUT THE ELEMENT DATA                     *
1450 REM  *                                                              *
1460 REM  * * * * * * * * * * * * * * * * * * * * * * * * * * * * * * * *
1470 REM
1480 PRINT : PRINT
1490 PRINT "ELEMENT "; K : PRINT
1500 INPUT "LOWER  NODE NUMBER   = ", IN
1510 INPUT "HIGHER NODE NUMBER   = ", JN
1520 PRINT
1530 PRINT         "+ + + + + + + + + + +"
1540 PRINT USING "+    ELEMENT ## ##   +"; IN, JN
1550 PRINT         "+ + + + + + + + + + +"
1560 PRINT
1570 INPUT "AREA                         = ", A
1580 INPUT "LENGTH                       = ", L
1590 INPUT "ELASTIC CONSTANT             = ", E
1600 INPUT "ANGLE WITH THE GLOBAL X-AXIS = ", ALPHA
1610 PRINT
1620 RETURN
1630 REM  * * * * * * * * * * * * * * * * * * * * * * * * * * * * * * * *
1640 REM  *                                                              *
1650 REM  *     SUBROUTINE TO CALCULATE AND OUTPUT THE ELEMENT STIFFNESS *
1660 REM  *     MATRIX IN THE MEMBER AXES SYSTEM                         *
1670 REM  *                                                              *
1680 REM  * * * * * * * * * * * * * * * * * * * * * * * * * * * * * * * *
1690 REM
1700 FOR I = 1 TO 4
1710 FOR J = 1 TO 4
1720 ESTIFF(I,J) = 0
1730 NEXT J
1740 NEXT I
1750 REM
1760 REM  --- CALCULATE THE STIFFNESS TERMS (UPPER TRIANGLE ONLY)
1770 REM
1780 ESTIFF(1,1) =   A * E / L
1790 ESTIFF(1,3) = - ESTIFF(1,1)
1800 ESTIFF(3,3) =   ESTIFF(1,1)
1810 REM
1820 REM  --- OUTPUT THE ELEMENT STIFFNESS MATRIX IN THE MEMBER AXES SYSTEM
1830 REM
1840 PRINT
1850 PRINT "* * * * * * * * * * * * * * *"
1860 PRINT "*   ELEMENT STIFFNESS MATRIX IN   *"
1870 PRINT "*     THE MEMBER AXES SYSTEM      *"
1880 PRINT "* * * * * * * * * * * * * * *"
1890 PRINT
1900 FOR I = 1 TO 4
1910 FOR J = 1 TO 4
1920 IF I <= J THEN PRINT USING "#.####^^^^    "; ESTIFF(I,J);
1930 IF I >  J THEN PRINT USING "#.####^^^^    "; ESTIFF(J,I);
1940 NEXT J
1950 PRINT
1960 NEXT I
1970 RETURN
1980 REM  * * * * * * * * * * * * * * * * * * * * * * * * * * * * * * * *
1990 REM  *                                                              *
2000 REM  *     SUBROUTINE TO CALCULATE AND OUTPUT THE ELEMENT STIFFNESS *
2010 REM  *     MATRIX IN THE GLOBAL AXES SYSTEM                         *
2020 REM  *                                                              *
2030 REM  * * * * * * * * * * * * * * * * * * * * * * * * * * * * * * * *
2040 REM
2050 REM  --- CONVERT THE ANGLE TO RADIANS
2060 REM
2070 AR = 4 * ATN(1) * ALPHA / 180
2080 C  = COS(AR)
2090 S  = SIN(AR)
2100 FOR I = 1 TO 4
2110 FOR J = 1 TO 4
2120 ESTIFF(I,J) = 0
2130 NEXT J
2140 NEXT I
```

PLANE TRUSSES

```
2150 REM
2160 REM --- CALCULATE THE STIFFNESS TERMS (UPPER TRIANGLE ONLY)
2170 REM
2180 ESTIFF(1,1) =     E * A * C * C / L
2190 ESTIFF(1,2) =     E * A * C * S / L
2200 ESTIFF(1,3) = -   ESTIFF(1,1)
2210 ESTIFF(1,4) = -   ESTIFF(1,2)
2220 ESTIFF(2,2) =     E * A * S * S / L
2230 ESTIFF(2,3) = -   ESTIFF(1,2)
2240 ESTIFF(2,4) = -   ESTIFF(2,2)
2250 ESTIFF(3,3) =     ESTIFF(1,1)
2260 ESTIFF(3,4) =     ESTIFF(1,2)
2270 ESTIFF(4,4) =     ESTIFF(2,2)
2280 REM
2290 REM --- OUTPUT THE ELEMENT STIFFNESS MATRIX IN THE GLOBAL AXES SYSTEM
2300 REM
2310 PRINT
2320 PRINT "* * * * * * * * * * * * * * * * *"
2330 PRINT "*    ELEMENT STIFFNESS MATRIX IN    *"
2340 PRINT "*       THE GLOBAL AXES SYSTEM      *"
2350 PRINT "* * * * * * * * * * * * * * * * *"
2360 PRINT
2370 FOR I = 1 TO 4
2380 FOR J = 1 TO 4
2390 IF I <= J THEN PRINT USING "#.####^^^^    "; ESTIFF(I,J);
2400 IF I >  J THEN PRINT USING "#.####^^^^    "; ESTIFF(J,I);
2410 NEXT J
2420 PRINT
2430 NEXT I
2440 RETURN
```

Sample Run The following run demonstrates program ESTIFF.PT being used to find the element stiffness matrices for the elements of the structure analysed in example 5.1. Input from the keyboard is shown in bold typeface.

```
--------------------------------------------------
PROGRAM     E S T I F F . P T    TO CALCULATE THE
ELEMENT STIFFNESS MATRICES IN THE MEMBER AND THE
GLOBAL AXES SYSTEMS FOR A PLANE TRUSS MEMBER
--------------------------------------------------

NUMBER OF ELEMENTS    =  3

ELEMENT  1

LOWER  NODE NUMBER    =  1
HIGHER NODE NUMBER    =  2

+ + + + + + + + + +
+    ELEMENT  1  2    +
+ + + + + + + + + +
AREA                           =  3000
LENGTH                         =  4243
ELASTIC CONSTANT               =  200
ANGLE WITH THE GLOBAL X-AXIS   =  45
```

COMPUTER ANALYSIS OF STRUCTURAL FRAMEWORKS

```
* * * * * * * * * * * * * * * * *
*   ELEMENT STIFFNESS MATRIX IN  *
*   THE MEMBER AXES SYSTEM       *
* * * * * * * * * * * * * * * * *

 0.1414E+03    0.0000E+00   -.1414E+03    0.0000E+00
 0.0000E+00    0.0000E+00    0.0000E+00    0.0000E+00
-.1414E+03     0.0000E+00    0.1414E+03    0.0000E+00
 0.0000E+00    0.0000E+00    0.0000E+00    0.0000E+00

* * * * * * * * * * * * * * * * *
*   ELEMENT STIFFNESS MATRIX IN  *
*   THE GLOBAL AXES SYSTEM       *
* * * * * * * * * * * * * * * * *

 0.7070E+02    0.7070E+02   -.7070E+02   -.7070E+02
 0.7070E+02    0.7070E+02   -.7070E+02   -.7070E+02
-.7070E+02    -.7070E+02    0.7070E+02    0.7070E+02
-.7070E+02    -.7070E+02    0.7070E+02    0.7070E+02
```

ELEMENT 2

LOWER NODE NUMBER   = 1
HIGHER NODE NUMBER   = 3

```
+ + + + + + + + + + +
+     ELEMENT  1  3  +
+ + + + + + + + + + +
```

AREA   = 3000
LENGTH   = 6000
ELASTIC CONSTANT   = 200
ANGLE WITH THE GLOBAL X-AXIS = 0

```
* * * * * * * * * * * * * * * * *
*   ELEMENT STIFFNESS MATRIX IN  *
*   THE MEMBER AXES SYSTEM       *
* * * * * * * * * * * * * * * * *

 0.1000E+03    0.0000E+00   -.1000E+03    0.0000E+00
 0.0000E+00    0.0000E+00    0.0000E+00    0.0000E+00
-.1000E+03     0.0000E+00    0.1000E+03    0.0000E+00
 0.0000E+00    0.0000E+00    0.0000E+00    0.0000E+00

* * * * * * * * * * * * * * * * *
*   ELEMENT STIFFNESS MATRIX IN  *
*   THE GLOBAL AXES SYSTEM       *
* * * * * * * * * * * * * * * * *

 0.1000E+03    0.0000E+00   -.1000E+03    0.0000E+00
 0.0000E+00    0.0000E+00    0.0000E+00    0.0000E+00
-.1000E+03     0.0000E+00    0.1000E+03    0.0000E+00
 0.0000E+00    0.0000E+00    0.0000E+00    0.0000E+00
```

ELEMENT 3

LOWER NODE NUMBER   = 2
HIGHER NODE NUMBER   = 3

```
+ + + + + + + + + + +
+     ELEMENT  2  3  +
+ + + + + + + + + + +
```

AREA   = 3000
LENGTH   = 4243
ELASTIC CONSTANT   = 200
ANGLE WITH THE GLOBAL X-AXIS = 315

PLANE TRUSSES

```
* * * * * * * * * * * * * * * *
*    ELEMENT STIFFNESS MATRIX IN   *
*       THE MEMBER AXES SYSTEM     *
* * * * * * * * * * * * * * * *

 0.1414E+03   0.0000E+00  -.1414E+03   0.0000E+00
 0.0000E+00   0.0000E+00   0.0000E+00   0.0000E+00
-.1414E+03    0.0000E+00   0.1414E+03   0.0000E+00
 0.0000E+00   0.0000E+00   0.0000E+00   0.0000E+00


* * * * * * * * * * * * * * * *
*    ELEMENT STIFFNESS MATRIX IN   *
*       THE GLOBAL AXES SYSTEM     *
* * * * * * * * * * * * * * * *

 0.7070E+02  -.7070E+02  -.7070E+02   0.7070E+02
-.7070E+02   0.7070E+02   0.7070E+02  -.7070E+02
-.7070E+02   0.7070E+02   0.7070E+02  -.7070E+02
 0.7070E+02  -.7070E+02  -.7070E+02   0.7070E+02
```

Program   ISTIFF.PT

This program illustrates how the program ESTIFF.PT can be developed to produce a program which will automatically generate the initial structure stiffness matrix.

Input   The number of nodes and the number of elements are input. The program then loops for each element in turn. From the element data the global element stiffness matrix is calculated and then added to the initial structure stiffness matrix.

Comment on the Algorithm   Only the upper triangle of the initial structure stiffness matrix is generated.

Output   The initial structure stiffness matrix is output, with advantage being taken of the symmetry of the matrix.

Listing

```
1000 REM  ***********************************************************************
1010 REM  *                                                                     *
1020 REM  *    PROGRAM   I S T I F F . P T    TO GENERATE THE INITIAL           *
1030 REM  *    STIFFNESS MATRIX FOR A PLANE TRUSS STRUCTURE                     *
1040 REM  *                                                                     *
1050 REM  *    J.A.D.BALFOUR                                                    *
1060 REM  *                                                                     *
1070 REM  ***********************************************************************
1080 REM
1090 REM  ---  NOTE THAT VARIABLES ARE DEFINED IN APPENDIX A
1100 REM
1110 OPTION BASE 1
1120 REM
1130 REM  ---  MAX PROBLEM SIZE SET BY THE FOLLOWING DIMENSION STATEMENT
1140 REM
1150 DIM ISTIFF(20,20), ESTIFF(2,2)
1160 PRINT : PRINT
```

COMPUTER ANALYSIS OF STRUCTURAL FRAMEWORKS

```
1170 PRINT "---------------------------------------------"
1180 PRINT
1190 PRINT "PROGRAM    I S T I F F . P T    TO GENERATE THE"
1200 PRINT "INITIAL STIFFNESS MATRIX FOR A PLANE TRUSS"
1210 PRINT
1220 PRINT "---------------------------------------------"
1230 PRINT : PRINT
1240 INPUT "NUMBER OF NODES      = ", NNODE
1250 INPUT "NUMBER OF ELEMENTS   = ", NMEMB
1260 REM
1270 REM  ---  ZERO THE INITIAL STIFFNESS MATRIX
1280 REM
1290 NDOF = 2 * NNODE
1300 FOR I = 1 TO NDOF
1310 FOR J = 1 TO NDOF
1320 ISTIFF(I,J) = 0
1330 NEXT J
1340 NEXT I
1350 REM
1360 REM  ---  LOOP FOR ALL ELEMENTS
1370 REM
1380 FOR K = 1 TO NMEMB
1390 REM
1400 REM  ---  INPUT ELEMENT DATA
1410 REM
1420 GOSUB 1530
1430 REM
1440 REM  ---  ADD THE ELEMENT STIFFNESS TO THE INITIAL STIFFNESS MATRIX
1450 REM
1460 GOSUB 1730
1470 NEXT K
1480 REM
1490 REM  ---  OUTPUT THE INITIAL STRUCTURE STIFFNESS MATRIX
1500 REM
1510 GOSUB 2060
1520 END
1530 REM  * * * * * * * * * * * * * * * * * * * * * * * * * * * * * * *
1540 REM  *                                                           *
1550 REM  *    SUBROUTINE TO INPUT THE ELEMENT DATA                   *
1560 REM  *                                                           *
1570 REM  * * * * * * * * * * * * * * * * * * * * * * * * * * * * * * *
1580 REM
1590 PRINT : PRINT
1600 PRINT "ELEMENT "; K : PRINT
1610 INPUT "LOWER  NODE NUMBER    = ", IN
1620 INPUT "HIGHER NODE NUMBER    = ", JN
1630 PRINT
1640 PRINT         "+ + + + + + + + + +"
1650 PRINT USING "+    ELEMENT ## ##   +"; IN, JN
1660 PRINT         "+ + + + + + + + + +"
1670 PRINT
1680 INPUT "AREA                        = ", A
1690 INPUT "LENGTH                      = ", L
1700 INPUT "ELASTIC CONSTANT            = ", E
1710 INPUT "ANGLE WITH THE GLOBAL X-AXIS = ", ALPHA
1720 RETURN
1730 REM  * * * * * * * * * * * * * * * * * * * * * * * * * * * * * * *
1740 REM  *                                                           *
1750 REM  *    SUBROUTINE TO ADD THE ELEMENT STIFFNESS TO THE INITIAL *
1760 REM  *    STRUCTURE STIFFNESS MATRIX                             *
1770 REM  *                                                           *
1780 REM  * * * * * * * * * * * * * * * * * * * * * * * * * * * * * * *
1790 REM
1800 REM  ---  CONVERT THE ANGLE TO RADIANS
1810 REM
1820 AR = 4 * ATN(1) * ALPHA / 180
1830 C  = COS(AR)
1840 S  = SIN(AR)
1850 REM
1860 REM  ---  CALCULATE THE UPPER TRIANGLE OF THE ELEMENT STIFFNESS MATRIX
```

```
1870 REM
1880 ESTIFF(1,1) = E * A * C * C / L
1890 ESTIFF(1,2) = E * A * C * S / L
1900 ESTIFF(2,2) = E * A * S * S / L
1910 REM
1920 REM  ---  ADD ELEMENT STIFFNESS TERMS TO THE UPPER TRIANGLE
1930 REM  ---  OF THE STRUCTURE STIFFNESS MATRIX
1940 REM
1950 ISTIFF(2*IN-1,2*IN-1) =    ISTIFF(2*IN-1,2*IN-1) + ESTIFF(1,1)
1960 ISTIFF(2*IN-1,2*IN  ) =    ISTIFF(2*IN-1,2*IN  ) + ESTIFF(1,2)
1970 ISTIFF(2*IN  ,2*IN  ) =    ISTIFF(2*IN  ,2*IN  ) + ESTIFF(2,2)
1980 ISTIFF(2*IN-1,2*JN-1) = -  ESTIFF(1,1)
1990 ISTIFF(2*IN-1,2*JN  ) = -  ESTIFF(1,2)
2000 ISTIFF(2*IN  ,2*JN-1) = -  ESTIFF(1,2)
2010 ISTIFF(2*IN  ,2*JN  ) = -  ESTIFF(2,2)
2020 ISTIFF(2*JN-1,2*JN-1) =    ISTIFF(2*JN-1,2*JN-1) + ESTIFF(1,1)
2030 ISTIFF(2*JN-1,2*JN  ) =    ISTIFF(2*JN-1,2*JN  ) + ESTIFF(1,2)
2040 ISTIFF(2*JN  ,2*JN  ) =    ISTIFF(2*JN  ,2*JN  ) + ESTIFF(2,2)
2050 RETURN
2060 REM * * * * * * * * * * * * * * * * * * * * * * * * * * * * *
2070 REM *                                                         *
2080 REM *    SUBROUTINE TO OUTPUT THE INITIAL STRUCTURE STIFFNESS MATRIX    *
2090 REM *                                                         *
2100 REM * * * * * * * * * * * * * * * * * * * * * * * * * * * * *
2110 REM
2120 PRINT : PRINT
2130 PRINT "* * * * * * * * * * *"
2140 PRINT "*   INITIAL STRUCTURE   *"
2150 PRINT "*   STIFFNESS MATRIX    *"
2160 PRINT "* * * * * * * * * * *"
2170 FOR I = 1 TO NDOF
2180 PRINT : PRINT
2190 FOR J = 1 TO NDOF
2200 IF I <= J THEN PRINT ISTIFF(I,J), ELSE PRINT ISTIFF(J,I),
2210 NEXT J
2220 NEXT I
2230 RETURN
```

Sample Run The following run demonstrates program ISTIFF.PT being used to find the initial structure stiffness for the structure analysed in example 5.1. Input from the keyboard is shown in bold typeface.

```
---------------------------------------------
PROGRAM   I S T I F F . P T   TO GENERATE THE
INITIAL STIFFNESS MATRIX FOR A PLANE TRUSS
---------------------------------------------


NUMBER OF NODES      =  3
NUMBER OF ELEMENTS   =  3


ELEMENT  1

LOWER  NODE NUMBER   =  1
HIGHER NODE NUMBER   =  2

+ + + + + + + + + + +
+   ELEMENT  1  2   +
+ + + + + + + + + + +

AREA                         =  3000
LENGTH                       =  4243
ELASTIC CONSTANT             =  200
ANGLE WITH THE GLOBAL X-AXIS =  45
```

ELEMENT 2

LOWER NODE NUMBER   = 1
HIGHER NODE NUMBER   = 3

```
+ + + + + + + + + + +
+   ELEMENT  1  3   +
+ + + + + + + + + + +
```

AREA   = 3000
LENGTH   = 6000
ELASTIC CONSTANT   = 200
ANGLE WITH THE GLOBAL X-AXIS   = 0

ELEMENT 3

LOWER NODE NUMBER   = 2
HIGHER NODE NUMBER   = 3

```
+ + + + + + + + + + +
+   ELEMENT  2  3   +
+ + + + + + + + + + +
```

AREA   = 3000
LENGTH   = 4243
ELASTIC CONSTANT   = 200
ANGLE WITH THE GLOBAL X-AXIS   = 315

```
* * * * * * * * * * * *
*   INITIAL STRUCTURE  *
*   STIFFNESS MATRIX   *
* * * * * * * * * * * *
```

170.705	70.7047	-70.7047	-70.7047	-100	0
70.7047	70.7047	-70.7047	-70.7047	0	0
-70.7047	-70.7047	141.409	-7.62939E-06	-70.7048	70.7047
-70.7047	-70.7047	-7.62939E-06	141.409	70.7047	-70.7046
-100	0	-70.7048	70.7047	170.705	-70.7047
0	0	70.7047	-70.7046	-70.7047	70.7046

5.7 APPLICATION OF THE BOUNDARY CONDITIONS

Prior to the application of the boundary conditions the initial structure stiffness equation is a mixed system of simultaneous equations (i.e. there are known and unknown quantities on each side of the equality). Consider, for example, the truss shown in fig 5.9.

The initial stiffness relationship is

$$\begin{bmatrix} K_{11} & K_{12} & K_{13} & K_{14} & K_{15} & K_{16} \\ K_{21} & K_{22} & K_{23} & K_{24} & K_{25} & K_{26} \\ K_{31} & K_{32} & K_{33} & K_{34} & K_{35} & K_{36} \\ K_{41} & K_{42} & K_{43} & K_{44} & K_{45} & K_{46} \\ K_{51} & K_{52} & K_{53} & K_{54} & K_{55} & K_{56} \\ K_{61} & K_{62} & K_{63} & K_{64} & K_{65} & K_{66} \end{bmatrix} \begin{bmatrix} \Delta_{1x} \\ \Delta_{1y} \\ \Delta_{2x} \\ \Delta_{2y} \\ \Delta_{3x} \\ \Delta_{3y} \end{bmatrix} = \begin{bmatrix} P_{1x} \\ P_{1y} \\ P_{2x} \\ P_{2y} \\ P_{3x} \\ P_{3y} \end{bmatrix} \quad (5.17)$$

PLANE TRUSSES

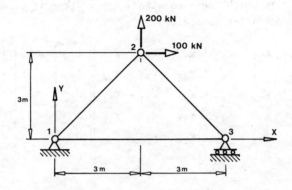

FIGURE 5.9   A THREE MEMBER TRUSS

The freebody diagram for the complete structure is shown in fig 5.10.

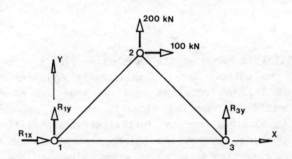

FIGURE 5.10   FREEBODY DIAGRAM FOR THE COMPLETE STRUCTURE

Hence the force vector **P** contains the known applied forces and the unknown reactive forces. Further, the boundary conditions demand that $\Delta_{1x} = \Delta_{1y} = \Delta_{3y} = 0$. Hence the vectors of force and displacement both contain known and unknown quantities, and equation (5.17) becomes

$$\begin{bmatrix} K_{11} & K_{12} & K_{13} & K_{14} & K_{15} & K_{16} \\ K_{21} & K_{22} & K_{23} & K_{24} & K_{25} & K_{26} \\ K_{31} & K_{32} & K_{33} & K_{34} & K_{35} & K_{36} \\ K_{41} & K_{42} & K_{43} & K_{44} & K_{45} & K_{46} \\ K_{51} & K_{52} & K_{53} & K_{54} & K_{55} & K_{56} \\ K_{61} & K_{62} & K_{63} & K_{64} & K_{65} & K_{66} \end{bmatrix} \begin{bmatrix} 0 \\ 0 \\ \Delta_{2x} \\ \Delta_{2y} \\ \Delta_{3x} \\ 0 \end{bmatrix} = \begin{bmatrix} R_{1x} \\ R_{1y} \\ 100 \\ 200 \\ 0 \\ R_{3y} \end{bmatrix}$$

As the order in which these equations are written is of no importance they can be rewritten as follows:

$$\begin{bmatrix} K_{31} & K_{32} & K_{33} & | & K_{34} & K_{35} & K_{36} \\ K_{41} & K_{42} & K_{43} & | & K_{44} & K_{45} & K_{46} \\ K_{51} & K_{52} & K_{53} & | & K_{54} & K_{55} & K_{56} \\ \hline K_{11} & K_{12} & K_{13} & | & K_{14} & K_{15} & K_{16} \\ K_{21} & K_{22} & K_{23} & | & K_{24} & K_{25} & K_{26} \\ K_{61} & K_{62} & K_{63} & | & K_{64} & K_{65} & K_{66} \end{bmatrix} \begin{bmatrix} \Delta_{2x} \\ \Delta_{2y} \\ \Delta_{3x} \\ 0 \\ 0 \\ 0 \end{bmatrix} = \begin{bmatrix} 100 \\ 200 \\ 0 \\ R_{1x} \\ R_{1y} \\ R_{3y} \end{bmatrix}$$

Or, in terms of the submatrices indicated by the broken lines:

$$\begin{bmatrix} K_{I,I} & | & K_{I,II} \\ \hline K_{II,I} & | & K_{II,II} \end{bmatrix} \begin{bmatrix} \Delta \\ 0 \end{bmatrix} = \begin{bmatrix} P \\ R \end{bmatrix} \quad (5.18)$$

hence

$$K_{I,I} \Delta = P \quad (5.19)$$

Equation (5.19) is known as the *final structure stiffness equation*. It relates the unknown nodal displacements and the known applied loads and it is, in fact, the equation that was solved when the stiffness method was applied directly in section 3.5. Equation (5.19) can be obtained from the initial structure stiffness equation by simply deleting the row and column centred on the diagonal for each boundary condition in turn, as shown below.

$$\begin{bmatrix} \cancel{K_{11}} & \cancel{K_{12}} & \cancel{K_{13}} & \cancel{K_{14}} & \cancel{K_{15}} & \cancel{K_{16}} \\ \cancel{K_{21}} & \cancel{K_{22}} & \cancel{K_{23}} & \cancel{K_{24}} & \cancel{K_{25}} & \cancel{K_{26}} \\ K_{31} & K_{32} & K_{33} & K_{34} & K_{35} & K_{36} \\ K_{41} & K_{42} & K_{43} & K_{44} & K_{45} & K_{46} \\ K_{51} & K_{52} & K_{53} & K_{54} & K_{55} & K_{56} \\ \cancel{K_{61}} & \cancel{K_{62}} & \cancel{K_{63}} & \cancel{K_{64}} & \cancel{K_{65}} & \cancel{K_{66}} \end{bmatrix} \begin{bmatrix} \cancel{0} \\ \cancel{0} \\ \Delta_{2x} \\ \Delta_{2y} \\ \Delta_{3x} \\ \cancel{0} \end{bmatrix} = \begin{bmatrix} \cancel{R_{1x}} \\ \cancel{R_{1y}} \\ 100 \\ 200 \\ 0 \\ \cancel{R_{3y}} \end{bmatrix}$$

Equation (5.18) also yields the following equation:

$$K_{II,I} \Delta = R \quad (5.20)$$

The solution of equation (5.19) yields the unknown structural displacements which can then be used in equation (5.20) to find the reactions.

Example 5.3

Find the displacements for the truss shown. As the initial stiffness matrix for this structure was found in example 5.2, the initial structure stiffness equation can be written down without calculation, as follows:

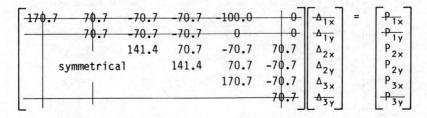

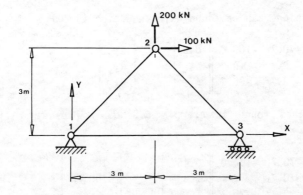

Applying the boundary conditions ($\Delta_{1x} = \Delta_{1y} = \Delta_{3y} = 0$) as shown, and inserting the values of the applied loads produces the final structure stiffness equation:

$$\begin{bmatrix} 141.4 & 0 & -70.7 \\ 0 & 141.4 & 70.7 \\ -70.7 & 70.7 & 170.7 \end{bmatrix} \begin{bmatrix} \Delta_{2x} \\ \Delta_{2y} \\ \Delta_{3x} \end{bmatrix} = \begin{bmatrix} 100 \\ 200 \\ 0 \end{bmatrix}$$

Solving these equations using any one of the equation solving programs given in Chapter 4 yields

$$\begin{bmatrix} \Delta_{2x} \\ \Delta_{2y} \\ \Delta_{3x} \end{bmatrix} = \begin{bmatrix} 0.457 \text{ mm} \\ 1.664 \text{ mm} \\ -0.500 \text{ mm} \end{bmatrix}$$

COMPUTER ANALYSIS OF STRUCTURAL FRAMEWORKS

Example 5.4

Find the reactions to the frame analysed in example 5.3.
If the boundary conditions are imposed by crossing out rows and columns in the initial structure stiffness matrix then the $K_{11,1}$ matrix comprises the elements eliminated by a horizontal line only. These elements are shown boxed in the following equation:

$$\begin{bmatrix} -170.7 & -70.7 & \boxed{-70.7 \quad -70.7 \quad -100.0} & -0 \\ & 70.7 & \boxed{-70.7 \quad -70.7 \quad -0} & 0 \\ & & 141.4 & 70.7 & -70.7 & 70.7 \\ & \text{symmetrical} & & 141.4 & 70.7 & -70.7 \\ & & & & 170.7 & -70.7 \\ & & \boxed{- \quad - \quad - \quad - \quad - \quad 70.7} \end{bmatrix} \begin{bmatrix} 0 \\ 0 \\ \Delta_{2x} \\ \Delta_{2y} \\ \Delta_{3x} \\ 0 \end{bmatrix} = \begin{bmatrix} R_{1x} \\ R_{1y} \\ P_{2x} \\ P_{2y} \\ P_{3x} \\ R_{3y} \end{bmatrix}$$

Hence the reactions can be found using equation (5.20):

$$\begin{bmatrix} -70.7 & -70.7 & -100.0 \\ -70.7 & -70.7 & 0 \\ 70.7 & -70.7 & -70.7 \end{bmatrix} \begin{bmatrix} \Delta_{2x} \\ \Delta_{2y} \\ \Delta_{3x} \end{bmatrix} = \begin{bmatrix} R_{1x} \\ R_{1y} \\ R_{3y} \end{bmatrix}$$

The displacements found in example 5.3 are now used to find the reactions.

$$\begin{bmatrix} -70.7 & -70.7 & -100.0 \\ -70.7 & -70.7 & 0 \\ 70.7 & -70.7 & -70.7 \end{bmatrix} \begin{bmatrix} 0.457 \\ 1.664 \\ -.500 \end{bmatrix} = \begin{bmatrix} -100 \text{ kN} \\ -150 \text{ kN} \\ -50 \text{ kN} \end{bmatrix}$$

Note that the $K_{11,1}$ matrix is not necessarily symmetrical.

5.8 THE FINAL STRUCTURE STIFFNESS MATRIX - Program FSTIFF.PT

In practice it is uneconomical to generate the initial structure stiffness matrix and apply the boundary conditions afterwards. It is far better to take account of the boundary conditions during the assembly of the stiffness matrix so that only the final structure stiffness matrix is generated. The $K_{11,1}$ submatrix in equation (5.18) need not be generated as the reactions can be found transforming the member end forces into the global axes system at each restrained node (see section 5.9).
To assemble the final structure stiffness matrix requires a knowledge of the boundary conditions. Restraints are conveniently described by the number of the node which is restrained, plus the direction of the restraint. Programming is simplified if the

directions of restraint are given numbers. For example, in the case of a space frame structure where the nodal degree of freedom is six, the restraint directions might be numbered thus:

```
1  -  translation restrained in the global x-direction
2  -  translation restrained in the global y-direction
3  -  translation restrained in the global z-direction
4  -  rotation restrained about the global x-axis
5  -  rotation restrained about the global y-axis
6  -  rotation restrained about the global z-axis.
```

In the case of a plane truss, restraint is only meaningful in directions 1 and 2. The boundary conditions for the truss shown in fig 5.11 would be input as follows:

```
NODE    DIRECTION
 1          1
 1          2
 3          2
```

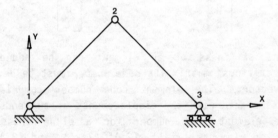

FIGURE 5.11   THREE-MEMBER TRUSS

Input can be reduced by entering the restraint directions at each node as a composite number, and then decoding the composite number within the program. This technique is particularly useful when each node has many possible directions of restraint (e.g. in space frame structures). Using composite numbers the restraints for the truss shown in fig 5.11 are

```
NODE    DIRECTION(S)
 1         12
 3          2
```

A one-dimensional array called the *freedom vector* will be used to facilitate the direct generation of the final structure stiffness matrix. The length of the freedom vector is equal to initial degree of freedom (i.e. the number of nodes times the nodal degree of freedom). It is developed from the boundary conditions as follows:

1. Zero all the elements of the freedom vector.

2. Replace the element corresponding to each restrained freedom by a one.

3. Starting from the beginning of the freedom vector replace ones with zeros, and zeros with consecutive integers.

For example, the freedom vector for the truss shown in fig 5.11 would be found as follows:

Zero the freedom vector. $\begin{bmatrix} 0 \\ 0 \\ 0 \\ 0 \\ 0 \\ 0 \end{bmatrix}$ Apply the boundary conditions. $\begin{bmatrix} 1 \\ 1 \\ 0 \\ 0 \\ 0 \\ 1 \end{bmatrix}$ Number the freedoms. $\begin{bmatrix} 0 \\ 0 \\ 1 \\ 2 \\ 3 \\ 0 \end{bmatrix}$

Before the stiffness of an element can be added to the final structure stiffness matrix its *code number* must be developed from the freedom vector. The element code number comprises the freedom numbers associated with the displacements at the ends of the member. Hence the element code number for a plane truss element "i,j" comprises the elements (2i-1), (2i), (2j-1), and (2j) of the freedom vector.

The truss shown in fig 5.11 will be used to illustrate how element code numbers are used to develop the final structure stiffness matrix. The freedom vector has already been shown to be

$$\{ \boxed{0 \quad 0} \quad 1 \quad 2 \quad \boxed{3 \quad 0} \}$$

The freedom numbers associated with the displacements at the ends of element 1,3 are shown boxed. Hence the code number for element 1,3 is [0 0 3 0].

$$\begin{array}{c} \quad 0 \quad\quad 0 \quad\quad 3 \quad\quad 0 \\ \begin{matrix} 0 \\ 0 \\ 3 \\ 0 \end{matrix} \begin{bmatrix} K_{11} & K_{12} & K_{13} & K_{14} \\ K_{21} & K_{22} & K_{23} & K_{24} \\ K_{31} & K_{32} & K_{33} & K_{34} \\ K_{41} & K_{42} & K_{43} & K_{44} \end{bmatrix} \end{array}$$

PLANE TRUSSES

When the code number is applied to the global element stiffness matrix as shown it is evident that the only stiffness term to contribute to the final stiffness matrix is K_{33} which should be added to location (3,3). The reader should note that the global element stiffness matrix comprises the four global element stiffness submatrices as defined in equations (5.9) to (5.12). Applying the code number for element 1,2 [0 0 1 2] to the global element stiffness matrix for element 1,2 gives

$$\begin{array}{c} \\ 0 \\ 0 \\ 1 \\ 2 \end{array} \begin{array}{cccc} 0 & 0 & 1 & 2 \\ \left[\begin{array}{cccc} K_{11} & K_{12} & K_{13} & K_{14} \\ K_{21} & K_{22} & K_{23} & K_{24} \\ K_{31} & K_{32} & K_{33} & K_{34} \\ K_{41} & K_{42} & K_{43} & K_{44} \end{array}\right] \end{array}$$

Hence element stiffness K_{33} should be added to location (1,1) in the final structure stiffness matrix, element stiffness K_{34} should be added to location (1,2), etc. Note that a zero in either the "horizontal" or the "vertical" code number means that that stiffness coefficient does not contribute to the final structure stiffness matrix.

Example 5.5

Generate the final structure stiffness matrix for the truss shown below using the structure freedom vector and element code numbers. Note the frame is identical to that analysed in example 5.2.

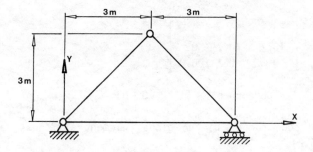

The freedom vector has already been developed for this structure (see fig 5.11) and is known to be {0 0 1 2 3 0}.

Member 1,2 - Code number [0 0 1 2]

Apply the code number to the global element stiffness matrix (from

125

example 5.2)

$$\begin{array}{c} & \begin{array}{cccc} 0 & 0 & 1 & 2 \end{array} \\ \begin{array}{c} 0 \\ 0 \\ 1 \\ 2 \end{array} & \left[\begin{array}{cccc} 70.7 & 70.7 & -70.7 & -70.7 \\ 70.7 & 70.7 & -70.7 & -70.7 \\ -70.7 & -70.7 & 70.7 & 70.7 \\ -70.7 & -70.7 & 70.7 & 70.7 \end{array} \right] \end{array}$$

Adding the element stiffness to the final structure stiffness matrix (currently set to zero) gives

$$\left[\begin{array}{ccc} 70.7 & 70.7 & 0 \\ 70.7 & 70.7 & 0 \\ 0 & 0 & 0 \end{array} \right]$$

<u>Member 1,3</u> - Code number [0 0 3 0]

Apply the code number to the global element stiffness matrix (from example 5.2)

$$\begin{array}{c} & \begin{array}{cccc} 0 & 0 & 3 & 0 \end{array} \\ \begin{array}{c} 0 \\ 0 \\ 3 \\ 0 \end{array} & \left[\begin{array}{cccc} 100.0 & 0 & -100.0 & 0 \\ 0 & 0 & 0 & 0 \\ -100.0 & 0 & 100.0 & 0 \\ 0 & 0 & 0 & 0 \end{array} \right] \end{array}$$

Adding the element stiffness to the final structure stiffness matrix gives

$$\left[\begin{array}{ccc} 70.7 & 70.7 & 0 \\ 70.7 & 70.7 & 0 \\ 0 & 0 & 100.0 \end{array} \right]$$

<u>Member 2,3</u> - Code number [1 2 3 0]

Apply the code number to the global element stiffness matrix (from example 5.2)

$$\begin{array}{c} & \begin{array}{cccc} 1 & 2 & 3 & 0 \end{array} \\ \begin{array}{c} 1 \\ 2 \\ 3 \\ 0 \end{array} & \left[\begin{array}{cccc} 70.7 & -70.7 & -70.7 & 70.7 \\ -70.7 & 70.7 & 70.7 & -70.7 \\ -70.7 & 70.7 & 70.7 & -70.7 \\ 70.7 & -70.7 & -70.7 & 70.7 \end{array} \right] \end{array}$$

PLANE TRUSSES

Adding this element's stiffness completes the generation of the final structure stiffness matrix, which is

$$\begin{bmatrix} 141.4 & 0 & -70.7 \\ 0 & 141.4 & 70.7 \\ -70.7 & 70.7 & 170.7 \end{bmatrix}$$

Program FSTIFF.PT

This program shows how the freedom vector is set up and then used to assemble the final structure stiffness matrix from the global element stiffness matrices using element code numbers. Note that no attempt has been made to save space by taking advantage of either the symmetry or the bandwidth of the structure stiffness matrix.

Input The program first requests the number of nodes, the number of elements, and the number of restrained nodes. It then requests details of the boundary conditions, followed by the data for each element in turn.

Comments on the Algorithm Once the boundary conditions have been established the freedom vector is developed. The program then loops for each element in turn. The global element stiffness matrix is evaluated and the element code number is used to add the element stiffness to the upper triangle of the structure stiffness matrix.

Output The final structure stiffness matrix is output, and the reader should observe that a special piece of code is required to output the complete array as only the upper triangle of the structure stiffness matrix is available.

Listing

```
1000 REM  ***********************************************************************
1010 REM  *                                                                     *
1020 REM  *      PROGRAM    F S T I F F . P T    TO GENERATE THE FINAL          *
1030 REM  *      STIFFNESS MATRIX FOR A PLANE TRUSS                             *
1040 REM  *                                                                     *
1050 REM  *      J.A.D.BALFOUR                                                  *
1060 REM  *                                                                     *
1070 REM  ***********************************************************************
1080 REM
1090 REM  ---  NOTE THAT VARIABLES ARE DEFINED IN APPENDIX A
1100 REM
1110 OPTION BASE 1
1120 REM
1130 REM  ---  PROBLEM SIZE SET BY THE FOLLOWING DIMENSION STATEMENT
1140 REM
1150 DIM ESTIFF(4,4), FSTIFF(20,20), EFREE(4), FREE(20)
1160 PRINT : PRINT
```

COMPUTER ANALYSIS OF STRUCTURAL FRAMEWORKS

```
1170 PRINT "----------------------------------------------"
1180 PRINT
1190 PRINT "PROGRAM      F S T I F F . P T     TO GENERATE"
1200 PRINT "THE FINAL STIFFNESS MATRIX FOR A PLANE TRUSS"
1210 PRINT
1220 PRINT "----------------------------------------------"
1230 PRINT : PRINT
1240 INPUT "NUMBER OF NODES              = ", NNODE
1250 INPUT "NUMBER OF ELEMENTS           = ", NMEMB
1260 INPUT "NUMBER OF RESTRAINED NODES   = ", NREST
1270 PRINT : PRINT
1280 REM
1290 REM  ---  INPUT THE BOUNDARY CONDITIONS AND GENERATE THE FREEDOM VECTOR
1300 REM
1310 GOSUB 1540
1320 REM
1330 REM  ---  ZERO THE STRUCTURE STIFFNESS MATRIX
1340 REM
1350 FOR I       = 1 TO NDOF
1360 FOR J       = 1 TO NDOF
1370 FSTIFF(I,J) = 0
1380 NEXT J
1390 NEXT I
1400 REM
1410 REM  ---  LOOP FOR ALL ELEMENTS
1420 REM
1430 FOR K = 1 TO NMEMB
1440 REM
1450 REM  ---  INPUT THE ELEMENT DATA AND ADD THE ELEMENT STIFFNESS
1460 REM
1470 GOSUB 1980
1480 NEXT K
1490 REM
1500 REM  ---  OUTPUT THE FINAL STRUCTURE STIFFNESS MATRIX
1510 REM
1520 GOSUB 2570
1530 END
1540 REM  * * * * * * * * * * * * * * * * * * * * * * * * * * * * * * * * *
1550 REM  *                                                                *
1560 REM  *    SUBROUTINE TO INPUT THE BOUNDARY CONDITIONS AND GENERATE    *
1570 REM  *    THE FREEDOM VECTOR                                          *
1580 REM  *                                                                *
1590 REM  * * * * * * * * * * * * * * * * * * * * * * * * * * * * * * * * *
1600 REM
1610 REM  ---  ZERO THE FREEDOM VECTOR
1620 REM
1630 FOR I     = 1 TO 2*NNODE
1640 FREE(I) = 0
1650 NEXT I
1660 PRINT "+ + + + + + + + + + + + +"
1670 PRINT "+  BOUNDARY   CONDITIONS    +"
1680 PRINT "+ + + + + + + + + + + + +"
1690 PRINT
1700 PRINT "INPUT BOUNDARY CONDITIONS"
1710 PRINT "RESTRAINT IN THE GLOBAL X-DIRECTION   =   1"
1720 PRINT "RESTRAINT IN THE GLOBAL Y-DIRECTION   =   2"
1730 PRINT "ENTER RESTRAINTS AS A COMPOSITE NUMBER"
1740 PRINT "(E.G. IF NODE IS RESTRAINED IN BOTH DIRECTIONS ENTER 12)"
1750 FOR I     = 1 TO NREST
1760 PRINT
1770 INPUT "NODE NUMBER     = ", IN
1780 INPUT "DIRECTION(S)    = ", DIRN
1790 REM
1800 REM  ---  EVALUATE THE FREEDOM TO BE RESTRAINED
1810 REM
1820 J        = 2*IN - 2 + (DIRN MOD 10)
1830 FREE(J) = 1
1840 DIRN     = INT(DIRN/10)
1850 IF DIRN > 0 THEN GOTO 1820
1860 NEXT I
1870 REM
```

PLANE TRUSSES

```
1880 REM --- NUMBER THE FREEDOMS (RESTRAINTS SET TO ZERO)
1890 REM
1900 FOR I     = 1 TO 2*NNODE
1910 IF FREE(I) = 1 THEN GOTO 1950
1920 NDOF     = NDOF + 1
1930 FREE(I) = NDOF
1940 GOTO 1960
1950 FREE(I) = 0
1960 NEXT I
1970 RETURN
1980 REM *******************************************
1990 REM *                                                             *
2000 REM *    SUBROUTINE TO INPUT THE ELEMENT DATA AND ADD THE ELEMENT *
2010 REM *    STIFFNESS TO THE FINAL STRUCTURE STIFFNESS MATRIX        *
2020 REM *                                                             *
2030 REM *******************************************
2040 REM
2050 PRINT : PRINT
2060 PRINT "ELEMENT "; K : PRINT
2070 INPUT "LOWER  NODE NUMBER  = ", IN
2080 INPUT "HIGHER NODE NUMBER  = ", JN
2090 PRINT
2100 PRINT         "+ + + + + + + + + + +"
2110 PRINT USING "+    ELEMENT ## ##   +"; IN, JN
2120 PRINT         "+ + + + + + + + + + +"
2130 PRINT
2140 INPUT "AREA                            = ", A
2150 INPUT "LENGTH                          = ", L
2160 INPUT "ELASTIC CONSTANT                = ", E
2170 INPUT "ANGLE WITH THE GLOBAL X-AXIS    = ", ALPHA
2180 REM
2190 REM --- CONVERT THE ANGLE TO RADIANS
2200 REM
2210 AR = 4 * ATN(1) * ALPHA / 180
2220 C  = COS(AR)
2230 S  = SIN(AR)
2240 REM
2250 REM --- GENERATE THE UPPER TRIANGLE OF THE ELEMENT STIFFNESS MATRIX
2260 REM
2270 ESTIFF(1,1) =    E * A * C * C / L
2280 ESTIFF(1,2) =    E * A * C * S / L
2290 ESTIFF(1,3) = -  ESTIFF(1,1)
2300 ESTIFF(1,4) = -  ESTIFF(1,2)
2310 ESTIFF(2,2) =    E * A * S * S / L
2320 ESTIFF(2,3) = -  ESTIFF(1,2)
2330 ESTIFF(2,4) = -  ESTIFF(2,2)
2340 ESTIFF(3,3) =    ESTIFF(1,1)
2350 ESTIFF(3,4) =    ESTIFF(1,2)
2360 ESTIFF(4,4) =    ESTIFF(2,2)
2370 REM
2380 REM --- SET UP THE ELEMENT CODE NUMBER
2390 REM
2400 EFREE(1) = FREE(2*IN-1)
2410 EFREE(2) = FREE(2*IN)
2420 EFREE(3) = FREE(2*JN-1)
2430 EFREE(4) = FREE(2*JN)
2440 REM
2450 REM --- ADD THE ELEMENT STIFFNESS TERMS
2460 REM
2470 FOR I       = 1 TO 4
2480 IF EFREE(I) = 0 THEN GOTO 2550
2490 FOR J       = 1 TO 4
2500 IF EFREE(J) = 0 THEN GOTO 2540
2510 L           = EFREE(I)
2520 M           = EFREE(J)
2530 FSTIFF(L,M) = FSTIFF(L,M) + ESTIFF(I,J)
2540 NEXT J
2550 NEXT I
2560 RETURN
```

COMPUTER ANALYSIS OF STRUCTURAL FRAMEWORKS

```
2570 REM * * * * * * * * * * * * * * * * * * * * * * * * * * * * * *
2580 REM *                                                          *
2590 REM *    SUBROUTINE TO OUTPUT THE FINAL STRUCTURE STIFFNESS MATRIX *
2600 REM *                                                          *
2610 REM * * * * * * * * * * * * * * * * * * * * * * * * * * * * * *
2620 REM
2630 PRINT : PRINT
2640 PRINT "* * * * * * * * * * *"
2650 PRINT "*    FINAL STRUCTURE    *"
2660 PRINT "*    STIFFNESS MATRIX   *"
2670 PRINT "* * * * * * * * * * *"
2680 FOR I = 1 TO NDOF
2690 PRINT : PRINT
2700 FOR J = 1 TO NDOF
2710 IF I <= J THEN PRINT FSTIFF(I,J), ELSE PRINT FSTIFF(J,I),
2720 NEXT J
2730 NEXT I
2740 RETURN
```

Sample Run The following run demonstrates program FSTIFF.PT being used to find the final structure stiffness for the structure analysed in example 5.5. Input from the keyboard is shown in bold typeface.

```
---------------------------------------------
PROGRAM    F S T I F F . P T     TO GENERATE
THE FINAL STIFFNESS MATRIX FOR A PLANE TRUSS
---------------------------------------------


NUMBER OF NODES             =  3
NUMBER OF ELEMENTS          =  3
NUMBER OF RESTRAINED NODES  =  2


+ + + + + + + + + + + + +
+   BOUNDARY  CONDITIONS +
+ + + + + + + + + + + + +

INPUT BOUNDARY CONDITIONS
RESTRAINT IN THE GLOBAL X-DIRECTION  =  1
RESTRAINT IN THE GLOBAL Y-DIRECTION  =  2
ENTER RESTRAINTS AS A COMPOSITE NUMBER
(E.G. IF NODE IS RESTRAINED IN BOTH DIRECTIONS ENTER 12)

NODE NUMBER    =  1
DIRECTION(S)   =  12

NODE NUMBER    =  3
DIRECTION(S)   =  2


ELEMENT  1

LOWER  NODE NUMBER  =  1
HIGHER NODE NUMBER  =  2

+ + + + + + + + + + +
+   ELEMENT  1  2   +
+ + + + + + + + + + +

AREA                          =  3000
LENGTH                        =  4243
ELASTIC CONSTANT              =  200
ANGLE WITH THE GLOBAL X-AXIS  =  45


ELEMENT  2
```

```
LOWER  NODE NUMBER    =  1
HIGHER NODE NUMBER    =  3

+ + + + + + + + + + +
+    ELEMENT  1  3   +
+ + + + + + + + + + +

AREA                             =  3000
LENGTH                           =  6000
ELASTIC CONSTANT                 =  200
ANGLE WITH THE GLOBAL X-AXIS     =  0

ELEMENT  3

LOWER  NODE NUMBER    =  2
HIGHER NODE NUMBER    =  3

+ + + + + + + + + + +
+    ELEMENT  2  3   +
+ + + + + + + + + + +

AREA                             =  3000
LENGTH                           =  4243
ELASTIC CONSTANT                 =  200
ANGLE WITH THE GLOBAL X-AXIS     =  315


* * * * * * * * * * * *
*    FINAL STRUCTURE   *
*    STIFFNESS MATRIX  *
* * * * * * * * * * * *

   141.409        -7.62939E-06    -70.7048

  -7.62939E-06    141.409          70.7047

   -70.7048        70.7047         170.705
```

5.9 MEMBER END FORCES AND REACTIONS - Program MFORCE.PT

The element stiffness submatrices in the member axes system are used to calculate the member end forces from the member end displacements using the following equation (see section 5.2):

$$f_{ij} = k_{ii}^{j} \delta_{ij} + k_{ij} \delta_{ji} \qquad (5.21)$$

Solution of the final structure stiffness equation yields the nodal displacements in terms of the nodal axes systems. To find the member end forces using equation (5.21) the nodal displacements must first be transformed into the member axes system.

hence
$$f_{ij} = k_{ii}^{j} T_{ij} \Delta_i + k_{ij} T_{ij} \Delta_j$$

$$\begin{bmatrix} f_{ijx} \\ f_{ijy} \end{bmatrix} = \begin{bmatrix} EA/L & 0 \\ 0 & 0 \end{bmatrix} \begin{bmatrix} \cos\alpha & \sin\alpha \\ -\sin\alpha & \cos\alpha \end{bmatrix} \begin{bmatrix} \Delta_{ix} \\ \Delta_{iy} \end{bmatrix} +$$

$$\begin{bmatrix} -EA/L & 0 \\ 0 & 0 \end{bmatrix} \begin{bmatrix} \cos\alpha & \sin\alpha \\ -\sin\alpha & \cos\alpha \end{bmatrix} \begin{bmatrix} \Delta_{jx} \\ \Delta_{jy} \end{bmatrix}$$

hence

$$f_{ijx} = EA/L(\cos\alpha\, \Delta_{ix} + \sin\alpha\, \Delta_{iy} - \cos\alpha\, \Delta_{jx} - \sin\alpha\, \Delta_{jy})$$
$$f_{ijy} = 0 \qquad (5.22)$$

Similarly it can be shown that f_{jix} is equal to $-f_{ijx}$ and that $f_{jiy} = 0$. Note also that if f_{ijx} is positive then the member is in *compression*.

Reactions

When loads are applied to a structure the reactive forces assume values that will keep the structure in equilibrium. A restrained node experiences forces from the following three sources:

1. The forces from the members

2. The reactive forces

3. Loads applied to the restrained node (if any).

Figure 5.12(a) shows a restrained node "i" which has members "i,a" and "i,b" framing into it. The freebody diagram for node "i" is shown in fig 5.12(b). Taking components of the axial forces in members "i,a" and "i,b" in the directions of the global axes system allows the equations of equilibrium for node "i" to be written as follows:

$$P_{ix} + R_{ix} + F'_{iax} + F'_{ibx} = 0$$

$$P_{iy} + R_{iy} + F'_{iay} + F'_{iby} = 0$$

Note that F'_{iax} is the component of f'_{ia} in the direction of the global x-axis.

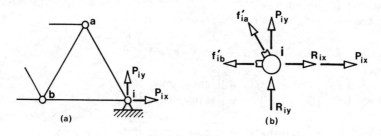

FIGURE 5.12   AN EXAMPLE OF A RESTRAINED NODE

PLANE TRUSSES

In section 5.5 the nodes at the ends of a member were shown to experience forces that were equal and opposite to the member end forces, i.e.

$$F'_{ijx} = -F_{ijx}$$

and

$$F'_{ijy} = -F_{ijy}$$

The equations of nodal equilibrium can, therefore, be written as follows:

$$R_{ix} = F_{iax} + F_{ibx} - P_{ix}$$

$$R_{iy} = F_{iay} + F_{iby} - P_{iy}$$

or, in matrix form

$$R_i = F_{ia} + F_{ib} - P_i$$

where

$$R_i = \begin{bmatrix} R_{ix} \\ R_{iy} \end{bmatrix}, \quad F_{ia} = \begin{bmatrix} F_{iax} \\ F_{iay} \end{bmatrix}, \quad F_{ib} = \begin{bmatrix} F_{ibx} \\ F_{iby} \end{bmatrix}, \quad P_i = \begin{bmatrix} P_{ix} \\ P_{iy} \end{bmatrix}$$

In general

$$R_i = \sum F_{ij} - P_i$$

where the summation is for all members connected to the restrained node "i". Note that the member end forces will have been found in terms of the member axes system and must be transformed into the global axes system prior to summation. Hence the above equation should be written as follows:

$$R_i = \sum T^{-1}_{ij} f_{ij} - P_i \qquad (5.23)$$

Program MFORCE.PT

This program calculates the force in a pin ended member from the member end displacements.

Input The number of members for which the force is to be found is input to allow the program to loop for all members. Then the member properties plus end displacements are input for each member in turn and the axial force in the member is calculated.

COMPUTER ANALYSIS OF STRUCTURAL FRAMEWORKS

Comments on the Algorithm Equation (5.22) is used to calculate the member forces.

Output The force in each member is output.

Listing

```
1000 REM ***********************************************************************
1010 REM *                                                                     *
1020 REM *      PROGRAM     M F O R C E . P T       TO CALCULATE THE MEMBER    *
1030 REM *      END FORCES FROM END DISPLACEMENTS FOR A PLANE TRUSS MEMBER     *
1040 REM *                                                                     *
1050 REM *      J.A.D.BALFOUR                                                  *
1060 REM *                                                                     *
1070 REM ***********************************************************************
1080 REM
1090 REM  ---  NOTE THAT VARIABLES ARE DEFINED IN APPENDIX A
1100 REM
1110 OPTION BASE 1
1120 PRINT : PRINT
1130 PRINT "------------------------------------------------------------"
1140 PRINT
1150 PRINT "PROGRAM    M F O R C E . P T      TO CALCULATE THE MEMBER"
1160 PRINT "END FORCES FROM END DISPLACEMENTS FOR A PLANE TRUSS MEMBER"
1170 PRNT
1180 PRINT "------------------------------------------------------------"
1190 PRINT : PRINT
1200 INPUT "NUMBER OF MEMBERS       = ", NMEMB
1210 REM
1220 REM  ---  LOOP FOR ALL MEMBERS
1230 REM
1240 FOR K = 1 TO NMEMB
1250 PRINT : PRINT
1260 PRINT "ELEMENT "; K : PRINT
1270 INPUT "LOWER  NODE NUMBER  = ", IN
1280 INPUT "HIGHER NODE NUMBER  = ", JN
1290 PRINT
1300 PRINT          "+ + + + + + + + + +"
1310 PRINT USING "+    ELEMENT ## ##    +"; IN, JN
1320 PRINT          "+ + + + + + + + + +"
1330 PRINT
1340 INPUT "AREA                          = ", A
1350 INPUT "LENGTH                        = ", L
1360 INPUT "ELASTIC CONSTANT              = ", E
1370 INPUT "ANGLE WITH THE GLOBAL X-AXIS  = ", ALPHA
1380 PRINT
1390 PRINT "ENTER THE NODAL DISPLACEMENTS (GLOBAL AXES)"
1400 PRINT
1410 PRINT USING "X-DISPLACEMENT AT NODE  ##    = "; IN;
1420 INPUT "", XI
1430 PRINT USING "Y-DISPLACEMENT AT NODE  ##    = "; IN;
1440 INPUT "", YI
1450 PRINT USING "X-DISPLACEMENT AT NODE  ##    = "; JN;
1460 INPUT "", XJ
1470 PRINT USING "Y-DISPLACEMENT AT NODE  ##    = "; JN;
1480 INPUT "", YJ
1490 REM
1500 REM  ---  CONVERT THE ANGLE TO RADIANS
1510 REM
1520 AR   = 4 * ATN(1) * ALPHA / 180
1530 REM
1540 REM  ---  CALCULATE AND OUTPUT THE MEMBER FORCE
1550 REM
1560 FORCE = (-E*A/L) * (COS(AR)*XI + SIN(AR)*YI - COS(AR)*XJ - SIN(AR)*YJ)
1570 PRINT
1580 PRINT USING "FORCE IN ELEMENT   ## ##     = "; IN, JN;
1590 PRINT ABS(FORCE);
1600 IF FORCE > 0 THEN PRINT "    (TENSION)"
```

PLANE TRUSSES

```
1610 IF FORCE < 0 THEN PRINT "   (COMPRESSION)"
1620 NEXT K
1630 END
```

Sample Run The following run demonstrates program MFORCE.PT being used to find the forces in the members of the structure analysed in example 5.3. Input from the keyboard is shown in bold typeface.

```
-------------------------------------------------------------
PROGRAM    M F O R C E . P T    TO CALCULATE THE MEMBER
END FORCES FROM END DISPLACEMENTS FOR A PLANE TRUSS MEMBER
-------------------------------------------------------------


NUMBER OF MEMBERS    =  3

ELEMENT  1

LOWER  NODE NUMBER   =  1
HIGHER NODE NUMBER   =  2

+ + + + + + + + + + +
+   ELEMENT  1  2   +
+ + + + + + + + + + +

AREA                         =  3000
LENGTH                       =  4243
ELASTIC CONSTANT             =  200
ANGLE WITH THE GLOBAL X-AXIS =  45
ENTER THE NODAL DISPLACEMENTS (GLOBAL AXES)

X-DISPLACEMENT AT NODE   1   =  0
Y-DISPLACEMENT AT NODE   1   =  0
X-DISPLACEMENT AT NODE   2   =  .457
Y-DISPLACEMENT AT NODE   2   =  1.66

FORCE IN ELEMENT     1  2    =  211.682    (TENSION)


ELEMENT  2

LOWER  NODE NUMBER   =  1
HIGHER NODE NUMBER   =  3

+ + + + + + + + + + +
+   ELEMENT  1  3   +
+ + + + + + + + + + +

AREA                         =  3000
LENGTH                       =  6000
ELASTIC CONSTANT             =  200
ANGLE WITH THE GLOBAL X-AXIS =  0

ENTER THE NODAL DISPLACEMENTS (GLOBAL AXES)

X-DISPLACEMENT AT NODE   1   =  0
Y-DISPLACEMENT AT NODE   1   =  0
X-DISPLACEMENT AT NODE   3   =  -.5
Y-DISPLACEMENT AT NODE   3   =  0

FORCE IN ELEMENT     1  3    =  50    (COMPRESSION)
```

```
ELEMENT  3

LOWER  NODE  NUMBER   =  2
HIGHER NODE  NUMBER   =  3

+ + + + + + + + + + +
+    ELEMENT  2  3   +
+ + + + + + + + + + +

AREA                            =  3000
LENGTH                          =  4243
ELASTIC CONSTANT                =  200
ANGLE WITH THE GLOBAL X-AXIS    =  315

ENTER THE NODAL DISPLACEMENTS (GLOBAL AXES)

X-DISPLACEMENT AT NODE    2    =   .457
Y-DISPLACEMENT AT NODE    2    =  1.66
X-DISPLACEMENT AT NODE    3    =  -.5
Y-DISPLACEMENT AT NODE    3    =   0

FORCE IN ELEMENT      2  3     =  70.2939    (TENSION)
```

5.10 PUTTING IT ALL TOGETHER - Program PTRUSS.PT

This chapter has now covered all of the procedures necessary for the analysis of plane trusses using the stiffness method. In this section the short computer programs to generate the final structure stiffness matrix (FSTIFF.PT) and to calculate member end forces (MFORCE.PT) are combined with the Gauss-Jordan equation solver (GAUSSJ.EQ) to produce a simple automatic truss analysis program PTRUSS.PT.

The volume of input required by all previous programs has been small enough for it to be entered from the keyboard without undue difficulty. This program requires the following information to be input:

1. Job title
2. Nodal coordinates
3. Member properties
4. Restraints
5. Loads.

The volume of input is likely to be so large that error-free data entry becomes difficult. Further, the program user may well wish to rerun the program using data that is only slightly different from the previous run. For these reasons it is desirable that the data should be read from a data file. This, of course, introduces another level of difficulty to the programming. The advantage of using data files will soon, however, become apparent. Finally it should be noted that commercial structural analysis software invariably makes use of data files.

PLANE TRUSSES

Program PTRUSS.PT

This program, which will automatically analyse plane trusses, is essentially an amalgamation of the programs FSTIFF.PT, MFORCE.PT and GAUSSJ.EQ.

Input The only piece of information the user supplies to the program from the keyboard is the name of the data file to be read by the program. This data file can be constructed using either a text editor, or the data preprocessor program PRE.MA/PRE.01/PRE.02 which is described in section 5.11. If using a text editor then the data file should be constructed to the following format:

```
<job title (not to exceed one line)>
1,2,4,<no. of nodes>,<no. of members>,<no. of restrained nodes>,
<no. of loads>
<x-coordinate node 1>,<y-coordinate node 1>
<x-coordinate node 2>,<y-coordinate node 2>
<x-coordinate node 3>,<y-coordinate node 3>
         .                      .
         .                      .
<x-coordinate node n>,<y-coordinate node n>
<node "i" member 1>,<node "j" member 1>,<area member 1>,<E member 1>
<node "i" member 2>,<node "j" member 2>,<area member 2>,<E member 2>
<node "i" member 3>,<node "j" member 3>,<area member 3>,<E member 3>
       .                  .                   .
    note "i" must be less then  "j"
       .                  .                   .
       .                  .                   .
<restrained node no. 1 node number>,<direction(s)>
<restrained node no. 2 node number>,<direction(s)>
       .                  .
    note - restraint directions are as follows:
           restraint in the global x-direction = 1
           restraint in the global y-direction = 2
       .                  .                   .
    Input restraints as a composite number (see text)
       .                  .                   .
       .                  .                   .
<restrained node no. n node number>,<direction(s)>
<nodal load no. 1  node number>,<direction>,<magnitude>
<nodal load no. 2  node number>,<direction>,<magnitude>
       .                  .                   .
       .                  .                   .
```

```
    note - load directions are as follows
        load in the global  x-direction = 1
        load in the global  y-direction = 2
            .               .               .
              .               .               .
<nodal load no. n  node number>,<direction>,<magnitude>
```

Comments on the Algorithm   The program first uses the boundary conditions to establish the freedom vector. It then adds the stiffness of each element to the final structure stiffness matrix in turn. The upper triangle only of the final structure stiffness matrix is generated. Symmetry is later used to complete the matrix. No attempt is made to minimise storage by taking advantage of either the symmetry or the bandwidth of the structure stiffness matrix. The loading vector is developed and then the equations are solved using Gauss-Jordan elimination. Once the unknown nodal displacements have been found they are used to calculate the member end forces and the reactions.

Output   The data from the data file is echoed and messages are output to indicate the beginning of each solution phase. After solution of the stiffness equation nodal displacements are output, followed by the axial forces in the members, and then the reactions.

Listing

```
1000 REM   ************************************************************************
1010 REM   *                                                                      *
1020 REM   *    PROGRAM    P T R U S S . P T    FOR THE AUTOMATIC ANALYSIS OF     *
1030 REM   *    PLANE TRUSSES.                                                    *
1040 REM   *                                                                      *
1050 REM   *    DATA FROM DATA FILE GENERATED BY A TEXT EDITOR OR BY THE          *
1060 REM   *    PREPROCESSOR PROGRAM    P R E . M A / P R E . 0 1 / P R E . 0 2   *
1070 REM   *                                                                      *
1080 REM   *    J.A.D.BALFOUR                                                     *
1090 REM   *                                                                      *
1100 REM   ************************************************************************
1110 REM
1120 REM   ---  NOTE THAT VARIABLES ARE DEFINED IN APPENDIX A
1130 REM
1140 OPTION BASE 1
1150 REM
1160 REM   ---  THE FOLLOWING DIMENSION STATEMENT SETS THE MAXIMUM PROBLEM SIZE
1170 REM
1180 COMMON FILE$
1190 DIM NODE(20,2),  MEMB(40,7), REST(20,4), NLOD(40,3), FREE(40)
1200 DIM ESTIFF(4,4), EFREE(4),   FSTIFF(40,41)
1210 PRINT : PRINT
1220 PRINT "-------------------------------------------------"
1230 PRINT
1240 PRINT "PROGRAM    P T R U S S . P T    FOR THE AUTOMATIC"
1250 PRINT "ANALYSIS OF PLANE TRUSSES"
1260 PRINT
1270 PRINT "-------------------------------------------------"
1280 REM
```

PLANE TRUSSES

```
1290 REM  ---  CHECK IF THIS PROGRAM HAS BEEN CHAINED FROM PRE.MA/PRE.01/PRE.02
1300 REM
1310 IF LEN(FILE$) <> 0 THEN GOTO 1340
1320 PRINT
1330 INPUT          "NAME OF THE DATA FILE =  ", FILE$
1340 OPEN "I", #1, FILE$
1350 PRINT
1360 PRINT          "READING DATA"
1370 GOSUB 1600
1380 REM
1390 REM                  PRINT DATA
1400 GOSUB 1960
1410 PRINT
1420 PRINT          "GENERATING THE FREEDOM VECTOR"
1430 GOSUB 2490
1440 PRINT
1450 PRINT          "GENERATING THE STRUCTURE STIFFNESS MATRIX"
1460 GOSUB 2850
1470 PRINT
1480 PRINT          "GENERATING THE LOADING VECTOR"
1490 GOSUB 3590
1500 PRINT
1510 PRINT          "SOLVING EQUATIONS  -  NO. OF EQUATIONS  = "; NDOF
1520 GOSUB 3740
1530 REM
1540 REM                  OUTPUT NODAL DISPLACEMENTS
1550 GOSUB 4080
1560 REM
1570 REM                  OUTPUT MEMBER FORCES AND REACTIONS
1580 GOSUB 4330
1590 END
1600 REM * * * * * * * * * * * * * * * * * * * * * * * * * * * * * * * *
1610 REM *                                                              *
1620 REM *     SUBROUTINE TO READ DATA FROM THE INPUT FILE              *
1630 REM *                                                              *
1640 REM * * * * * * * * * * * * * * * * * * * * * * * * * * * * * * * *
1650 REM
1660 REM  ---  READ BASIC PROBLEM DATA
1670 REM
1680 INPUT #1, TITLE$
1690 INPUT #1, PTYPE, NCORD, NPROP, NNODE, NMEMB, NREST, NNLOD
1700 IF PTYPE = 1 THEN GOTO 1750
1710 PRINT : PRINT "NOT A PLANE TRUSS DATA FILE" : END
1720 REM
1730 REM  ---  READ THE NODAL COORDINATES
1740 REM
1750 FOR I = 1 TO NNODE
1760 INPUT #1, NODE(I,1), NODE(I,2)
1770 NEXT I
1780 REM  ---  READ MEMBER DATA
1790 REM
1800 FOR I = 1 TO NMEMB
1810 INPUT #1, MEMB(I,1), MEMB(I,2), MEMB(I,3), MEMB(I,4)
1820 NEXT I
1830 REM
1840 REM  ---  READ RESTRAINT DATA
1850 REM
1860 FOR I = 1 TO NREST
1870 INPUT #1, REST(I,1), REST(I,2)
1880 NEXT I
1890 REM
1900 REM  ---  READ LOADING DATA
1910 REM
1920 FOR I = 1 TO NNLOD
1930 INPUT #1, NLOD(I,1), NLOD(I,2), NLOD(I,3)
1940 NEXT I
1950 RETURN
1960 REM * * * * * * * * * * * * * * * * * * * * * * * * * * * * * * * *
1970 REM *                                                              *
1980 REM *     SUBROUTINE TO PRINT DATA                                 *
1990 REM *                                                              *
2000 REM * * * * * * * * * * * * * * * * * * * * * * * * * * * * * * * *
2010 REM
```

```
2020 PRINT
2030 PRINT "+ + + + + + + + +"
2040 PRINT "+    JOB TITLE   +   -    "; TITLE$
2050 PRINT "+ + + + + + + + +"
2060 PRINT
2070 PRINT "+ + + + + + + + + + + + +"
2080 PRINT "+   NODAL COORDINATES   +"
2090 PRINT "+ + + + + + + + + + + + +"
2100 PRINT
2110 PRINT "NODE           X             Y"
2120 PRINT "NUMBER
2130 FOR I = 1 TO NNODE
2140 PRINT " "; I, NODE(I,1), NODE(I,2)
2150 NEXT I
2160 PRINT
2170 PRINT "+ + + + + + + + + + + + +"
2180 PRINT "+   MEMBER PROPERTIES   +"
2190 PRINT "+ + + + + + + + + + + + +"
2200 PRINT
2210 PRINT "MEMBER          A             E"
2220 PRINT "NUMBER"
2230 FOR I = 1 TO NMEMB
2240 PRINT USING "## ##    "; MEMB(I,1), MEMB(I,2);
2250 PRINT USING "  #.###^^^^"; MEMB(I,3), MEMB(I,4)
2260 NEXT I
2270 PRINT
2280 PRINT "+ + + + + + + + +"
2290 PRINT "+   RESTRAINTS  +"
2300 PRINT "+ + + + + + + + +"
2310 PRINT
2320 PRINT "NODE     DIRECTION(S)"
2330 PRINT "NUMBER"
2340 FOR I = 1 TO NREST
2350 PRINT USING "  ##         ##"; REST(I,1), REST(I,2)
2360 NEXT I
2370 PRINT
2380 PRINT "+ + + + + + + + +"
2390 PRINT "+   NODAL LOADS +"
2400 PRINT "+ + + + + + + + +"
2410 PRINT
2420 PRINT       "NODE    DIRECTION   VALUE"
2430 PRINT       "NUMBER"
2440 FOR I = 1 TO NNLOD
2450 PRINT USING " ##         #"; NLOD(I,1), NLOD(I,2);
2460 PRINT USING "   #.###^^^^"; NLOD(I,3)
2470 NEXT I
2480 RETURN
2490 REM * * * * * * * * * * * * * * * * * * * * * * * * * * * * * * * *
2500 REM *                                                              *
2510 REM *    SUBROUTINE TO GENERATE THE FREEDOM VECTOR                 *
2520 REM *                                                              *
2530 REM * * * * * * * * * * * * * * * * * * * * * * * * * * * * * * * *
2540 REM
2550 REM --- ZERO THE FREEDOM VECTOR
2560 REM
2570 FOR I   = 1 TO 2*NNODE
2580 FREE(I) = 0
2590 NEXT I
2600 REM
2610 REM --- LOOP FOR ALL RESTRAINED NODES
2620 REM
2630 FOR I   = 1 TO NREST
2640 K       = REST(I,2)
2650 REM
2660 REM --- EVALUATE THE FREEDOM TO BE RESTRAINED FROM RIGHTMOST DIGIT OF K
2670 REM
2680 J       = 2*REST(I,1) - 2 + (K MOD 10)
2690 FREE(J) = 1
2700 K       = INT(K/10)
2710 IF K    > 0 THEN GOTO 2680
2720 NEXT I
2730 REM
```

PLANE TRUSSES

```
2740 REM --- NUMBER THE FREEDOMS (RESTRAINTS SET TO ZERO)
2750 REM
2760 NDOF     = 0
2770 FOR I    = 1 TO 2*NNODE
2780 IF FREE(I) = 1 THEN GOTO 2820
2790 NDOF     = NDOF + 1
2800 FREE(I) = NDOF
2810 GOTO 2830
2820 FREE(I) = 0
2830 NEXT I
2840 RETURN
2850 REM * * * * * * * * * * * * * * * * * * * * * * * * * * * * * * *
2860 REM *                                                             *
2870 REM *    SUBROUTINE TO GENERATE THE STRUCTURE STIFFNESS MATRIX    *
2880 REM *                                                             *
2890 REM * * * * * * * * * * * * * * * * * * * * * * * * * * * * * * *
2900 REM
2910 REM --- ZERO THE AUGMENTED MATRIX
2920 REM
2930 FOR I    = 1 TO NDOF
2940 FOR J    = 1 TO NDOF + 1
2950 FSTIFF(I,J) = 0
2960 NEXT J
2970 NEXT I
2980 REM
2990 REM --- LOOP FOR EACH ELEMENT
3000 REM
3010 PRINT "ADDING ELEMENT    :-";
3020 FOR K = 1 TO NMEMB
3030 PRINT K;
3040 IN    = MEMB(K,1)
3050 JN    = MEMB(K,2)
3060 REM
3070 REM --- CALCULATE ELEMENT LENGTH (STORE IN MEMB(K,5))
3080 REM
3090 MEMB(K,5) = SQR((NODE(JN,1)-NODE(IN,1))^2 + (NODE(JN,2)-NODE(IN,2))^2)
3100 REM
3110 REM --- CALCULATE THE COSINE AND SINE OF THE ANGLE THE MEMBER MAKES
3120 REM --- WITH THE COORDINATE X-AXIS (STORE IN MEMB(K,6) AND MEMB(K,7))
3130 REM
3140 MEMB(K,6) = (NODE(JN,1) - NODE(IN,1)) / MEMB(K,5)
3150 MEMB(K,7) = (NODE(JN,2) - NODE(IN,2)) / MEMB(K,5)
3160 REM
3170 REM --- GENERATE THE UPPER TRIANGLE OF THE ELEMENT STIFFNESS MATRIX
3180 REM
3190 ESTIFF(1,1) =     MEMB(K,3) * MEMB(K,4) * MEMB(K,6) * MEMB(K,6) / MEMB(K,5)
3200 ESTIFF(1,2) =     MEMB(K,3) * MEMB(K,4) * MEMB(K,6) * MEMB(K,7) / MEMB(K,5)
3210 ESTIFF(1,3) = - ESTIFF(1,1)
3220 ESTIFF(1,4) = - ESTIFF(1,2)
3230 ESTIFF(2,2) =     MEMB(K,3) * MEMB(K,4) * MEMB(K,7) * MEMB(K,7) / MEMB(K,5)
3240 ESTIFF(2,3) = - ESTIFF(1,2)
3250 ESTIFF(2,4) = - ESTIFF(2,2)
3260 ESTIFF(3,3) =     ESTIFF(1,1)
3270 ESTIFF(3,4) =     ESTIFF(1,2)
3280 ESTIFF(4,4) =     ESTIFF(2,2)
3290 REM
3300 REM --- SET UP THE ELEMENT CODE NUMBER
3310 REM
3320 EFREE(1) = FREE(2*IN-1)
3330 EFREE(2) = FREE(2*IN)
3340 EFREE(3) = FREE(2*JN-1)
3350 EFREE(4) = FREE(2*JN)
3360 REM
3370 REM --- ADD THE ELEMENT STIFFNESS
3380 REM
3390 FOR I    = 1 TO 4
3400 IF EFREE(I) = 0 THEN GOTO 3470
3410 FOR J    = 1 TO 4
3420 IF EFREE(J) = 0 THEN GOTO 3460
3430 L       = EFREE(I)
3440 M       = EFREE(J)
3450 FSTIFF(L,M) = FSTIFF(L,M) + ESTIFF(I,J)
3460 NEXT J
```

```
3470 NEXT I
3480 NEXT K
3490 PRINT
3500 REM
3510 REM   ---   USE SYMMETRY TO FILL OUT THE STIFFNESS ARRAY
3520 REM
3530 FOR I          = 1 TO NDOF-1
3540 FOR J          = I+1 TO NDOF
3550 FSTIFF(J,I) = FSTIFF(I,J)
3560 NEXT J
3570 NEXT I
3580 RETURN
3590 REM  * * * * * * * * * * * * * * * * * * * * * * * * * * * * * * * *
3600 REM  *                                                              *
3610 REM  *    SUBROUTINE TO SET UP THE LOADING VECTOR (IN THE AUGMENTED *
3620 REM  *    MATRIX)                                                   *
3630 REM  *                                                              *
3640 REM  * * * * * * * * * * * * * * * * * * * * * * * * * * * * * * * *
3650 REM
3660 REM   ---   LOOP FOR ALL LOADS
3670 REM
3680 FOR I          = 1 TO NNLOD
3690 J              = 2*NLOD(I,1) - 2 + NLOD(I,2)
3700 J              = FREE(J)
3710 FSTIFF(J,NDOF+1) = NLOD(I,3)
3720 NEXT I
3730 RETURN
3740 REM  * * * * * * * * * * * * * * * * * * * * * * * * * * * * * * * *
3750 REM  *                                                              *
3760 REM  *    SUBROUTINE TO SOLVE LINEAR SIMULTANEOUS EQUATIONS BY GAUSS-*
3770 REM  *    JORDAN ELIMINATION WITH NO ROW INTERCHANGE                *
3780 REM  *                                                              *
3790 REM  * * * * * * * * * * * * * * * * * * * * * * * * * * * * * * * *
3800 REM
3810 REM   ---   LOOP FOR ALL PIVOTS
3820 REM
3830 PRINT "USING PIVOT NUMBER :- ";
3840 FOR I = 1 TO NDOF
3850 PRINT I;
3860 TEMP      = FSTIFF(I,I)
3870 REM
3880 REM   ---   NORMALISE
3890 REM
3900 FOR J         = 1 TO NDOF+1
3910 FSTIFF(I,J) = FSTIFF(I,J) / TEMP
3920 NEXT J
3930 REM
3940 REM   ---   LOOP FOR ALL ROWS USING THE CURRENT PIVOT
3950 REM
3960 FOR J  = 1 TO NDOF
3970 IF J   = I THEN GOTO 4050
3980 TEMP   = FSTIFF(J,I)
3990 REM
4000 REM   ---   LOOP FOR ALL ELEMENTS IN THE ROW
4010 REM
4020 FOR K         = 1 TO NDOF+1
4030 FSTIFF(J,K) = FSTIFF(J,K) - TEMP*FSTIFF(I,K)
4040 NEXT K
4050 NEXT J
4060 NEXT I
4070 RETURN
4080 REM  * * * * * * * * * * * * * * * * * * * * * * * * * * * * * * * *
4090 REM  *                                                              *
4100 REM  *    SUBROUTINE TO OUTPUT THE DISPLACEMENTS                    *
4110 REM  *                                                              *
4120 REM  * * * * * * * * * * * * * * * * * * * * * * * * * * * * * * * *
4130 REM
4140 PRINT : PRINT
4150 PRINT "* * * * * * * * * * *"
4160 PRINT "*   DISPLACEMENTS   *          * :- RESTRAINT"
4170 PRINT "* * * * * * * * * * *"
4180 PRINT
4190 PRINT "NODE     X-DISPLACEMENT  Y-DISPLACEMENT"
```

PLANE TRUSSES

```
4200 PRINT
4210 FOR I = 1 TO NNODE
4220 PRINT USING "###      "; I;
4230 FOR J = 1 TO 2
4240 K       = FREE(2*I+J-2)
4250 IF K    = 0 THEN GOTO 4280
4260 PRINT USING "     #.####^^^^"; FSTIFF(K,NDOF+1);
4270 GOTO 4290
4280 PRINT   "          *      ";
4290 NEXT J
4300 PRINT
4310 NEXT I
4320 RETURN
4330 REM * * * * * * * * * * * * * * * * * * * * * * * * * * * * * *
4340 REM *                                                         *
4350 REM *    SUBROUTINE TO CALCULATE AND OUTPUT THE MEMBER FORCES AND *
4360 REM *    THE REACTIONS                                        *
4370 REM *                                                         *
4380 REM * * * * * * * * * * * * * * * * * * * * * * * * * * * * * *
4390 REM
4400 REM --- ZERO THE REACTION COMPONENTS
4410 REM
4420 FOR I = 1 TO NREST
4430 REST(I,3) = 0
4440 REST(I,4) = 0
4450 NEXT I
4460 PRINT : PRINT
4470 PRINT "* * * * * * * * * *"
4480 PRINT "*   MEMBER FORCES  *           TENSION POSITIVE"
4490 PRINT "* * * * * * * * * *"
4500 PRINT
4510 PRINT "MEMBER            FORCE"
4520 PRINT
4530 FOR I = 1 TO NMEMB
4540 II    = 2*MEMB(I,1) - 1
4550 JJ    = 2*MEMB(I,2) - 1
4560 K     = FREE(II)
4570 IF K  = 0 THEN XI = 0 ELSE XI = FSTIFF(K,NDOF+1)
4580 K     = FREE(II+1)
4590 IF K  = 0 THEN YI = 0 ELSE YI = FSTIFF(K,NDOF+1)
4600 K     = FREE(JJ)
4610 IF K  = 0 THEN XJ = 0 ELSE XJ = FSTIFF(K,NDOF+1)
4620 K     = FREE(JJ+1)
4630 IF K  = 0 THEN YJ = 0 ELSE YJ = FSTIFF(K,NDOF+1)
4640 FORCE = MEMB(I,3)*MEMB(I,4)*(MEMB(I,6)*(XJ-XI) + MEMB(I,7)*(YJ-YI))
4650 FORCE = FORCE / MEMB(I,5)
4660 PRINT USING "## ##         #.####^^^^"; MEMB(I,1); MEMB(I,2); FORCE
4670 REM
4680 REM --- ADD ANY CONTRIBUTION FROM THE MEMBER FORCES TO THE REACTIONS
4690 REM
4700 FOR J = 1 TO NREST
4710 K = 0
4720 IF MEMB(I,1) = REST(J,1) THEN K = -1
4730 IF MEMB(I,2) = REST(J,1) THEN K =  1
4740 IF K = 0 THEN GOTO 4770
4750 REST(J,3) = REST(J,3) + K*MEMB(I,6)*FORCE
4760 REST(J,4) = REST(J,4) + K*MEMB(I,7)*FORCE
4770 NEXT J
4780 NEXT I
4790 REM
4800 REM --- ADD ANY APPLIED LOADS
4810 REM
4820 FOR I = 1 TO NREST
4830 FOR J = 1 TO NNLOD
4840 IF REST(I,1) <> NLOD(J,1) THEN GOTO 4870
4850 K    = NLOD(J,2)
4860 REST(I,2+K) = REST(I,2+K) - NLOD(J,3)
4870 NEXT J
4880 NEXT I
4890 REM
4900 REM --- OUTPUT THE REACTIONS
4910 REM
4920 PRINT : PRINT
```

```
4930 PRINT "* * * * * * * *"
4940 PRINT "*   REACTIONS   *"
4950 PRINT "* * * * * * * *"
4960 PRINT
4970 PRINT "              X-FORCE         Y-FORCE"
4980 PRINT
4990 FOR I   = 1 TO NREST
5000 PRINT USING "NODE ##"; REST(I,1);
5010 PRINT USING "       #.####^^^^"; REST(I,3), REST(I,4)
5020 NEXT I
5030 RETURN
5030 RETURN
```

Sample Run 1 The following run demonstrates program PTRUSS.PT being used to analyse the truss in example 5.3. Input from the keyboard is shown in bold typeface.

```
-------------------------------------------------
PROGRAM    P T R U S S . P T    FOR THE AUTOMATIC
ANALYSIS OF PLANE TRUSSES
-------------------------------------------------

NAME OF THE DATA FILE = DATA1.PT

READING DATA

+ + + + + + + + +
+  JOB TITLE   +      -    EXAMPLE 5.3 - DATA1.PT
+ + + + + + + + +


+ + + + + + + + + + + +
+  NODAL COORDINATES  +
+ + + + + + + + + + + +

NODE              X                 Y
NUMBER
  1               0                 0
  2              3000              3000
  3              6000               0

+ + + + + + + + + + + +
+  MEMBER PROPERTIES  +
+ + + + + + + + + + + +

MEMBER            A                 E
NUMBER
  1   2        0.300E+04         0.200E+03
  1   3        0.300E+04         0.200E+03
  2   3        0.300E+04         0.200E+03

+ + + + + + + + +
+  RESTRAINTS   +
+ + + + + + + + +

NODE      DIRECTION(S)
NUMBER
  1           12
  3            2

+ + + + + + + + +
+  NODAL LOADS  +
+ + + + + + + + +

NODE      DIRECTION    VALUE
NUMBER
  2           1       0.100E+03
  2           2       0.200E+03
```

PLANE TRUSSES

```
GENERATING THE FREEDOM VECTOR

GENERATING THE STRUCTURE STIFFNESS MATRIX
ADDING ELEMENT     :-  1  2  3

GENERATING THE LOADING VECTOR

SOLVING EQUATIONS - NO. OF EQUATIONS  =  3
USING PIVOT NUMBER :-   1   2   3
```

```
* * * * * * * * * *
*   DISPLACEMENTS   *           * :-  RESTRAINT
* * * * * * * * * *

NODE      X-DISPLACEMENT    Y-DISPLACEMENT

  1              *                  *
  2           0.4571E+00         0.1664E+01
  3          -.5000E+00             *
```

```
* * * * * * * * * *
*   MEMBER FORCES   *           TENSION POSITIVE
* * * * * * * * * *

MEMBER              FORCE

 1  2            0.2121E+03
 1  3           -.5000E+02
 2  3            0.7071E+02
```

```
* * * * * * * * *
*   REACTIONS    *
* * * * * * * * *

                X-FORCE          Y-FORCE

NODE  1       -.1000E+03       -.1500E+03
NODE  3        0.3815E-05      -.5000E+02
```

Sample Run 2 This run shows PTRUSS.PT being used to solve the truss analysed in example 3.2.

```
--------------------------------------------------
PROGRAM     P T R U S S . P T     FOR THE AUTOMATIC
ANALYSIS OF PLANE TRUSSES
--------------------------------------------------


NAME OF THE DATA FILE  =  DATA2.PT

READING DATA

+ + + + + + + +
+   JOB TITLE    +      -     EXAMPLE 3.2 - DATA2.PT
+ + + + + + + +

+ + + + + + + + + + + +
+   NODAL COORDINATES    +
+ + + + + + + + + + + +

NODE            X              Y
NUMBER
  1             0              0
  2             0              2000
  3             2000           2000
  4             2000           0
```

```
+ + + + + + + + + + + +
+   MEMBER PROPERTIES   +
+ + + + + + + + + + + +

MEMBER           A              E
NUMBER
 1   2       0.400E+03      0.100E+02
 1   3       0.400E+03      0.100E+02
 1   4       0.400E+03      0.100E+02
 2   3       0.400E+03      0.100E+02
 2   4       0.400E+03      0.100E+02
 3   4       0.400E+03      0.100E+02


+ + + + + + + + + +
+   RESTRAINTS    +
+ + + + + + + + + +

NODE         DIRECTION(S)
NUMBER
  1              12
  4               2


+ + + + + + + + + +
+   NODAL LOADS   +
+ + + + + + + + + +

NODE      DIRECTION     VALUE
NUMBER
  3           1       0.100E+02
  3           2       0.500E+01

GENERATING THE FREEDOM VECTOR

GENERATING THE STRUCTURE STIFFNESS MATRIX
ADDING ELEMENT       :-  1  2  3  4  5  6

GENERATING THE LOADING VECTOR

SOLVING EQUATIONS  -  NO. OF EQUATIONS  =  5
USING PIVOT NUMBER :-  1  2  3  4  5

* * * * * * * * * * *
*   DISPLACEMENTS   *           * :-  RESTRAINT
* * * * * * * * * * *

NODE       X-DISPLACEMENT    Y-DISPLACEMENT

  1              *                 *
  2          0.8321E+01        0.1723E+01
  3          0.1004E+02       -.7767E+00
  4          0.1723E+01            *


* * * * * * * * * * *
*   MEMBER FORCES   *           TENSION POSITIVE
* * * * * * * * * * *

MEMBER            FORCE

 1   2         0.3447E+01
 1   3         0.9268E+01
 1   4         0.3447E+01
 2   3         0.3447E+01
 2   4        -.4874E+01
 3   4        -.1553E+01
```

```
        * * * * * * * * *
        *   REACTIONS    *
        * * * * * * * * *
                    X-FORCE         Y-FORCE
        NODE  1    -.1000E+02      -.1000E+02
        NODE  4    -.2384E-06       0.5000E+01
```

5.11 A DATA PREPROCESSOR   -   Program   PRE.MA/PRE.01/PRE.02

This program preprocesses the data for the plane truss analysis program PTRUSS.PT. It will also preprocess data for the programs STRUSS.ST, PFRAME.PF, GRID.GD, and SFRAME.SF which are described in subsequent chapters. The program allows either the creation of a new data file or the modification of an existing data file. It is interesting to note that the preprocessor is far longer than PTRUSS.PT itself. This is perhaps testimony to the compactness and efficiency of structural analysis programs based upon the stiffness method. To reduce the amount of memory necessary to run the program it has been segmented into three parts:

1. PRE.MA - Chains program PRE.01 if a new data file is to be created or chains PRE.02 if an existing data file is to be modified.

2. PRE.01 - Creates a new data file, requests data, and then chains PRE.02.

3. PRE.02 - This program gives the user the option to list all of the data, modify the data, run the data file with the appropriate analysis program, or to stop. If chained from PRE.01 the data is first tidied.

PRE.MA, PRE.01 and PRE.02 occupy 3, 15 and 20 kilobytes of memory respectively (all figures include space for variables). Should the available computer have too little memory then there are two obvious courses of action. If the shortage of memory is not very severe then the omission of comments and/or extra spaces from the code may create enough space. If this fails then the programs can be further segmented and the resulting program segments chained. The program operates by a question and answer sequence that is intended to be self-explanatory.

COMPUTER ANALYSIS OF STRUCTURAL FRAMEWORKS

Program Organisation The main program, PRE.MA, first establishes the name of the data file and whether the file exists or is to be created. If an existing data file is to modified PRE.MA chains program PRE.02. If a new file is to be created then PRE.MA chains PRE.01 which requests information about the job title, the nodal locations, the member properties, the restraints, and the loads. Once this information is complete the program PRE.02 is chained. The member, restraint and loading data is then tidied into order. PRE.02 then treats the new data file exactly as a file to modified and allows the user to add, delete, or modify the data. Prior to modifying a block of data (e.g. loads) the block is displayed. Once the modifications have been made, that block of data is tidied and the revised data displayed. When the data is complete the user can either exit the program or run PTRUSS.PT with the current data file. In either case the current data is copied to disk.

Listing - Program PRE.MA

```
1000 REM   **********************************************************************
1010 REM   *                                                                    *
1020 REM   *     PROGRAM    P R E . M A    TO PREPROCESS DATA FOR PROGRAMS      *
1030 REM   *     PTRUSS.PT, STRUSS.ST, PFRAME.PF, GRID.GD, AND SFRAME.SF        *
1040 REM   *                                                                    *
1050 REM   *     CHAINS PROGRAMS    P R E . 0 1   TO CREATE NEW DATA FILE       *
1060 REM   *                        P R E . 0 2   TO MODIFY/TIDY/PRINT DATA     *
1070 REM   *                                                                    *
1080 REM   *     J.A.D.BALFOUR                                                  *
1090 REM   *                                                                    *
1100 REM   **********************************************************************
1110 REM
1120 REM   --- NOTE THAT VARIABLES ARE DEFINED IN APPENDIX A
1130 REM
1140 OPTION BASE 1
1150 COMMON FILE$
1160 PRINT : PRINT
1170 PRINT "---------------------------------------------------------------"
1180 PRINT
1190 PRINT "PROGRAM    P R E . M A    TO PREPROCESS DATA FOR PROGRAMS"
1200 PRINT "PTRUSS.PT, STRUSS.ST, PFRAME.PF, GRID.GD, AND SFRAME.SF
1210 PRINT
1220 PRINT "---------------------------------------------------------------"
1230 PRINT
1240 PRINT "DO YOU WISH TO"
1250 PRINT "(1) CREATE A NEW DATA FILE"
1260 PRINT "(2) MODIFY AN EXISTING DATA FILE"
1270 INPUT ANS$
1280 PRINT
1290 INPUT "NAME OF THE DATA FILE = "; FILE$
1300 REM
1310 REM   --- CHAIN APPROPRIATE PROGRAM
1320 REM
1330 PRINT
1340 PRINT "WAIT - CHAINING PRE.0"; ANS$
1350 CHAIN ("PRE.0" + ANS$)
1360 END
```

PLANE TRUSSES

Listing - Program PRE.01

```
1000 REM **********************************************************************
1010 REM *                                                                    *
1020 REM *      PROGRAM     P R E . 0 1     TO CREATE A DATA FILE FOR PROGRAMS *
1030 REM *      PTRUSS.PT, STRUSS.ST, PFRAME.PF, GRID.GD, AND SFRAME.SF         *
1040 REM *                                                                    *
1050 REM *      CHAINED FROM PROGRAM     P R E . M A                          *
1060 REM *      CHAINS   PROGRAM         P R E . 0 2                          *
1070 REM *                                                                    *
1080 REM *      J.A.D.BALFOUR                                                 *
1090 REM *                                                                    *
1100 REM **********************************************************************
1110 REM
1120 REM --- NOTE THAT VARIABLES ARE DEFINED IN APPENDIX A
1130 REM
1140 OPTION BASE 1
1150 REM
1160 REM --- PROBLEM SIZE IS SET BY THE FOLLOWING DIMENSION STATEMENT
1170 REM
1180 COMMON FILE$
1190 DIM NODE(30,3), MEMB(50,9), REST(20,2), NLOD(50,3), MLOD(20,5), TEMP(10)
1200 REM
1210 REM --- SET PROBLEM TYPE
1220 REM
1230 GOSUB 1570
1240 REM
1250 REM --- READ JOB TITLE AND NODAL COORDINATES
1260 REM
1270 GOSUB 1780
1280 REM
1290 REM --- INPUT MEMBER DATA
1300 REM
1310 GOSUB 2010
1320 REM
1330 REM --- INPUT RESTRAINTS
1340 REM
1350 GOSUB 2500
1360 REM
1370 REM --- INPUT NODAL LOADS
1380 REM
1390 GOSUB 2920
1400 REM
1410 REM --- INPUT MEMBER LOADS
1420 REM
1430 GOSUB 3340
1440 REM
1450 REM --- OUTPUT TO DISK FILE
1460 REM
1470 GOSUB 3680
1480 CLOSE #1
1490 REM
1500 REM --- CHAIN PROGRAM PRE.02 TO TIDY/PRINT/MODIFY THE DATA
1510 REM
1520 PRINT
1530 PRINT    "WAIT - CHAINING PRE.02"
1540 REM
1550 CHAIN "PRE.02"
1560 END
1570 REM * * * * * * * * * * * * * * * * * * * * * * * * * * * * * * * * *
1580 REM *                                                                *
1590 REM *      SUBROUTINE TO SET THE PROBLEM TYPE                        *
1600 REM *                                                                *
1610 REM * * * * * * * * * * * * * * * * * * * * * * * * * * * * * * * * *
1620 REM
1630 PRINT : PRINT
1640 PRINT "SELECT PROBLEM TYPE"
1650 PRINT "(1) PLANE TRUSS"
1660 PRINT "(2) SPACE TRUSS"
1670 PRINT "(3) PLANE FRAME"
1680 PRINT "(4) GRILLAGE"
1690 PRINT "(5) SPACE FRAME"
```

```
1700 INPUT PTYPE
1710 NCORD    = 2
1720 IF PTYPE = 2 THEN NCORD = 3
1730 IF PTYPE = 5 THEN NCORD = 3
1740 NPROP    = 6
1750 IF PTYPE < 3 THEN NPROP = 4
1760 IF PTYPE = 5 THEN NPROP = 9
1770 RETURN
1780 REM ******************************************
1790 REM *                                        *
1800 REM *   SUBROUTINE TO INPUT JOB TITLE AND NODAL COORDINATES  *
1810 REM *                                        *
1820 REM ******************************************
1830 REM
1840 PRINT
1850 INPUT "JOB TITLE  = "; TITLE$
1860 REM
1870 REM --- INPUT NODAL COORDINATES
1880 REM
1890 PRINT : PRINT
1900 INPUT "NUMBER OF NODES  = "; NNODE
1910 FOR I = 1 TO NNODE
1920 PRINT
1930 PRINT "NODE"; I
1940 PRINT
1950 INPUT "X-COORDINATE      = "; NODE(I,1)
1960 INPUT "Y-COORDINATE      = "; NODE(I,2)
1970 IF NCORD = 2 THEN GOTO 1990
1980 INPUT "Z-COORDINATE      = "; NODE(I,3)
1990 NEXT I
2000 RETURN
2010 REM ******************************************
2020 REM *                                        *
2030 REM *   SUBROUTINE TO INPUT MEMBER DATA      *
2040 REM *                                        *
2050 REM ******************************************
2060 REM
2070 PRINT : PRINT
2080 INPUT "NUMBER OF MEMBERS        = ", NMEMB
2090 IF PTYPE <> 3 THEN GOTO 2180
2100 PRINT
2110 PRINT "ELEMENT     LOWER       HIGHER"
2120 PRINT " TYPE       NODE        NODE"
2130 PRINT
2140 PRINT "   1        FIXED       FIXED"
2150 PRINT "   2        FIXED       PINNED"
2160 PRINT "   3        PINNED      FIXED"
2170 PRINT "   4        PINNED      PINNED"
2180 PRINT
2190 PRINT "HIT CARRIAGE RETURN IF THE VALUE IS "
2200 PRINT "THE SAME AS THE PREVIOUS MEMBER"
2210 FOR I = 1 TO NMEMB
2220 PRINT
2230 PRINT "MEMBER"; I
2240 PRINT
2250 PRINT "LOWER NODE NUMBER       = "; : J = 1 : GOSUB 4170
2260 PRINT "HIGHER NODE NUMBER      = "; : J = 2 : GOSUB 4170
2270 IF PTYPE = 4 THEN GOTO 2350
2280 PRINT "CROSS-SECTIONAL AREA    = "; : J = 3 : GOSUB 4170
2290 IF PTYPE < 3 THEN GOTO 2440
2300 IF PTYPE = 5 THEN GOTO 2390
2310 PRINT "SECOND MOMENT OF AREA   = "; : J = 4 : GOSUB 4170
2320 IF PTYPE = 4 THEN GOTO 2440
2330 PRINT "ELEMENT TYPE            = "; : J = 5 : GOSUB 4170
2340 GOTO 2440
2350 PRINT "SECOND MOMENT OF AREA   = "; : J = 3 : GOSUB 4170
2360 PRINT "TORSIONAL CONSTANT      = "; : J = 4 : GOSUB 4170
2370 PRINT "MODULUS OF RIGIDITY     = "; : J = 5 : GOSUB 4170
2380 GOTO 2440
2390 PRINT "2ND MOMENT OF AREA Y-Y  = "; : J = 4 : GOSUB 4170
2400 PRINT "2ND MOMENT OF AREA Z-Z  = "; : J = 5 : GOSUB 4170
2410 PRINT "TORSIONAL 2ND MT. AREA  = "; : J = 6 : GOSUB 4170
2420 PRINT "BETA                    = "; : J = 7 : GOSUB 4170
```

```
2430 PRINT "MODULUS OF RIGIDITY      = "; : J = 8 : GOSUB 4170
2440 PRINT "ELASTIC CONSTANT         = "; : J = NPROP : GOSUB 4170
2450 FOR J = 1 TO NPROP
2460 MEMB(I,J) = TEMP(J)
2470 NEXT J
2480 NEXT I
2490 RETURN
2500 REM **********************************************
2510 REM *                                              *
2520 REM *      SUBROUTINE TO INPUT RESTRAINTS          *
2530 REM *                                              *
2540 REM **********************************************
2550 REM
2560 PRINT : PRINT
2570 INPUT "NUMBER OF RESTRAINED NODES = "; NREST
2580 PRINT
2590 PRINT "RESTRAINT DIRECTIONS ARE"
2600 IF PTYPE = 4 THEN GOTO 2730
2610 PRINT "X-TRANSLATION    = 1"
2620 PRINT "Y-TRANSLATION    = 2"
2630 IF NCORD = 2 THEN GOTO 2700
2640 PRINT "Z-TRANSLATION    = 3"
2650 IF PTYPE <> 5 THEN GOTO 2760
2660 PRINT "X-ROTATION       = 4"
2670 PRINT "Y-ROTATION       = 5"
2680 PRINT "Z-ROTATION       = 6"
2690 GOTO 2760
2700 IF PTYPE = 1 THEN GOTO 2760
2710 PRINT "ROTATION         = 3"
2720 GOTO 2760
2730 PRINT "Z-TRANSLATION    = 1"
2740 PRINT "X-ROTATION       = 2"
2750 PRINT "Y-ROTATION       = 3"
2760 PRINT
2770 PRINT "ENTER RESTRAINED DIRECTIONS AS A COMPOSITE NUMBER"
2780 PRINT "E.G.  IF NODE IS RESTRAINED IN DIRECTIONS 1 AND 2 ENTER 12"
2790 PRINT
2800 PRINT "HIT CARRIAGE RETURN IF THE VALUE IS "
2810 PRINT "THE SAME AS THE PREVIOUS RESTRAINT"
2820 FOR I = 1 TO NREST
2830 PRINT
2840 PRINT "RESTRAINED NODE NUMBER "; I
2850 PRINT
2860 PRINT "NODE TO BE RESTRAINED   = "; : J = 1 : GOSUB 4170
2870 PRINT "DIRECTION(S)            = "; : J = 2 : GOSUB 4170
2880 REST(I,1) = TEMP(1)
2890 REST(I,2) = TEMP(2)
2900 NEXT I
2910 RETURN
2920 REM **********************************************
2930 REM *                                              *
2940 REM *      SUBROUTINE TO INPUT NODAL LOADS         *
2950 REM *                                              *
2960 REM **********************************************
2970 REM
2980 PRINT : PRINT
2990 INPUT "NUMBER OF NODAL LOADS   = "; NNLOD
3000 IF NNLOD = 0 THEN GOTO 3330
3010 PRINT
3020 PRINT "LOAD DIRECTIONS ARE"
3030 IF PTYPE = 4 THEN GOTO 3160
3040 PRINT "FORCE IN THE X-DIRECTION   = 1"
3050 PRINT "FORCE IN THE Y-DIRECTION   = 2"
3060 IF NCORD = 2 THEN GOTO 3130
3070 PRINT "FORCE IN THE Z-DIRECTION   = 3"
3080 IF PTYPE = 2 THEN GOTO 3190
3090 PRINT "MOMENT ABOUT THE X-AXIS    = 4"
3100 PRINT "MOMENT ABOUT THE Y-AXIS    = 5"
3110 PRINT "MOMENT ABOUT THE Z-AXIS    = 6"
3120 GOTO 3190
3130 IF PTYPE = 1 THEN GOTO 3190
3140 PRINT "MOMENT                     = 3"
3150 GOTO 3190
```

```
3160 PRINT "FORCE IN THE Z-DIRECTION  =  1"
3170 PRINT "MOMENT ABOUT THE X-AXIS   =  2"
3180 PRINT "MOMENT ABOUT THE Y-AXIS   =  3"
3190 PRINT
3200 PRINT "HIT CARRIAGE RETURN IF THE VALUE IS "
3210 PRINT "THE SAME AS THE PREVIOUS LOAD"
3220 FOR I = 1 TO NNLOD
3230 PRINT
3240 PRINT "NODAL LOAD NUMBER "; I
3250 PRINT
3260 PRINT "NODE CARRYING LOAD      = "; : J = 1 : GOSUB 4170
3270 PRINT "DIRECTION               = "; : J = 2 : GOSUB 4170
3280 PRINT "MAGNITUDE OF LOAD       = "; : J = 3 : GOSUB 4170
3290 NLOD(I,1) = TEMP(1)
3300 NLOD(I,2) = TEMP(2)
3310 NLOD(I,3) = TEMP(3)
3320 NEXT I
3330 RETURN
3340 REM * * * * * * * * * * * * * * * * * * * * * * * * * * * * * * * * * *
3350 REM *                                                                  *
3360 REM *      SUBROUTINE TO INPUT MEMBER LOADS                            *
3370 REM *                                                                  *
3380 REM * * * * * * * * * * * * * * * * * * * * * * * * * * * * * * * * * *
3390 REM
3400 PRINT : PRINT
3410 IF PTYPE < 3 THEN GOTO 3670
3420 IF PTYPE > 4 THEN GOTO 3670
3430 INPUT "NUMBER OF MEMBER LOADS  = ", NMLOD
3440 IF NMLOD = 0 THEN GOTO 3670
3450 PRINT
3460 PRINT "LOAD TYPES ARE"
3470 PRINT "UDL             = 1"
3480 PRINT "POINT LOAD      = 2"
3490 PRINT
3500 PRINT "HIT CARRIAGE RETURN IF THE VALUE IS "
3510 PRINT "THE SAME AS THE PREVIOUS LOAD"
3520 FOR I = 1 TO NMLOD
3530 PRINT
3540 PRINT "MEMBER LOAD NUMBER "; I
3550 PRINT
3560 PRINT "LOWER  NODE NUMBER        = "; : J = 1 : GOSUB 4170
3570 PRINT "HIGHER NODE NUMBER        = "; : J = 2 : GOSUB 4170
3580 PRINT "LOAD TYPE                 = "; : J = 3 : GOSUB 4170
3590 PRINT "MAGNITUDE OF LOAD         = "; : J = 4 : GOSUB 4170
3600 IF TEMP(3) = 1 THEN GOTO 3630
3610 PRINT USING "DIST OF LOAD FROM NODE ## = "; MLOD(I,3);
3620 J = 5 : GOSUB 4170
3630 FOR J     = 1 TO 5
3640 MLOD(I,J) = TEMP(J)
3650 NEXT J
3660 NEXT I
3670 RETURN
3680 REM * * * * * * * * * * * * * * * * * * * * * * * * * * * * * * * * * *
3690 REM *                                                                  *
3700 REM *      SUBROUTINE TO OUTPUT THE DATA TO DISK FILE                  *
3710 REM *                                                                  *
3720 REM * * * * * * * * * * * * * * * * * * * * * * * * * * * * * * * * * *
3730 REM
3740 OPEN "O", #1, FILE$
3750 WRITE #1, TITLE$
3760 REM
3770 REM ---  PTYPE SET -VE TO SHOW DATA NEEDS TIDYING
3780 REM
3790 TEMP(1) = - PTYPE
3800 WRITE #1, TEMP(1), NCORD, NPROP, NNODE, NMEMB, NREST, NNLOD
3810 IF PTYPE = 3 THEN WRITE #1, NMLOD
3820 IF PTYPE = 4 THEN WRITE #1, NMLOD
3830 REM
3840 REM ---  WRITE NODAL COORDINATES
3850 REM
3860 FOR I = 1 TO NNODE
3870 FOR J = 1 TO NCORD
3880 WRITE #1, NODE(I,J)
```

PLANE TRUSSES

```
3890 NEXT J
3900 NEXT I
3910 REM --- WRITE MEMBER DATA
3920 REM
3930 FOR I = 1 TO NMEMB
3940 FOR J = 1 TO NPROP
3950 WRITE #1, MEMB(I,J)
3960 NEXT J
3970 NEXT I
3980 REM
3990 REM --- WRITE RESTRAINT DATA
4000 REM
4010 FOR I = 1 TO NREST
4020 WRITE #1, REST(I,1), REST(I,2)
4030 NEXT I
4040 REM
4050 REM --- WRITE NODAL LOADS
4060 REM
4070 FOR I = 1 TO NNLOD
4080 WRITE #1, NLOD(I,1), NLOD(I,2), NLOD(I,3)
4090 NEXT I
4100 REM
4110 REM --- WRITE MEMBER LOADS
4120 REM
4130 FOR I = 1 TO NMLOD
4140 WRITE #1, MLOD(I,1), MLOD(I,2), MLOD(I,3), MLOD(I,4), MLOD(I,5)
4150 NEXT I
4160 RETURN
4170 REM * * * * * * * * * * * * * * * * * * * * * * * * * * * * * * *
4180 REM *                                                             *
4190 REM *    SUBROUTINE TO ALLOW A CARRIAGE RETURN TO BE USED TO SET  *
4200 REM *    CURRENT VALUE EQUAL TO THE PREVIOUS VALUE                *
4210 REM *                                                             *
4220 REM * * * * * * * * * * * * * * * * * * * * * * * * * * * * * * *
4230 REM
4240 INPUT; "", T$
4250 IF LEN(T$) = 0 THEN GOTO 4290
4260 TEMP(J)  = VAL(T$)
4270 PRINT
4280 GOTO 4300
4290 PRINT CHR$(8); TEMP(J)
4300 RETURN
```

Listing - Program PRE.02

```
1000 REM ******************************************************************
1010 REM *                                                                *
1020 REM *   PROGRAM   P R E . 0 2   TO PRINT/TIDY/MODIFY DATA FOR PROGRAMS *
1030 REM *   PTRUSS.PT, STRUSS.PT, PFRAME.PF, GRID.GD, AND SFRAME.SF       *
1040 REM *                                                                *
1050 REM *   CHAINED FROM PROGRAMS      P R E . M A    AND    P R E . 0 1  *
1060 REM *                                                                *
1070 REM *   J.A.D.BALFOUR                                                *
1080 REM *                                                                *
1090 REM ******************************************************************
1100 REM
1110 REM --- NOTE THAT VARIABLES ARE DEFINED IN APPENDIX A
1120 REM
1130 OPTION BASE 1
1140 REM
1150 REM --- PROBLEM SIZE IS SET BY THE FOLLOWING DIMENSION STATEMENT
1160 REM
1170 COMMON FILE$
1180 DIM NODE(30,3), MEMB(50,9), REST(20,2), NLOD(50,3), MLOD(20,5), TEMP(10)
1190 REM
1200 REM --- READ DATA FROM FILE
1210 REM
1220 GOSUB 1800
1230 REM
1240 REM --- CHECK IF DATA NEEDS TO BE TIDIED (I.E. PTYPE < 0)
```

```
1250 REM
1260 IF PTYPE > 0 THEN GOTO 1400
1270 GOSUB 2280
1280 GOSUB 2460
1290 GOSUB 2860
1300 GOSUB 3890
1310 GOSUB 4470
1320 GOSUB 5130
1330 REM
1340 REM  ---  IF NEWLY CREATED FILE LIST TIDIED DATA
1350 REM
1360 IF PTYPE > 0 THEN GOTO 1400
1370 PTYPE    = ABS(PTYPE)
1380 ANS      = 2
1390 GOTO 1270
1400 PRINT
1410 PRINT "DO YOU WISH TO    (CR = CARRIAGE RETURN)"
1420 PRINT "(1)   MODIFY DATA
1430 PRINT "(2)   LIST ALL THE CURRENT DATA"
1440 PRINT "(3)   RUN DATA FILE WITH APPROPRIATE ANALYSIS PROGRAM"
1450 PRINT "(CR) STOP"
1460 INPUT ANS
1470 IF ANS = 0 THEN ANS = 4
1480 ON ANS GOTO 1490, 1270, 1690, 1690
1490 PRINT
1500 PRINT "DO YOU WISH TO    (CR = CARRIAGE RETURN)"
1510 PRINT "(1)   CHANGE JOB TITLE"
1520 PRINT "(2)   MODIFY NODAL DATA"
1530 PRINT "(3)   MODIFY MEMBER DATA"
1540 PRINT "(4)   MODIFY RESTRAINTS"
1550 PRINT "(5)   MODIFY NODAL LOADS"
1560 IF PTYPE < 3 THEN GOTO 1590
1570 IF PTYPE > 4 THEN GOTO 1590
1580 PRINT "(6)   MODIFY MEMBER LOADS"
1590 PRINT "(CR) STOP MODIFYING DATA"
1600 INPUT ANS
1610 IF ANS = 0 THEN ANS = 7
1620 ON      ANS     GOSUB 2280, 2460, 2860, 3890, 4470, 5130, 1400
1630 ANS = -ANS
1640 PRINT
1650 ON ABS(ANS) GOSUB 2280, 2460, 2860, 3890, 4470, 5130, 1400
1660 PRINT
1670 INPUT "AGAIN -   (Y)ES OR (CR)"; T$
1680 IF T$ = "Y" THEN GOTO 1640 ELSE GOTO 1490
1690 GOSUB 5880
1700 IF ANS    = 4 THEN GOTO 1790
1710 IF PTYPE = 1 THEN PROG$ = "PTRUSS.PT"
1720 IF PTYPE = 2 THEN PROG$ = "STRUSS.ST"
1730 IF PTYPE = 3 THEN PROG$ = "PFRAME.PF"
1740 IF PTYPE = 4 THEN PROG$ = "GRID.GD"
1750 IF PTYPE = 5 THEN PROG$ = "SFRAME.SF"
1760 PRINT
1770 PRINT "WAIT - CHAINING "; PROG$
1780 CHAIN PROG$
1790 END
1800 REM  * * * * * * * * * * * * * * * * * * * * * * * * * * * * * * * *
1810 REM  *                                                              *
1820 REM  *    SUBROUTINE TO INPUT DATA GENERATED BY THE PREPROCESSOR    *
1830 REM  *    PROGRAMS    P R E . M A / P R E . 0 1                     *
1840 REM  *                                                              *
1850 REM  * * * * * * * * * * * * * * * * * * * * * * * * * * * * * * * *
1860 REM
1870 OPEN "I", #1, FILE$
1880 REM
1890 REM  ---  TITLE AND NO. OF NODES, MEMBERS, RESTRAINTS AND LOADS
1900 REM
1910 INPUT #1, TITLE$
1920 INPUT #1, PTYPE, NCORD, NPROP, NNODE, NMEMB, NREST, NNLOD
1930 IF ABS(PTYPE) = 3 THEN INPUT #1, NMLOD
1940 IF ABS(PTYPE) = 4 THEN INPUT #1, NMLOD
1950 FOR I   = 1 TO NNODE
1960 FOR J   = 1 TO NCORD
1970 INPUT #1, NODE(I,J)
```

PLANE TRUSSES

```
1980 NEXT J
1990 NEXT I
2000 REM
2010 REM  ---  READ MEMBER DATA
2020 REM
2030 FOR I = 1 TO NMEMB
2040 FOR J = 1 TO NPROP
2050 INPUT #1, MEMB(I,J)
2060 NEXT J
2070 NEXT I
2080 REM
2090 REM  ---  READ RESTRAINT DATA
2100 REM
2110 FOR I = 1 TO NREST
2120 INPUT #1, REST(I,1), REST(I,2)
2130 NEXT I
2140 REM
2150 REM  ---  READ NODAL LOADS
2160 REM
2170 FOR I = 1 TO NNLOD
2180 INPUT #1, NLOD(I,1), NLOD(I,2), NLOD(I,3)
2190 NEXT I
2200 REM
2210 REM  ---  READ MEMBER LOADS (PLANE FRAMES AND GRILLAGES ONLY)
2220 REM
2230 FOR I   = 1 TO NMLOD
2240 INPUT #1, MLOD(I,1), MLOD(I,2), MLOD(I,3), MLOD(I,4), MLOD(I,5)
2250 NEXT I
2260 CLOSE #1
2270 RETURN
2280 REM  * * * * * * * * * * * * * * * * * * * * * * * * * * * * * * * *
2290 REM  *                                                              *
2300 REM  *     SUBROUTINE TO MODIFY/PRINT JOB TITLE                     *
2310 REM  *                                                              *
2320 REM  * * * * * * * * * * * * * * * * * * * * * * * * * * * * * * * *
2330 REM
2340 IF PTYPE < 0 THEN GOTO 2450
2350 IF ANS   > 0 THEN GOTO 2400
2360 REM
2370 REM  ---  CHANGE JOB TITLE
2380 REM
2390 INPUT "NEW JOB TITLE  = ", TITLE$
2400 PRINT
2410 PRINT "+ + + + + + + +"
2420 PRINT "+   JOB TITLE    +  -  "; TITLE$
2430 PRINT "+ + + + + + + +"
2440 PRINT
2450 RETURN
2460 REM  * * * * * * * * * * * * * * * * * * * * * * * * * * * * * * * *
2470 REM  *                                                              *
2480 REM  *     SUBROUTINE TO MODIFY/PRINT THE NODAL COORDINATES         *
2490 REM  *                                                              *
2500 REM  * * * * * * * * * * * * * * * * * * * * * * * * * * * * * * * *
2510 REM
2520 IF PTYPE < 0 THEN GOTO 2850
2530 IF ANS   > 0 THEN GOTO 2730
2540 REM
2550 REM  ---  MODIFY NODAL DATA
2560 REM
2570 INPUT "NODE TO BE REVISED/ADDED  = ", N
2580 INPUT "X-COORDINATE               = ", X
2590 INPUT "Y-COORDINATE               = ", Y
2600 IF NCORD = 2 THEN GOTO 2660
2610 INPUT "Z-COORDINATE               = ", Z
2620 REM
2630 REM  ---  THIS IS THE    O N L Y    ERROR CHECK IN THE PROGRAM
2640 REM  ---  THE USER SHOULD INTRODUCE MORE
2650 REM
2660 IF (N - NNODE) <= 1 THEN GOTO 2700
2670 PRINT "NODE NUMBERS MUST BE SEQUENTIAL"
2680 PRINT
2690 GOTO 2570
2700 IF N > NNODE  THEN NNODE     = N
```

COMPUTER ANALYSIS OF STRUCTURAL FRAMEWORKS

```
2710 NODE(N,1) = X  :    NODE(N,2) = Y
2720 IF NCORD = 3 THEN NODE(N,3) = Z
2730 PRINT
2740 PRINT "+ + + + + + + + + + + +"
2750 PRINT "+    NODAL COORDINATES   +"
2760 PRINT "+ + + + + + + + + + + +"
2770 PRINT
2780 PRINT "NODE            X                Y";
2790 IF NCORD = 3 THEN PRINT "             Z" ELSE PRINT
2800 PRINT "NUMBER
2810 FOR I = 1 TO NNODE
2820 PRINT " "; I, NODE(I,1), NODE(I,2),
2830 IF NCORD = 3 THEN PRINT NODE(I,3) ELSE PRINT
2840 NEXT I
2850 RETURN
2860 REM * * * * * * * * * * * * * * * * * * * * * * * * * * * * * * * * *
2870 REM *                                                                *
2880 REM *   SUBROUTINE TO MODIFY/TIDY/PRINT THE MEMBER PROPERTIES        *
2890 REM *                                                                *
2900 REM * * * * * * * * * * * * * * * * * * * * * * * * * * * * * * * * *
2910 REM
2920 IF PTYPE < 0 THEN GOTO 3410
2930 IF ANS   > 0 THEN GOTO 3670
2940 REM
2950 REM --- MODIFY MEMBER DATA
2960 REM
2970 PRINT "ENTER DATA FOR ELEMENT TO BE MODIFIED/ADDED"
2980 PRINT "SET CROSS-SECTIONAL AREA TO ZERO TO CANCEL AN ELEMENT"
2990 IF PTYPE <> 3 THEN GOTO 3070 ELSE PRINT
3000 PRINT "ELEMENT         NODE I       NODE J"
3010 PRINT " TYPE"
3020 PRINT
3030 PRINT"    1           FIXED        FIXED"
3040 PRINT"    2           FIXED        PINNED"
3050 PRINT"    3           PINNED       FIXED"
3060 PRINT"    4           PINNED       PINNED"
3070 PRINT
3080 INPUT "LOWER NODE NUMBER         = ", TEMP(1)
3090 INPUT "HIGHER NODE NUMBER        = ", TEMP(2)
3100 IF PTYPE =  4 THEN GOTO 3180
3110 INPUT "CROSS-SECTIONAL AREA      = ", TEMP(3)
3120 IF PTYPE <  3 THEN GOTO 3270
3130 IF PTYPE =  5 THEN GOTO 3220
3140 INPUT "SECOND MOMENT OF AREA     = ", TEMP(4)
3150 IF PTYPE <> 3 THEN GOTO 3270
3160 INPUT "ELEMENT TYPE              = ", TEMP(5)
3170 GOTO 3270
3180 INPUT "SECOND MOMENT OF AREA     = ", TEMP(3)
3190 INPUT "TORSIONAL CONSTANT        = ", TEMP(4)
3200 INPUT "MODULUS OF RIGIDITY       = ", TEMP(5)
3210 GOTO 3270
3220 INPUT "2ND MOMENT OF AREA Y-Y    = ", TEMP(4)
3230 INPUT "2ND MOMENT OF AREA Z-Z    = ", TEMP(5)
3240 INPUT "TORSIONAL 2ND MT. AREA    = ", TEMP(6)
3250 INPUT "BETA                      = ", TEMP(7)
3260 INPUT "MODULUS OF RIGIDITY       = ", TEMP(8)
3270 INPUT "ELASTIC CONSTANT          = ", TEMP(NPROP)
3280 K = 0
3290 FOR I         = 1 TO NMEMB
3300 IF MEMB(I,1) <> TEMP(1) THEN GOTO 3320
3310 IF MEMB(I,2) =  TEMP(2) THEN K = I
3320 NEXT I
3330 IF K = 0 THEN NMEMB = NMEMB + 1
3340 IF K = 0 THEN K     = NMEMB
3350 FOR J         = 1 TO NPROP
3360 MEMB(K,J) = TEMP(J)
3370 NEXT J
3380 REM
3390 REM --- REMOVE ANY DELETED MEMBERS
3400 REM
3410 COUNT         = 0
3420 FOR I         = 1 TO NMEMB
3430 IF MEMB(I,3)  = 0 THEN GOTO 3480
```

```
3440 COUNT         = COUNT + 1
3450 FOR J = 1 TO NPROP
3460 MEMB(COUNT,J) = MEMB(I,J)
3470 NEXT J
3480 NEXT I
3490 NMEMB         = COUNT
3500 REM
3510 REM   ---   SORT MEMBERS INTO ORDER
3520 REM
3530 FLAG  = 0
3540 FOR I = 1 TO NMEMB-1
3550 IF MEMB(I,1) >  MEMB(I+1,1) THEN GOTO 3580
3560 IF MEMB(I,1) <  MEMB(I+1,1) THEN GOTO 3640
3570 IF MEMB(I,2) <= MEMB(I+1,2) THEN GOTO 3640
3580 FLAG          = 1
3590 FOR J         = 1 TO NPROP
3600 TEMP(1)       = MEMB(I,J)
3610 MEMB(I,J)     = MEMB(I+1,J)
3620 MEMB(I+1,J)   = TEMP(1)
3630 NEXT J
3640 NEXT I
3650 IF FLAG       = 1 THEN GOTO 3530
3660 IF PTYPE      < 0 THEN GOTO 3880
3670 PRINT
3680 PRINT "+ + + + + + + + + + + + +"
3690 PRINT "+    MEMBER PROPERTIES    +"
3700 PRINT "+ + + + + + + + + + + + +"
3710 PRINT
3720 PRINT "MEMBER";
3730 IF PTYPE <> 4 THEN PRINT "          A";
3740 IF PTYPE =  4 THEN PRINT "          I";
3750 IF PTYPE <  3 THEN PRINT "          E"
3760 IF PTYPE =  3 THEN PRINT "     I       PTYPE       E"
3770 IF PTYPE =  4 THEN PRINT "     J         G         E"
3780 IF PTYPE =  5 THEN PRINT "   IYY       IZZ        J";
3790 IF PTYPE =  5 THEN PRINT "   BETA        G         E"
3800 PRINT "NUMBER"
3810 FOR I = 1 TO NMEMB
3820 PRINT USING "## ##      "; MEMB(I,1), MEMB(I,2);
3830 FOR J = 3 TO NPROP
3840 PRINT USING " #.###^^^^"; MEMB(I,J);
3850 NEXT J
3860 PRINT
3870 NEXT I
3880 RETURN
3890 REM * * * * * * * * * * * * * * * * * * * * * * * * * * * * * * *
3900 REM *                                                            *
3910 REM *    SUBROUTINE TO MODIFY/TIDY/PRINT THE RESTRAINT DATA      *
3920 REM *                                                            *
3930 REM * * * * * * * * * * * * * * * * * * * * * * * * * * * * * * *
3940 REM
3950 IF PTYPE < 0 THEN GOTO 4140
3960 IF ANS   > 0 THEN GOTO 4360
3970 REM
3980 REM   ---   MODIFY RESTRAINT DATA
3990 REM
4000 PRINT "SET DIRECTION(S) EQUAL TO ZERO TO CANCEL RESTRAINT"
4010 PRINT
4020 INPUT "NODE NUMBER    = ", TEMP(1)
4030 INPUT "DIRECTION(S)   = ", TEMP(2)
4040 K = 0
4050 FOR I = 1 TO NREST
4060 IF REST(I,1) = TEMP(1) THEN K = I
4070 NEXT I
4080 IF K = 0 THEN NREST = NREST + 1
4090 IF K = 0 THEN K     = NREST
4100 REST(K,1) = TEMP(1) : REST(K,2) = TEMP(2)
4110 REM
4120 REM   ---   REMOVE ANY DELETED RESTRAINTS
4130 REM
4140 COUNT         = 0
4150 FOR I         = 1 TO NREST
4160 IF REST(I,2)  = 0 THEN GOTO 4190
```

```
4170 COUNT           = COUNT + 1
4180 REST(COUNT,1) = REST(I,1) : REST(COUNT,2) = REST(I,2)
4190 NEXT I
4200 NREST           = COUNT
4210 REM
4220 REM  ---  PUT RESTRAINTS INTO ORDER
4230 REM
4240 FLAG            = 0
4250 FOR I           = 1 TO NREST-1
4260 IF REST(I,1) <= REST(I+1,1) THEN GOTO 4330
4270 FLAG            = 1
4280 FOR J           = 1 TO 2
4290 TEMP(1)         = REST(I,J)
4300 REST(I,J)       = REST(I+1,J)
4310 REST(I+1,J)     = TEMP(1)
4320 NEXT J
4330 NEXT I
4340 IF FLAG         = 1 THEN GOTO 4240
4350 IF PTYPE        < 0 THEN GOTO 4460
4360 PRINT
4370 PRINT "+ + + + + + + + + +"
4380 PRINT "+    RESTRAINTS    +"
4390 PRINT "+ + + + + + + + + +"
4400 PRINT
4410 PRINT          "NODE      DIRECTION(S)"
4420 PRINT          "NUMBER"
4430 FOR I = 1 TO NREST
4440 PRINT USING " ##         ######"; REST(I,1),REST(I,2)
4450 NEXT I
4460 RETURN
4470 REM * * * * * * * * * * * * * * * * * * * * * * * * * * * * * * * *
4480 REM *                                                              *
4490 REM *     SUBROUTINE TO MODIFY/TIDY/PRINT THE NODAL LOADS          *
4500 REM *                                                              *
4510 REM * * * * * * * * * * * * * * * * * * * * * * * * * * * * * * * *
4520 REM
4530 IF ANS    = -5 THEN GOTO 4600
4540 IF ANS    =  1 THEN GOTO 5010
4550 IF PTYPE  <  0 THEN GOTO 4740
4560 IF ANS    >  0 THEN GOTO 5000
4570 REM
4580 REM  ---  MODIFY NODAL LOADS
4590 REM
4600 INPUT "NODE WHERE LOAD IS TO MODIFIED  = ", TEMP(1)
4610 INPUT "DIRECTION                       = ", TEMP(2)
4620 INPUT "REVISED VALUE OF LOAD           = ", TEMP(3)
4630 K = 0
4640 FOR I           = 1 TO NNLOD
4650 IF NLOD(I,1) <> TEMP(1) THEN GOTO 4670
4660 IF NLOD(I,2) =  TEMP(2) THEN K = I
4670 NEXT I
4680 IF K = 0 THEN NNLOD = NNLOD + 1
4690 IF K = 0 THEN K     = NNLOD
4700 NLOD(K,1) = TEMP(1) : NLOD(K,2) = TEMP(2) : NLOD(K,3) = TEMP(3)
4710 REM
4720 REM  ---  REMOVE ANY ZERO NODAL LOADS
4730 REM
4740 COUNT           = 0
4750 FOR I           = 1 TO NNLOD
4760 IF NLOD(I,3)    = 0 THEN GOTO 4810
4770 COUNT           = COUNT + 1
4780 FOR J = 1 TO 3
4790 NLOD(COUNT,J) = NLOD(I,J)
4800 NEXT J
4810 NEXT I
4820 NNLOD           = COUNT
4830 REM
4840 REM  ---  PUT NODAL LOADS INTO ORDER
4850 REM
4860 FLAG            = 0
4870 FOR I           = 1 TO NNLOD-1
4880 IF NLOD(I,1) >  NLOD(I+1,1) THEN GOTO 4910
4890 IF NLOD(I,1) <  NLOD(I+1,1) THEN GOTO 4970
```

PLANE TRUSSES

```
4900 IF NLOD(I,2) <= NLOD(I+1,2) THEN GOTO 4970
4910 FLAG           = 1
4920 FOR J          = 1 TO 3
4930 TEMP(1)        = NLOD(I,J)
4940 NLOD(I,J)      = NLOD(I+1,J)
4950 NLOD(I+1,J)    = TEMP(1)
4960 NEXT J
4970 NEXT I
4980 IF FLAG        = 1 THEN GOTO 4860
4990 IF PTYPE       < 0 THEN GOTO 5120
5000 IF NNLOD       = 0 THEN GOTO 5120
5010 PRINT
5020 PRINT "+ + + + + + + + +"
5030 PRINT "+    NODAL LOADS    +"
5040 PRINT "+ + + + + + + + +"
5050 PRINT
5060 PRINT          "NODE    DIRECTION    VALUE"
5070 PRINT          "NUMBER"
5080 FOR I = 1 TO NNLOD
5090 PRINT USING " ##          #"; NLOD(I,1), NLOD(I,2);
5100 PRINT USING "     #.###^^^^"; NLOD(I,3)
5110 NEXT I
5120 RETURN
5130 REM * * * * * * * * * * * * * * * * * * * * * * * * * * * * * * * *
5140 REM *                                                              *
5150 REM *   SUBROUTINE TO MODIFY/TIDY/PRINT THE MEMBER LOADS           *
5160 REM *                                                              *
5170 REM * * * * * * * * * * * * * * * * * * * * * * * * * * * * * * * *
5180 REM
5190 IF ANS    = -6 THEN GOTO 5260
5200 IF ANS    = 1 THEN GOTO 5740
5210 IF PTYPE  < 0 THEN GOTO 5460
5220 IF ANS    > 0 THEN GOTO 5730
5230 REM
5240 REM --- MODIFY MEMBER LOADS
5250 REM
5260 INPUT "LOWER  NODE NUMBER              = ", TEMP(1)
5270 INPUT "HIGHER NODE NUMBER              = ", TEMP(2)
5280 INPUT "LOAD TYPE (UDL = 1, PT LOAD = 2) = ", TEMP(3)
5290 INPUT "MAGNITUDE OF LOAD               = ", TEMP(4)
5300 IF TEMP(3)     = 1 THEN GOTO 5340
5310 PRINT USING "DISTANCE OF LOAD FROM NODE ##"; TEMP(1);
5320 INPUT "   = ", TEMP(5)
5330 K = 0
5340 FOR I          = 1 TO NMLOD
5350 IF MLOD(I,1) <> TEMP(1) THEN GOTO 5370
5360 IF MLOD(I,2) = TEMP(2) THEN K = I
5370 NEXT I
5380 IF K = 0 THEN NMLOD = NMLOD + 1
5390 IF K = 0 THEN K      = NMLOD
5400 FOR J          = 1 TO 5
5410 MLOD(K,J)      = TEMP(J)
5420 NEXT J
5430 REM
5440 REM --- REMOVE ANY ZERO MEMBER LOADS
5450 REM
5460 IF NMLOD       = 0 THEN GOTO 5870
5470 COUNT          = 0
5480 FOR I          = 1 TO NMLOD
5490 IF MLOD(I,3)   = 0 THEN GOTO 5540
5500 COUNT          = COUNT + 1
5510 FOR J = 1 TO 5
5520 MLOD(COUNT,J) = MLOD(I,J)
5530 NEXT J
5540 NEXT I
5550 NMLOD          = COUNT
5560 REM
5570 REM --- PUT MEMBER LOADS INTO ORDER
5580 REM
5590 FLAG           = 0
5600 FOR I          = 1 TO NMLOD-1
5610 IF MLOD(I,1) >  MLOD(I+1,1) THEN GOTO 5640
5620 IF MLOD(I,1) <  MLOD(I+1,1) THEN GOTO 5700
```

```
5630 IF MLOD(I,2) <= MLOD(I+1,2) THEN GOTO 5700
5640 FLAG         = 1
5650 FOR J        = 1 TO 5
5660 TEMP(1)      = MLOD(I,J)
5670 MLOD(I,J)    = MLOD(I+1,J)
5680 MLOD(I+1,J)  = TEMP(1)
5690 NEXT J
5700 NEXT I
5710 IF FLAG      = 1 THEN GOTO 5590
5720 IF PTYPE     < 0 THEN GOTO 5870
5730 IF NMLOD     = 0 THEN GOTO 5870
5740 PRINT
5750 PRINT "+ + + + + + + + + + +"
5760 PRINT "+    MEMBER LOADS    +   -   UDL = 1,    POINT LOAD = 2"
5770 PRINT "+ + + + + + + + + + +"
5780 PRINT
5790 PRINT "MEMBER     LOAD       MAGNITUDE      DISTANCE FROM"
5800 PRINT "           TYPE                      LOWER NODE"
5810 FOR I = 1 TO NMLOD
5820 PRINT USING "## ##          #"; MLOD(I,1), MLOD(I,2), MLOD(I,3);
5830 PRINT USING "    #.###^^^^"; MLOD(I,4);
5840 IF MLOD(I,3) =  2 THEN PRINT USING "     #.###^^^^"; MLOD(I,5)
5850 IF MLOD(I,3) <> 2 THEN PRINT "          N/A"
5860 NEXT I
5870 RETURN
5880 REM *********************************************
5890 REM *                                           *
5900 REM *   SUBROUTINE TO WRITE THE DATA TO THE DATA FILE   *
5910 REM *                                           *
5920 REM *********************************************
5930 REM
5940 OPEN "O", #1, FILE$
5950 WRITE #1, TITLE$
5960 WRITE #1, PTYPE, NCORD, NPROP, NNODE, NMEMB, NREST, NNLOD
5970 IF PTYPE = 3 THEN WRITE #1, NMLOD
5980 IF PTYPE = 4 THEN WRITE #1, NMLOD
5990 REM
6000 REM --- WRITE NODAL COORDINATES
6010 REM
6020 FOR I = 1 TO NNODE
6030 FOR J = 1 TO NCORD
6040 WRITE #1, NODE(I,J)
6050 NEXT J
6060 NEXT I
6070 REM --- WRITE MEMBER DATA
6080 REM
6090 FOR I = 1 TO NMEMB
6100 FOR J = 1 TO NPROP
6110 WRITE #1, MEMB(I,J)
6120 NEXT J
6130 NEXT I
6140 REM
6150 REM --- WRITE RESTRAINT DATA
6160 REM
6170 FOR I = 1 TO NREST
6180 WRITE #1, REST(I,1), REST(I,2)
6190 NEXT I
6200 REM
6210 REM --- WRITE NODAL LOADS
6220 REM
6230 FOR I = 1 TO NNLOD
6240 WRITE #1, NLOD(I,1), NLOD(I,2), NLOD(I,3)
6250 NEXT I
6260 REM
6270 REM --- WRITE MEMBER LOADS
6280 REM
6290 FOR I = 1 TO NMLOD
6300 WRITE #1, MLOD(I,1), MLOD(I,2), MLOD(I,3), MLOD(I,4), MLOD(I,5)
6310 NEXT I
6320 CLOSE #1
6330 RETURN
```

PLANE TRUSSES

Sample Run The following run demonstrates programs PRE.MA/PRE.01/PRE.02 being used to generate the data necessary to solve example 5.3 using the program PTRUSS.PT. Input from the keyboard is shown in bold typeface.

```
---------------------------------------------------------
PROGRAM    P R E . M A   TO PREPROCESS DATA FOR PROGRAMS
PTRUSS.PT, STRUSS.PT, PFRAME.PF, GRID.GD, AND SFRAME.SF
---------------------------------------------------------

DO YOU WISH TO
(1) CREATE A NEW DATA FILE
(2) MODIFY AN EXISTING DATA FILE
? 1

NAME OF THE DATA FILE  =  DATA1.PT

WAIT - CHAINING PRE.01


SELECT PROBLEM TYPE
(1) PLANE TRUSS
(2) SPACE TRUSS
(3) PLANE FRAME
(4) GRILLAGE
(5) SPACE FRAME
? 1

JOB TITLE =  EXAMPLE 5.3 - DATA1.PT

NUMBER OF NODES  =  3

NODE 1

X-COORDINATE       =  0
Y-COORDINATE       =  0

NODE 2

X-COORDINATE       =  3
Y-COORDINATE       =  3

NODE 3

X-COORDINATE       =  6000
Y-COORDINATE       =  0


NUMBER OF MEMBERS       =  3

HIT CARRIAGE RETURN IF THE VALUE IS
THE SAME AS THE PREVIOUS MEMBER

MEMBER 1

LOWER NODE NUMBER       =  1
HIGHER NODE NUMBER      =  3
CROSS-SECTIONAL AREA    =  200
ELASTIC CONSTANT        =  200

MEMBER 2

LOWER NODE NUMBER       =  2
HIGHER NODE NUMBER      =  3
CROSS-SECTIONAL AREA    =  3000
ELASTIC CONSTANT        =  200
```

Note that input that is not shown in boldface type has been entered simply by hitting carriage return.

161

MEMBER 3

```
LOWER NODE NUMBER      = 1
HIGHER NODE NUMBER     = 2
CROSS-SECTIONAL AREA   = 3000
ELASTIC CONSTANT       = 200
```

NUMBER OF RESTRAINED NODES = 2

RESTRAINT DIRECTIONS ARE
X-TRANSLATION = 1
Y-TRANSLATION = 2

ENTER RESTRAINED DIRECTIONS AS A COMPOSITE NUMBER
E.G. IF NODE IS RESTRAINED IN DIRECTIONS 1 AND 2 ENTER 12

HIT CARRIAGE RETURN IF THE VALUE IS
THE SAME AS THE PREVIOUS RESTRAINT

RESTRAINED NODE NUMBER 1

```
NODE TO BE RESTRAINED  = 3
DIRECTION(S)           = 2
```

RESTRAINED NODE NUMBER 2

```
NODE TO BE RESTRAINED  = 1
DIRECTION(S)           = 12
```

NUMBER OF NODAL LOADS = 2

LOAD DIRECTIONS ARE
FORCE IN THE X-DIRECTION = 1
FORCE IN THE Y-DIRECTION = 2

HIT CARRIAGE RETURN IF THE VALUE IS
THE SAME AS THE PREVIOUS LOAD
NODAL LOAD NUMBER 1

```
NODE CARRYING LOAD  = 2
DIRECTION           = 2
MAGNITUDE OF LOAD   = 200
```

NODAL LOAD NUMBER 2

```
NODE CARRYING LOAD  = 2
DIRECTION           = 1
MAGNITUDE OF LOAD   = 100
```

WAIT - CHAINING PRE.02


```
+ + + + + + + +
+  JOB TITLE  +   -   EXAMPLE 5.3  -  DATA1.PT
+ + + + + + + +


+ + + + + + + + + + + +
+  NODAL COORDINATES  +
+ + + + + + + + + + + +
```

NODE NUMBER	X	Y
1	0	0
2	3	3
3	6000	0

PLANE TRUSSES

```
+ + + + + + + + + + + +
+    MEMBER PROPERTIES   +
+ + + + + + + + + + + +

MEMBER            A          E
NUMBER
  1   2      0.300E+04  0.200E+03
  1   3      0.200E+03  0.200E+03
  2   3      0.300E+04  0.200E+03


+ + + + + + + + + +
+    RESTRAINTS    +
+ + + + + + + + + +

NODE      DIRECTION(S)
NUMBER
  1           12
  3            2


+ + + + + + + + + +
+    NODAL LOADS   +
+ + + + + + + + + +

NODE      DIRECTION    VALUE
NUMBER
  2           1       0.100E+03
  2           2       0.200E+03

DO YOU WISH TO    (CR = CARRIAGE RETURN)
(1)  MODIFY DATA
(2)  LIST ALL THE CURRENT DATA
(3)  RUN DATA FILE WITH APPROPRIATE ANALYSIS PROGRAM
(CR) STOP
? 1

DO YOU WISH TO    (CR = CARRIAGE RETURN)
(1)  CHANGE JOB TITLE
(2)  MODIFY NODAL DATA
(3)  MODIFY MEMBER DATA
(4)  MODIFY RESTRAINTS
(5)  MODIFY NODAL LOADS
(CR) STOP MODIFYING DATA
? 2

+ + + + + + + + + + + +
+    NODAL COORDINATES   +
+ + + + + + + + + + + +

NODE          X             Y
NUMBER
  1           0             0
  2           3             3
  3          6000           0

NODE TO BE REVISED/ADDED  =  2
X-COORDINATE              =  3000
Y-COORDINATE              =  3000

+ + + + + + + + + + + +
+    NODAL COORDINATES   +
+ + + + + + + + + + + +

NODE          X             Y
NUMBER
  1           0             0
  2          3000          3000
  3          6000           0

AGAIN  -  (Y)ES OR (CR) ?              Carriage return entered
```

COMPUTER ANALYSIS OF STRUCTURAL FRAMEWORKS

```
DO YOU WISH TO    (CR = CARRIAGE RETURN)
(1)   CHANGE JOB TITLE
(2)   MODIFY NODAL DATA
(3)   MODIFY MEMBER DATA
(4)   MODIFY RESTRAINTS
(5)   MODIFY NODAL LOADS
(CR)  STOP MODIFYING DATA
? 3

+ + + + + + + + + + + + +
+    MEMBER PROPERTIES   +
+ + + + + + + + + + + + +

MEMBER           A         E
NUMBER
  1  2      0.300E+04 0.200E+03
  1  3      0.200E+03 0.200E+03
  2  3      0.300E+04 0.200E+03

ENTER DATA FOR ELEMENT TO BE MODIFIED/ADDED
SET CROSS-SECTIONAL AREA TO ZERO TO CANCEL AN ELEMENT

LOWER NODE NUMBER       =  1
HIGHER NODE NUMBER      =  3
CROSS-SECTIONAL AREA    =  3000
ELASTIC CONSTANT        =  200


+ + + + + + + + + + + + +
+    MEMBER PROPERTIES   +
+ + + + + + + + + + + + +

MEMBER           A         E
NUMBER
  1  2      0.300E+04 0.200E+03
  1  3      0.300E+04 0.200E+03
  2  3      0.300E+04 0.200E+03

AGAIN  -  (Y)ES OR (CR) ?              Carriage return entered

DO YOU WISH TO (CR = CARRIAGE RETURN)
(1)   CHANGE JOB TITLE
(2)   MODIFY NODAL DATA
(3)   MODIFY MEMBER DATA
(4)   MODIFY RESTRAINTS
(5)   MODIFY NODAL LOADS
(CR)  STOP MODIFYING DATA
?                                      Carriage return entered

DO YOU WISH TO (CR = CARRIAGE RETURN)
(1)   MODIFY DATA
(2)   LIST ALL THE CURRENT DATA
(3)   RUN DATA FILE WITH APPROPRIATE ANALYSIS PROGRAM
(CR)  STOP
?                                      Carriage return entered
```

6 Space Trusses

6.1 INTRODUCTION

The previous chapter considered only plane trusses. This chapter shows how the theory for plane trusses is easily extended to deal with three-dimensional trusses. The principles underlying the theory developed for plane trusses are equally applicable to space trusses. The implementation is, however, complicated by the introduction of an additional freedom at each node. Consequently the nodal force and displacement vectors now have three components and the transformation matrix and element stiffness submatrices are 3 x 3 arrays. This chapter has been written assuming the reader to be familiar with the material presented in the previous chapter because, for compactness, only the essential differences between the analysis of space trusses and plane trusses are covered.

6.2 THE ELEMENT STIFFNESS MATRIX IN THE MEMBER AXES SYSTEM

Figure 6.1 shows the member axes and freedoms for a space truss element.

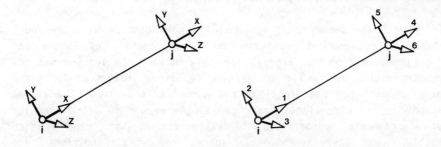

FIGURE 6.1 MEMBER AXES AND FREEDOMS FOR A SPACE TRUSS ELEMENT

In Chapter 5 the relationship between the member end forces and the member end displacements for a plane truss member was shown to be:

$$\begin{bmatrix} f_{ijx} \\ f_{ijy} \\ f_{jix} \\ f_{jiy} \end{bmatrix} = \begin{bmatrix} EA/L & 0 & -EA/L & 0 \\ 0 & 0 & 0 & 0 \\ -EA/L & 0 & EA/L & 0 \\ 0 & 0 & 0 & 0 \end{bmatrix} \begin{bmatrix} \delta_{ijx} \\ \delta_{ijy} \\ \delta_{jix} \\ \delta_{jiy} \end{bmatrix}$$

The principal difference between space trusses and plane trusses is that space trusses have an additional freedom at each node (in the z-direction). Displacements in the z-direction (member axes) generate no force in the member (on the basis of small displacement theory). Hence the element stiffness equation for a space truss element is as follows:

$$\begin{bmatrix} f_{ijx} \\ f_{ijy} \\ f_{ijz} \\ f_{jix} \\ f_{jiy} \\ f_{jiz} \end{bmatrix} = \begin{bmatrix} EA/L & 0 & 0 & -EA/L & 0 & 0 \\ 0 & 0 & 0 & 0 & 0 & 0 \\ 0 & 0 & 0 & 0 & 0 & 0 \\ -EA/L & 0 & 0 & EA/L & 0 & 0 \\ 0 & 0 & 0 & 0 & 0 & 0 \\ 0 & 0 & 0 & 0 & 0 & 0 \end{bmatrix} \begin{bmatrix} \delta_{ijx} \\ \delta_{ijy} \\ \delta_{ijz} \\ \delta_{jix} \\ \delta_{jiy} \\ \delta_{jiz} \end{bmatrix}$$

or, in terms of the submatrices indicated by the broken lines:

$$\mathbf{f}_{ij} = \mathbf{k}_{ii}^j \delta_{ij} + \mathbf{k}_{ij} \delta_{ji} \tag{6.1}$$

and

$$\mathbf{f}_{ji} = \mathbf{k}_{ji} \delta_{ij} + \mathbf{k}_{jj}^i \delta_{ji} \tag{6.2}$$

The reader should note that equations (6.1) and (6.2) are exactly the same as the corresponding equations for a plane truss member.

6.3 TRANSFORMATION OF FORCE AND DISPLACEMENT

In general the member axes systems will not coincide with the nodal axes systems, making necessary the transformation of forces and displacements from one axes system to another. In section 5.3 the transformation matrix T_{ij} was used to transform vectors of force and displacement from the global to the member axes system. Transformations in two-dimensional space are simple enough to allow adoption of a pragmatic approach to the evaluation of the transformation matrix. In three-dimensional space the transformations are sufficiently complex to warrant a more rigorous approach.

SPACE TRUSSES

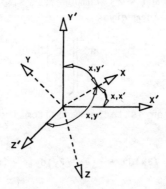

FIGURE 6.2   ROTATION OF AXES IN THREE-DIMENSIONAL SPACE

Figure 6.2 shows two, right handed, three-dimensional cartesian coordinate axes systems with a common origin, but oriented at different angles. The problem in hand is to find a matrix that will transform a vector of force or displacement from the unprimed to the primed axes system. Consider first the "x" component which is, of course, a function of "x'", "y'", and "z'",

i.e.
$$x = F(x', y', z')$$
hence
$$dx = (\partial x/\partial x')dx' + (\partial x/\partial y')dy' + (\partial x/\partial z')dz' \qquad (6.3)$$
and
$$\partial x/\partial x' = \cos(x, x') = l_x$$
$$\partial x/\partial y' = \cos(x, y') = m_x$$
$$\partial x/\partial z' = \cos(x, z') = n_x$$

where $\cos(x,x')$ is the cosine of the angle between the "x" and "x'" axes, $\cos(x,y')$ is the cosine of the angle between the "x" and "y'" axes, etc. Note that the direction in which the angle is measured is immaterial as $\cos(360 - \alpha) = \cos(\alpha)$. "$l_x$", "$m_x$", and "$n_x$" are known as the *direction cosines* of the "x" axis with respect to the "x',y', z'" axes system. Equation (6.3) can be rewritten:

$$dx = l_x\, dx' + m_x\, dy' + n_x\, dz'$$

Integrating

$$x = l_x\, x' + m_x\, y' + n_x\, z'$$

Treating the "y" and "z" components similarly and writing the equations in matrix form gives

$$\begin{bmatrix} x \\ y \\ z \end{bmatrix} = \begin{bmatrix} l_x & m_x & n_x \\ l_y & m_y & n_y \\ l_z & m_z & n_z \end{bmatrix} \begin{bmatrix} x' \\ y' \\ z' \end{bmatrix} \qquad (6.4)$$

If the transformation is to be made from the unprimed to the primed axes system then

$$dx' = (\partial x'/\partial x)dx + (\partial x'/\partial y)dy + (\partial x'/\partial z)dz \qquad (6.5)$$

Figure 6.3 shows the "x" and "x'" axes from the axes systems shown in fig 6.2, only the paper represents the "x,x'" plane. Note that, in general, no other axis will lie in this plane.

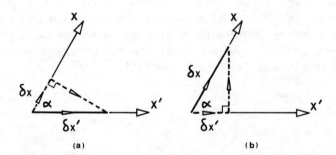

FIGURE 6.3   EVALUATION OF DIRECTION COSINES

The partial derivative $\partial x/\partial x'$ represents the rate of change of "x" with "x'", both "y'" and "z'" remaining constant. Figure 6.3(a) shows a change "$\delta x'$" being made to "x'". As the change takes place along the "x'" axis, "y'" and "z'" are, by definition, constant. The corresponding change in "x" is "δx" as shown,

hence

$$\partial x/\partial x' = \lim_{x \to 0}(\delta x/\delta x') = \cos\alpha \qquad (6.6)$$

Now consider the partial derivative $\partial x'/\partial x$ which represents the rate of change of "x'" with "x", both "y" and "z" remaining constant. Figure 6.3(b) shows "x" being varied by "δx" and the corresponding change in "x'". By inspection

$$\partial x'/\partial x = \lim_{x \to 0}(\delta x'/\delta x) = \cos\alpha \qquad (6.7)$$

SPACE TRUSSES

Hence by comparison of equations (6.6) and (6.7):

$$\partial x'/\partial x = \partial x/\partial x' \quad (\text{or } l_{x'} = l_x)$$

similarly

$$\partial x'/\partial y = \partial y/\partial x' \quad (\text{or } m_{x'} = l_y)$$
$$\partial x'/\partial z = \partial z/\partial x' \quad (\text{or } n_{x'} = l_z)$$

hence equation (6.5) reduces to

$$dx' = l_x\, dx + l_y\, dy + l_z\, dz$$

Integrating

$$x' = l_x\, x + l_y\, y + l_z\, z$$

Treating the "y'" and "z'" components similarly yields the following matrix equation:

$$\begin{bmatrix} x' \\ y' \\ z' \end{bmatrix} = \begin{bmatrix} l_x & l_y & l_z \\ m_x & m_y & m_z \\ n_x & n_y & n_z \end{bmatrix} \begin{bmatrix} x \\ y \\ z \end{bmatrix} \qquad (6.8)$$

By definition the transformation matrices in equations (6.4) and (6.8) must be the inverse of each other, which means that

$$\begin{bmatrix} l_x & m_x & n_x \\ l_y & m_y & n_y \\ l_z & m_z & n_z \end{bmatrix} \begin{bmatrix} l_x & l_y & l_z \\ m_x & m_y & m_z \\ n_x & n_y & n_z \end{bmatrix} = \begin{bmatrix} 1 & 0 & 0 \\ 0 & 1 & 0 \\ 0 & 0 & 1 \end{bmatrix}$$

or

$$l_i\, l_j + m_i\, m_j + n_i\, n_j = 0, 1 \qquad i \neq j,\ i = j$$
$$(i,\ j\ =\ x,\ y,\ z)$$

This is the orthogonality property of direction cosines. The six orthogonality equations mean that only three direction cosines out of the total of nine are required to define the position of one axes system relative to the other, provided that not all three known direction cosines lie in one row or one column. For instance if l_x, m_x, and n_x only were known (see fig 6.2) then the position of the "x" axis would be uniquely defined, but the positions of the "y" and "z" axes would not, as the "x,y,z" axes system could be rotated about the "x" axis.

6.4 THE ELEMENT STIFFNESS MATRIX IN THE GLOBAL AXES SYSTEM

Taking the primed axes in fig 6.2 as the global axes system and the unprimed axes system as the member axes system, then equation (6.9) gives the transformation of displacement from the global to the member axes system:

$$\begin{bmatrix} \delta_{ijx} \\ \delta_{ijy} \\ \delta_{ijz} \end{bmatrix} = \begin{bmatrix} l_x & m_x & n_x \\ l_y & m_y & n_y \\ l_z & m_z & n_z \end{bmatrix} \begin{bmatrix} \Delta_{ix} \\ \Delta_{iy} \\ \Delta_{iz} \end{bmatrix}$$

or

$$\delta_{ij} = T_{ij}\Delta_i \qquad (6.9)$$

Similarly the transformation of force from the global to the member axes system is

$$\begin{bmatrix} f_{ijx} \\ f_{ijy} \\ f_{ijz} \end{bmatrix} = \begin{bmatrix} l_x & m_x & n_x \\ l_y & m_y & n_y \\ l_z & m_z & n_z \end{bmatrix} \begin{bmatrix} F_{ijx} \\ F_{ijy} \\ F_{ijz} \end{bmatrix}$$

or

$$f_{ij} = T_{ij} F_{ij} \qquad (6.10)$$

The inverse of the transformation matrix effects the reverse transformation, i.e.

$$\Delta_i = T_{ij}^{-1} \delta_{ij} \qquad (6.11)$$

and

$$F_{ij} = T_{ij}^{-1} f_{ij} \qquad (6.12)$$

The equation of nodal equilibrium developed in the previous chapter for plane trusses (equation (5.15)) is perfectly general. Equation (5.15) has been copied as equation (6.13) for convenience.

$$P_i = F_{ia} + F_{ib} + F_{ic} \ldots + F_{in} \qquad (6.13)$$

P_i is the vector of external applied loads at node "i" and F_{ij} is the vector of member end forces at end "i" of member "i,j". All vectors are expressed in terms of the nodal axes system. Now, from equations (6.12) and (6.1)

$$F_{ij} = T_{ij}^{-1} f_{ij}$$
$$= T_{ij}^{-1} (k_{ii}^j \delta_{ij} + k_{ij} \delta_{ji})$$

and from equation (6.9)

hence
$$\delta_{ij} = T_{ij} \Delta_i$$
$$F_{ij} = T_{ij}^{-1} k_{ii}^{j} T_{ij} \Delta_i + T_{ij}^{-1} k_{ij} T_{ij} \Delta_j$$

This allows the equation of nodal equilibrium (equation (6.13)) to be written as follows:

$$P_i = K_{ii} \Delta_i + K_{ia} \Delta_a + K_{ib} \Delta_b + \ldots + K_{in} \Delta_n \tag{6.14}$$

where
$$K_{ii} = \sum K_{ii}^{j} = \sum T_{ij}^{-1} k_{ii}^{j} T_{ij}$$
$$K_{ij} = T_{ij}^{-1} k_{ij} T_{ij}$$

The summation is for all elements connected to node "i".

The reader should note that equations (6.13) and (6.14) are generally applicable as they simply state that each node of a structure must be in equilibrium under the action of the external applied forces (which may include reactive forces) and the internal member end forces. In the previous chapter the vectors of nodal force and displacement for a plane truss were seen to have two components and consequently the element stiffness submatrices had dimension 2 x 2. Three-dimensional trusses have an additional freedom at each node. Hence the vectors of nodal force and displacement have three components and the element stiffness submatrices are 3 x 3. Writing equation (6.14) for each node of the structure in turn produces a system of simultaneous equations relating the nodal displacements to the nodal loads. That system of equations is, as was seen in the previous chapter, the initial structure stiffness equation.

In practice it is convenient to explicitly evaluate the global element stiffness submatrices as follows:

$$K_{ii}^{j} = T_{ij}^{-1} k_{ii}^{j} T_{ij}$$

$$= \begin{bmatrix} l_x & l_y & l_z \\ m_x & m_y & m_z \\ n_x & n_y & n_z \end{bmatrix} \begin{bmatrix} EA/L & 0 & 0 \\ 0 & 0 & 0 \\ 0 & 0 & 0 \end{bmatrix} \begin{bmatrix} l_x & m_x & n_x \\ l_y & m_y & n_y \\ l_z & m_z & n_z \end{bmatrix}$$

$$= EA/L \begin{bmatrix} l_x^2 & l_x m_x & l_x n_x \\ l_x m_x & m_x^2 & m_x n_x \\ l_x n_x & m_x n_x & n_x^2 \end{bmatrix}$$

Similarly

$$K_{ij} = T_{ij}^{-1} k_{ij} T_{ij}$$

$$= -EA/L \begin{bmatrix} l_x^2 & l_x m_x & l_x n_x \\ l_x m_x & m_x^2 & m_x n_x \\ l_x n_x & m_x n_x & n_x^2 \end{bmatrix}$$

Evaluation of the K_{ji} and the K_{jj}^i submatrices shows that

and
$$K_{jj}^i = K_{ii}^j$$
$$K_{ji} = K_{ij}$$

To evaluate the element stiffness submatrices developed above requires a knowledge of the direction cosines of the member "x" axis. These are easily obtained from the coordinates of the nodes at the ends of the member. If the length of member "i,j" is L_{ij} and the coordinates of nodes "i" and "j" are (x_i, y_i, z_i) and (x_j, y_j, z_j) as illustrated in fig 6.4, then

$$l_x = (x_j - x_i) / L_{ij}$$
$$m_x = (y_j - y_i) / L_{ij}$$
$$n_x = (z_j - z_i) / L_{ij}$$

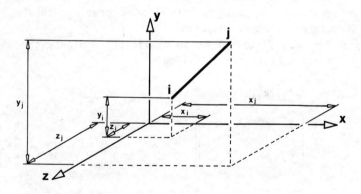

FIGURE 6.4   CALCULATION OF DIRECTION COSINES FOR MEMBER "i,j"

SPACE TRUSSES

6.5 THE STRUCTURE STIFFNESS MATRIX

Chapter 5 showed that, prior to the application of the boundary conditions, the equations of nodal equilibrium form a mixed set of simultaneous equations. The coefficient matrix of these equations is known as the initial structure stiffness matrix. The procedure to generate the initial structure stiffness matrix for a three-dimensional truss is identical to that used for a two-dimensional truss. However, because the submatrices K_{ii}^j, K_{ij}, K_{ji}, and K_{jj}^i are now 3 x 3 they must be added to the initial structure stiffness matrix such that element (1,1) is added to locations (3i-2, 3i-2), (3i-2, 3j-2), (3j-2, 3i-2), and (3j-2, 3j-2) respectively. Figure 6.5 shows the locations where the four global element stiffness submatrices for member 2,5 would be added to the initial structure stiffness matrix. The initial structure stiffness matrix is 18 x 18, indicating that the space truss has 6 nodes.

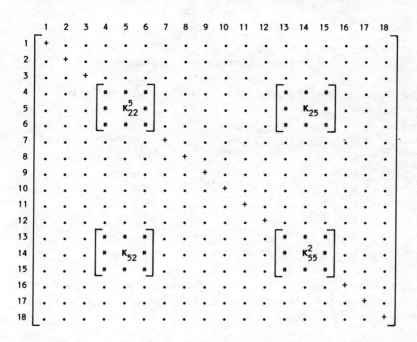

FIGURE 6.5   ADDING ELEMENT 2,5 TO THE INITIAL STRUCTURE STIFFNESS MATRIX

Chapter 5 showed that it is more efficient to generate the final structure stiffness matrix directly, rather than via the initial structure stiffness matrix.

To assemble the final structure stiffness matrix requires a knowledge of the boundary conditions. Restraints are conveniently

COMPUTER ANALYSIS OF STRUCTURAL FRAMEWORKS

described by the number of the restrained node, plus the direction of the restraint. Programming is simplified if the directions of restraint are given numbers. For space trusses the number system that follows will be adopted:

1 - translation restrained in the global x-direction
2 - translation restrained in the global y-direction
3 - translation restrained in the global z-direction.

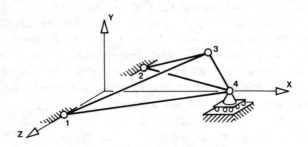

FIGURE 6.6   A SIMPLE THREE-DIMENSIONAL TRUSS

The boundary conditions for the truss shown in fig 6.6 would be input as follows:

NODE	DIRECTION
1	1
1	2
1	3
2	1
2	2
2	3
4	2

Input can be reduced by entering the restraint directions as a composite number (see section 5.8). Using composite numbers the above restraints become

NODE	DIRECTION(S)
1	123
2	123
4	2

The procedure to generate the final stiffness matrix for a space truss is identical to that used to generate the final stiffness matrix for a plane truss. Hence the freedom vector for the truss shown in fig 6.6 is generated as follows. Note that the length of the freedom vector is equal to the initial degree of freedom (i.e. 3 times the number of nodes).

Zero the freedom vector $\begin{bmatrix} 0\\0\\0\\0\\0\\0\\0\\0\\0\\0\\0\\0 \end{bmatrix}$ Apply the boundary conditions $\begin{bmatrix} 1\\1\\1\\1\\1\\1\\0\\0\\0\\0\\1\\0 \end{bmatrix}$ Number the freedoms $\begin{bmatrix} 0\\0\\0\\0\\0\\0\\1\\2\\3\\4\\0\\5 \end{bmatrix}$

The code number for element "i,j" comprises of the elements (3i-2), (3i-1), (3i), (3j-2), (3j-1) and (3j) of the freedom vector.

Consider again the truss shown in fig 6.6. The freedom vector has already been shown to be

$$\{ 0 \quad 0 \quad 0 \quad 0 \quad 0 \quad 0 \quad 1 \quad 2 \quad 3 \quad 4 \quad 0 \quad 5 \}$$

Hence the code number for element 1,3 is [0 0 0 1 2 3] which is applied to the global element stiffness matrix as shown:

$$\begin{array}{c} \\ 0\\ 0\\ 0\\ 1\\ 2\\ 3 \end{array} \begin{array}{cccccc} 0 & 0 & 0 & 1 & 2 & 3 \\ \begin{bmatrix} K_{11} & K_{12} & K_{13} & K_{14} & K_{15} & K_{16}\\ K_{21} & K_{22} & K_{23} & K_{24} & K_{25} & K_{26}\\ K_{31} & K_{32} & K_{33} & K_{34} & K_{35} & K_{36}\\ K_{41} & K_{42} & K_{43} & K_{44} & K_{45} & K_{46}\\ K_{51} & K_{52} & K_{53} & K_{54} & K_{55} & K_{56}\\ K_{61} & K_{62} & K_{63} & K_{64} & K_{65} & K_{66} \end{bmatrix} \end{array}$$

The code number indicates that stiffness term K_{44} should be added to the final structure stiffness matrix at location (1,1), the stiffness term K_{45} should be added at location (1,2), ... the stiffness term K_{66} should be added at location (3,3). Note that if a zero element appears in either the "horizontal" or the "vertical" code number then that stiffness term is ignored. The final stiffness matrix is complete when the stiffness contributions of all elements have been added using this technique.

6.6 MEMBER END FORCES AND REACTIONS

Once the final structure stiffness matrix has been assembled and the equations solved to find the unknown nodal displacements, equation (6.1) is used to find the forces in the members.

As the nodal displacements are found in the global axes system they must first be transformed into the member axes system.

Equation (6.1) is

$$f_{ij} = k_{ii}^{j} \delta_{ij} + k_{ij} \delta_{ji}$$

Expressing the member end displacements as transformed nodal displacements gives

$$f_{ij} = k_{ii}^{j} T_{ij} \Delta_i + k_{ij} T_{ij} \Delta_j$$

$$\begin{bmatrix} f_{ijx} \\ f_{ijy} \\ f_{ijz} \end{bmatrix} = \begin{bmatrix} EA/L & 0 & 0 \\ 0 & 0 & 0 \\ 0 & 0 & 0 \end{bmatrix} \begin{bmatrix} l_x & m_x & n_x \\ l_y & m_y & n_y \\ l_z & m_z & n_z \end{bmatrix} \begin{bmatrix} \Delta_{ix} \\ \Delta_{iy} \\ \Delta_{iz} \end{bmatrix} +$$

$$\begin{bmatrix} -EA/L & 0 & 0 \\ 0 & 0 & 0 \\ 0 & 0 & 0 \end{bmatrix} \begin{bmatrix} l_x & m_x & n_x \\ l_y & m_y & n_y \\ l_z & m_z & n_z \end{bmatrix} \begin{bmatrix} \Delta_{jx} \\ \Delta_{jy} \\ \Delta_{jz} \end{bmatrix}$$

or
$$f_{ijx} = EA/L\{l_x(\Delta_{ix} - \Delta_{jx}) + m_x(\Delta_{iy} - \Delta_{jy}) + n_x(\Delta_{iz} - \Delta_{jz})\}$$
$$f_{ijy} = 0$$
$$f_{ijz} = 0$$

Consideration of the equilibrium of the element (or a similar matrix expansion) shows that the forces at end "j" of the element are as follows:

$$f_{jix} = -f_{ijx}$$
$$f_{jiy} = 0$$
$$f_{jiz} = 0$$

Note that if f_{ijx} is positive then the member is in compression.

Reactions

If node "i" is restrained, the reactive forces (in the global axes system) are found using equation (5.23) which was developed in the previous chapter.

SPACE TRUSSES

$$R_i = \sum T_{ij}^{-1} f_{ij} - P_i$$

$$\begin{bmatrix} R_{ix} \\ R_{iy} \\ R_{iz} \end{bmatrix} = \sum \begin{bmatrix} l_x & l_y & l_z \\ m_x & m_y & m_z \\ n_x & n_y & n_z \end{bmatrix} \begin{bmatrix} f_{ijx} \\ f_{ijy} \\ f_{ijz} \end{bmatrix} - \begin{bmatrix} P_{ix} \\ P_{iy} \\ P_{iz} \end{bmatrix}$$

$$= \sum \begin{bmatrix} l_x f_{ijx} \\ m_x f_{ijx} \\ n_x f_{ijx} \end{bmatrix} - \begin{bmatrix} P_{ix} \\ P_{iy} \\ P_{iz} \end{bmatrix}$$

6.7 SPACE TRUSS PROGRAM - STRUSS.ST

This program is essentially the plane truss analysis program (PTRUSS.PT) modified to deal with three-dimensional trusses. The reader should note how little the code has had to be changed during the conversion, which illustrates, once again, the generality of the stiffness method.

Input The only piece of information the user supplies to the program from the keyboard is the name of the data file to be read by the program. This data file can be constructed using either a text editor, or the data preprocessor program PRE.MA/PRE.01/PRE.02 that is described in section 5.11. If using a text editor the data file should be constructed to the following format:

```
<job title (not to exceed one line)>
2,3,4,<no. of nodes>,<no. of members>,<no. of restrained nodes>,
<no. of loads>
<x-coordinate node 1>,<y-coordinate node 1>,<z-coordinate node 1>
<x-coordinate node 2>,<y-coordinate node 2>,<z-coordinate node 2>
<x-coordinate node 3>,<y-coordinate node 3>,<z-coordinate node 3>
         .                  .                    .
         .                  .                    .
         .                  .                    .
<x-coordinate node n>,<y-coordinate node n>,<z-coordinate node n>
<node "i" member 1>,<node "j" member 1>,<area member 1>,<E member 1>
<node "i" member 2>,<node "j" member 2>,<area member 2>,<E member 2>
<node "i" member 3>,<node "j" member 3>,<area member 3>,<E member 3>
         .                  .                    .
         .                  .                    .
         note - "i" must be less than "j"
         .                  .                    .
         .                  .                    .
```

```
<node "i" member n>,<node "j" member n>,<area member n>,<E member n>
<restrained node no. 1 node number>,<direction(s)>
<restrained node no. 2 node number>,<direction(s)>
         .                .                 .
         .                .                 .
   note - restraint directions are as follows:
          restraint in the global "x" direction = 1
          restraint in the global "y" direction = 2
          restraint in the global "z" direction = 3
         .                .                 .
   Input restraints as a composite number (see text)
         .                .                 .
         .                .                 .
<restrained node no. n node number>,<direction(s)>
<nodal load no. 1  node number>,<direction>,<magnitude>
<nodal load no. 2  node number>,<direction>,<magnitude>
         .                .                 .
         .                .                 .
   note - load directions are as follows:
          load in the global  x-direction = 1
          load in the global  y-direction = 2
          load in the global  z-direction = 3
         .                .                 .
         .                .                 .
<nodal load no. n  node number>,<direction>,<magnitude>
```

Comments on the Algorithm The program first uses the boundary conditions to establish the freedom vector. The stiffness of each element is then added to the structure stiffness matrix in turn. The upper triangle only of the final structure stiffness matrix is generated and its symmetry is later used to complete the matrix. No attempt is made to minimise storage by taking advantage of either the symmetry or the bandwidth of the structure stiffness matrix. The loading vector is developed and then the equations are solved using Gauss-Jordan elimination. Once the nodal displacements have been found they are used to calculate the member forces and reactions.

Output The data from the data file is echoed and messages are output to indicate the beginning of each solution phase. After solution of the stiffness equation nodal displacements are output, followed by the axial forces in the members and the reactions.

SPACE TRUSSES

Listing

```
1000 REM ***********************************************************************
1010 REM *                                                                     *
1020 REM *    PROGRAM      S T R U S S . S T      FOR THE AUTOMATIC ANALYSIS   *
1030 REM *    OF SPACE TRUSSES                                                 *
1040 REM *                                                                     *
1050 REM *    DATA FROM DATA FILE GENERATED BY A TEXT EDITOR OR BY THE         *
1060 REM *    PREPROCESSOR PROGRAM  P R E . M A / P R E . 0 1 / P R E . 0 2    *
1070 REM *                                                                     *
1080 REM *    J.A.D.BALFOUR                                                    *
1090 REM *                                                                     *
1100 REM ***********************************************************************
1110 REM
1120 REM --- NOTE THAT VARIABLES ARE DEFINED IN APPENDIX A
1130 REM
1140 OPTION BASE 1
1150 REM
1160 REM --- THE FOLLOWING DIMENSION STATEMENT SETS THE PROBLEM SIZE
1170 REM
1180 COMMON FILE$
1190 DIM NODE(20,3), MEMB(40,8), REST(20,5), NLOD(40,3), FSTIFF(40,41)
1200 DIM FREE(40),   EFREE(6),    ESTIFF(6,6)
1210 PRINT : PRINT
1220 PRINT "------------------------------------------------"
1230 PRINT
1240 PRINT "PROGRAM    S T R U S S . S T    FOR THE AUTOMATIC"
1250 PRINT "ANALYSIS OF SPACE TRUSSES"
1260 PRINT
1270 PRINT "------------------------------------------------"
1280 REM
1290 REM --- CHECK IF THIS PROGRAM HAS BEEN CHAINED FROM PRE.MA/PRE.01/PRE.02
1300 REM
1310 IF LEN(FILE$) <> 0 THEN GOTO 1340
1320 PRINT
1330 INPUT           "NAME OF THE DATA FILE  =  ", FILE$
1340 OPEN "I", #1, FILE$
1350 PRINT
1360 PRINT           "READING DATA"
1370 GOSUB 1600
1380 REM
1390 REM            PRINT DATA
1400 GOSUB 1960
1410 PRINT
1420 PRINT           "GENERATING THE FREEDOM VECTOR"
1430 GOSUB 2490
1440 PRINT
1450 PRINT           "GENERATING THE STRUCTURE STIFFNESS MATRIX"
1460 GOSUB 2850
1470 PRINT
1480 PRINT           "GENERATING THE LOADING VECTOR"
1490 GOSUB 3750
1500 PRINT
1510 PRINT           "SOLVING EQUATIONS  -  NO. OF EQUATIONS  = "; NDOF
1520 GOSUB 3870
1530 REM
1540 REM            OUTPUT NODAL DISPLACEMENTS
1550 GOSUB 4210
1560 REM
1570 REM            OUTPUT MEMBER FORCES AND REACTIONS
1580 GOSUB 4460
1590 END
1600 REM * * * * * * * * * * * * * * * * * * * * * * * * * * * * * * * * * *
1610 REM *                                                                 *
1620 REM *   SUBROUTINE TO READ DATA FROM THE INPUT FILE                   *
1630 REM *                                                                 *
1640 REM * * * * * * * * * * * * * * * * * * * * * * * * * * * * * * * * * *
1650 REM
1660 REM --- READ BASIC PROBLEM DATA
1670 REM
1680 INPUT #1, TITLE$
1690 INPUT #1, PTYPE, NCORD, NPROP, NNODE, NMEMB, NREST, NNLOD
```

COMPUTER ANALYSIS OF STRUCTURAL FRAMEWORKS

```
1700 IF PTYPE = 2 THEN GOTO 1750
1710 PRINT : PRINT "NOT A SPACE TRUSS DATA FILE" : END
1720 REM
1730 REM    ---   READ NODAL COORDINATES
1740 REM
1750 FOR I = 1 TO NNODE
1760 INPUT #1, NODE(I,1), NODE(I,2), NODE(I,3)
1770 NEXT I
1780 REM    ---   READ MEMBER DATA
1790 REM
1800 FOR I = 1 TO NMEMB
1810 INPUT #1, MEMB(I,1), MEMB(I,2), MEMB(I,3), MEMB(I,4)
1820 NEXT I
1830 REM
1840 REM    ---   READ RESTRAINT DATA
1850 REM
1860 FOR I = 1 TO NREST
1870 INPUT #1, REST(I,1), REST(I,2)
1880 NEXT I
1890 REM
1900 REM    ---   READ LOADING DATA
1910 REM
1920 FOR I = 1 TO NNLOD
1930 INPUT #1, NLOD(I,1), NLOD(I,2), NLOD(I,3)
1940 NEXT I
1950 RETURN
1960 REM * * * * * * * * * * * * * * * * * * * * * * * * * * * * * * * *
1970 REM *                                                              *
1980 REM *       SUBROUTINE TO PRINT DATA                               *
1990 REM *                                                              *
2000 REM * * * * * * * * * * * * * * * * * * * * * * * * * * * * * * * *
2010 REM
2020 PRINT
2030 PRINT "+ + + + + + + + "
2040 PRINT "+   JOB TITLE   +   -    "; TITLE$
2050 PRINT "+ + + + + + + + +"
2060 PRINT
2070 PRINT "+ + + + + + + + + + + + +"
2080 PRINT "+    NODAL COORDINATES   +"
2090 PRINT "+ + + + + + + + + + + + +"
2100 PRINT
2110 PRINT "NODE              X              Y              Z
2120 PRINT "NUMBER
2130 FOR I = 1 TO NNODE
2140 PRINT " "; I, NODE(I,1), NODE(I,2), NODE(I,3)
2150 NEXT I
2160 PRINT
2170 PRINT "+ + + + + + + + + + + + +"
2180 PRINT "+    MEMBER PROPERTIES   +"
2190 PRINT "+ + + + + + + + + + + + +"
2200 PRINT
2210 PRINT "MEMBER            A              E"
2220 PRINT "NUMBER"
2230 FOR I = 1 TO NMEMB
2240 PRINT USING "## ##      "; MEMB(I,1), MEMB(I,2);
2250 PRINT USING "    #.###^^^^"; MEMB(I,3), MEMB(I,4)
2260 NEXT I
2270 PRINT
2280 PRINT "+ + + + + + + + + "
2290 PRINT "+   RESTRAINTS   +"
2300 PRINT "+ + + + + + + + + "
2310 PRINT
2320 PRINT "NODE     DIRECTION(S)"
2330 PRINT "NUMBER"
2340 FOR I = 1 TO NREST
2350 PRINT USING "   ##          ###"; REST(I,1), REST(I,2)
2360 NEXT I
2370 PRINT
2380 PRINT "+ + + + + + + + + +"
2390 PRINT "+   NODAL LOADS   +"
2400 PRINT "+ + + + + + + + + +"
2410 PRINT
```

SPACE TRUSSES

```
2420 PRINT "NODE    DIRECTION    VALUE"
2430 PRINT "NUMBER"
2440 FOR I = 1 TO NNLOD
2450 PRINT USING "  ##          #"; NLOD(I,1), NLOD(I,2);
2460 PRINT USING "     #.###^^^^"; NLOD(I,3)
2470 NEXT I
2480 RETURN
2490 REM ********************************
2500 REM *                                *
2510 REM *  SUBROUTINE TO GENERATE THE FREEDOM VECTOR     *
2520 REM *                                *
2530 REM ********************************
2540 REM
2550 REM --- ZERO THE FREEDOM VECTOR
2560 REM
2570 FOR I     = 1 TO 3*NNODE
2580 FREE(I) = 0
2590 NEXT I
2600 REM
2610 REM --- LOOP FOR ALL RESTRAINED NODES
2620 REM
2630 FOR I     = 1 TO NREST
2640 K         = REST(I,2)
2650 REM
2660 REM --- EVALUATE THE FREEDOM TO BE RESTRAINED FROM RIGHT-MOST DIGIT OF K
2670 REM
2680 J         = 3*REST(I,1) - 3 + (K MOD 10)
2690 FREE(J) = 1
2700 K         = INT(K/10)
2710 IF K      > 0 THEN GOTO 2680
2720 NEXT I
2730 REM
2740 REM --- NUMBER THE FREEDOMS (RESTRAINTS SET TO ZERO)
2750 REM
2760 NDOF      = 0
2770 FOR I     = 1 TO 3*NNODE
2780 IF FREE(I) = 1 THEN GOTO 2820
2790 NDOF      = NDOF + 1
2800 FREE(I) = NDOF
2810 GOTO 2830
2820 FREE(I) = 0
2830 NEXT I
2840 RETURN
2850 REM ********************************
2860 REM *                                *
2870 REM *  SUBROUTINE TO GENERATE THE STRUCTURE STIFFNESS MATRIX  *
2880 REM *                                *
2890 REM ********************************
2900 REM
2910 REM --- ZERO THE AUGMENTED MATRIX
2920 REM
2930 FOR I     = 1 TO NDOF
2940 FOR J     = 1 TO NDOF + 1
2950 FSTIFF(I,J) = 0
2960 NEXT J
2970 NEXT I
2980 REM
2990 REM --- LOOP FOR EACH ELEMENT
3000 REM
3010 PRINT : PRINT "ADDING ELEMENT :- ";
3020 FOR K = 1 TO NMEMB
3030 PRINT K;
3040 IN    = MEMB(K,1)
3050 JN    = MEMB(K,2)
3060 REM
3070 REM --- CALCULATE ELEMENT LENGTH (STORE IN MEMB(K,5))
3080 REM
3090 TEMP      = (NODE(JN,1)-NODE(IN,1))^2 + (NODE(JN,2)-NODE(IN,2))^2
3100 TEMP      = SQR(TEMP + (NODE(JN,3)-NODE(IN,3))^2)
3110 MEMB(K,5) = TEMP
3120 REM
3130 REM --- CALCULATE THE DIRECTION COSINES THE MEMBER X-AXIS MAKES WITH
3140 REM --- THE GLOBAL X-AXES, STORE IN MEMB(I,6), MEMB(I,7), MEMB(I,8)
```

```
3150 REM
3160 MEMB(K,6) = (NODE(JN,1) - NODE(IN,1)) / TEMP
3170 MEMB(K,7) = (NODE(JN,2) - NODE(IN,2)) / TEMP
3180 MEMB(K,8) = (NODE(JN,3) - NODE(IN,3)) / TEMP
3190 REM
3200 REM    --- GENERATE THE UPPER TRIANGLE OF THE ELEMENT STIFFNESS MATRIX
3210 REM
3220 ESTIFF(1,1) =     MEMB(K,3) * MEMB(K,4) * MEMB(K,6) * MEMB(K,6) / MEMB(K,5)
3230 ESTIFF(1,2) =     MEMB(K,3) * MEMB(K,4) * MEMB(K,6) * MEMB(K,7) / MEMB(K,5)
3240 ESTIFF(1,3) =     MEMB(K,3) * MEMB(K,4) * MEMB(K,6) * MEMB(K,8) / MEMB(K,5)
3250 ESTIFF(1,4) = -   ESTIFF(1,1)
3260 ESTIFF(1,5) = -   ESTIFF(1,2)
3270 ESTIFF(1,6) = -   ESTIFF(1,3)
3280 ESTIFF(2,2) =     MEMB(K,3) * MEMB(K,4) * MEMB(K,7) * MEMB(K,7) / MEMB(K,5)
3290 ESTIFF(2,3) =     MEMB(K,3) * MEMB(K,4) * MEMB(K,7) * MEMB(K,8) / MEMB(K,5)
3300 ESTIFF(2,4) = -   ESTIFF(1,2)
3310 ESTIFF(2,5) = -   ESTIFF(2,2)
3320 ESTIFF(2,6) = -   ESTIFF(2,3)
3330 ESTIFF(3,3) =     MEMB(K,3) * MEMB(K,4) * MEMB(K,8) * MEMB(K,8) / MEMB(K,5)
3340 ESTIFF(3,4) = -   ESTIFF(1,3)
3350 ESTIFF(3,5) = -   ESTIFF(2,3)
3360 ESTIFF(3,6) = -   ESTIFF(3,3)
3370 ESTIFF(4,4) =     ESTIFF(1,1)
3380 ESTIFF(4,5) =     ESTIFF(1,2)
3390 ESTIFF(4,6) =     ESTIFF(1,3)
3400 ESTIFF(5,5) =     ESTIFF(2,2)
3410 ESTIFF(5,6) =     ESTIFF(2,3)
3420 ESTIFF(6,6) =     ESTIFF(3,3)
3430 REM
3440 REM    --- SET UP THE ELEMENT CODE NUMBER
3450 REM
3460 EFREE(1) = FREE(3*IN-2)
3470 EFREE(2) = FREE(3*IN-1)
3480 EFREE(3) = FREE(3*IN)
3490 EFREE(4) = FREE(3*JN-2)
3500 EFREE(5) = FREE(3*JN-1)
3510 EFREE(6) = FREE(3*JN)
3520 REM
3530 REM    --- ADD THE ELEMENT STIFFNESS
3540 REM
3550 FOR I          = 1 TO 6
3560 IF EFREE(I) = 0 THEN GOTO 3630
3570 FOR J          = 1 TO 6
3580 IF EFREE(J) = 0 THEN GOTO 3620
3590 L              = EFREE(I)
3600 M              = EFREE(J)
3610 FSTIFF(L,M) = FSTIFF(L,M) + ESTIFF(I,J)
3620 NEXT J
3630 NEXT I
3640 NEXT K
3650 PRINT
3660 REM
3670 REM    --- USE SYMMETRY TO COMPLETE THE STIFFNESS MATRIX
3680 REM
3690 FOR I          = 1 TO NDOF-1
3700 FOR J          = I+1 TO NDOF
3710 FSTIFF(J,I) = FSTIFF(I,J)
3720 NEXT J
3730 NEXT I
3740 RETURN
3750 REM   * * * * * * * * * * * * * * * * * * * * * * * * * * * * * * * *
3760 REM   *                                                              *
3770 REM   *    SUBROUTINE TO SET UP THE LOADING VECTOR                   *
3780 REM   *                                                              *
3790 REM   * * * * * * * * * * * * * * * * * * * * * * * * * * * * * * * *
3800 REM
3810 FOR I              = 1 TO NNLOD
3820 J                  = 3*NLOD(I,1) - 3 + NLOD(I,2)
3830 J                  = FREE(J)
3840 FSTIFF(J,NDOF+1) = NLOD(I,3)
3850 NEXT I
3860 RETURN
```

SPACE TRUSSES

```
3870 REM ******************************************
3880 REM *                                                              *
3890 REM *     SUBROUTINE TO SOLVE LINEAR SIMULTANEOUS EQUATIONS BY     *
3900 REM *     GAUSS-JORDAN ELIMINATION WITH NO ROW INTERCHANGE         *
3910 REM *                                                              *
3920 REM ******************************************
3930 REM
3940 REM --- LOOP FOR ALL PIVOTS
3950 REM
3960 PRINT "USING PIVOT    :- ";
3970 FOR I = 1 TO NDOF
3980 PRINT I;
3990 TEMP    = FSTIFF(I,I)
4000 REM
4010 REM --- NORMALISE
4020 REM
4030 FOR J       = 1 TO NDOF+1
4040 FSTIFF(I,J) = FSTIFF(I,J) / TEMP
4050 NEXT J
4060 REM
4070 REM --- LOOP FOR ALL ROWS USING THE CURRENT PIVOT
4080 REM
4090 FOR J = 1 TO NDOF
4100 IF  J = I THEN GOTO 4180
4110 TEMP    = FSTIFF(J,I)
4120 REM
4130 REM --- LOOP FOR ALL ELEMENTS IN THE ROW
4140 REM
4150 FOR K       = 1 TO NDOF+1
4160 FSTIFF(J,K) = FSTIFF(J,K) - TEMP*FSTIFF(I,K)
4170 NEXT K
4180 NEXT J
4190 NEXT I
4200 RETURN
4210 REM ******************************************
4220 REM *                                                              *
4230 REM *     SUBROUTINE TO OUTPUT THE DISPLACEMENTS                   *
4240 REM *                                                              *
4250 REM ******************************************
4260 REM
4270 PRINT : PRINT
4280 PRINT "* * * * * * * * * * *"
4290 PRINT "*   DISPLACEMENTS   *          * :-  RESTRAINT"
4300 PRINT "* * * * * * * * * * *"
4310 PRINT
4320 PRINT "NODE     X-DISPLACEMENT  Y-DISPLACEMENT  Z-DISPLACEMENT"
4330 PRINT
4340 FOR I = 1 TO NNODE
4350 PRINT USING "###     "; I;
4360 FOR J = 1 TO 3
4370 K     = FREE(3*I+J-3)
4380 IF K = 0 THEN GOTO 4410
4390 PRINT USING "    #.####^^^^"; FSTIFF(K,NDOF+1);
4400 GOTO 4420
4410 PRINT "          *     ";
4420 NEXT J
4430 PRINT
4440 NEXT I
4450 RETURN
4460 REM ******************************************
4470 REM *                                                              *
4480 REM *     SUBROUTINE TO CALCULATE AND OUTPUT THE MEMBER FORCES AND *
4490 REM *     THE REACTIONS                                            *
4500 REM *                                                              *
4510 REM ******************************************
4520 REM
4530 PRINT : PRINT
4540 PRINT "* * * * * * * * * * *"
4550 PRINT "*   MEMBER FORCES   *          TENSION POSITIVE"
4560 PRINT "* * * * * * * * * * *"
4570 PRINT: PRINT "MEMBER           FORCE"
4580 PRINT
```

```
4590 FOR I = 1 TO NMEMB
4600 II   = 3*MEMB(I,1) - 2
4610 JJ   = 3*MEMB(I,2) - 2
4620 K    = FREE(II)
4630 IF K = 0 THEN XI = 0 ELSE XI = FSTIFF(K,NDOF+1)
4640 K    = FREE(II+1)
4650 IF K = 0 THEN YI = 0 ELSE YI = FSTIFF(K,NDOF+1)
4660 K    = FREE(II+2)
4670 IF K = 0 THEN ZI = 0 ELSE ZI = FSTIFF(K,NDOF+1)
4680 K    = FREE(JJ)
4690 IF K = 0 THEN XJ = 0 ELSE XJ = FSTIFF(K,NDOF+1)
4700 K    = FREE(JJ+1)
4710 IF K = 0 THEN YJ = 0 ELSE YJ = FSTIFF(K,NDOF+1)
4720 K    = FREE(JJ+2)
4730 IF K = 0 THEN ZJ = 0 ELSE ZJ = FSTIFF(K,NDOF+1)
4740 TEMP = MEMB(I,6)*(XJ-XI) + MEMB(I,7)*(YJ-YI) + MEMB(I,8)*(ZJ-ZI)
4750 TEMP = MEMB(I,3) * MEMB(I,4) * TEMP / MEMB(I,5)
4760 PRINT USING "## ##          #.####^^^^"; MEMB(I,1); MEMB(I,2); TEMP
4770 REM
4780 REM  ---  ADD ANY CONTRIBUTION FROM THE MEMBER FORCES TO THE REACTIONS
4790 REM
4800 FOR J = 1 TO NREST
4810 K = 0
4820 IF MEMB(I,1) = REST(J,1) THEN K = -1
4830 IF MEMB(I,2) = REST(J,1) THEN K =  1
4840 IF K = 0 THEN GOTO 4880
4850 REST(J,3) = REST(J,3) + K*MEMB(I,6)*TEMP
4860 REST(J,4) = REST(J,4) + K*MEMB(I,7)*TEMP
4870 REST(J,5) = REST(J,5) + K*MEMB(I,8)*TEMP
4880 NEXT J
4890 NEXT I
4900 REM
4910 REM  ---  ADD ANY APPLIED LOADS
4920 REM
4930 FOR I = 1 TO NREST
4940 FOR J = 1 TO NNLOD
4950 IF REST(I,1) <> NLOD(J,1) THEN GOTO 4980
4960 K    = NLOD(J,2)
4970 REST(I,2+K) = REST(I,2+K) - NLOD(J,2)
4980 NEXT J
4990 NEXT I
5000 REM
5010 REM  ---  OUTPUT THE REACTIONS
5020 REM
5030 PRINT : PRINT
5040 PRINT "* * * * * * * * *"
5050 PRINT "*    REACTIONS  *"
5060 PRINT "* * * * * * * * *"
5070 PRINT
5080 PRINT "              X-FORCE          Y-FORCE          Z-FORCE"
5090 PRINT
5100 FOR I    = 1 TO NREST
5110 PRINT USING "NODE ##"; REST(I,1);
5120 PRINT USING "     #.####^^^^"; REST(I,3), REST(I,4), REST(I,5)
5130 NEXT I
5140 RETURN
```

SPACE TRUSSES

Sample Run There follows a sample run in which the programs PRE.MA/ PRE.01/PRE.02 are used to generate the data for the three-dimensional framework shown below. The structure is then analysed using the space truss analysis program STRUSS.ST. $E = 200$ kN/mm^2 and $A = 5000$ mm^2 for all members.

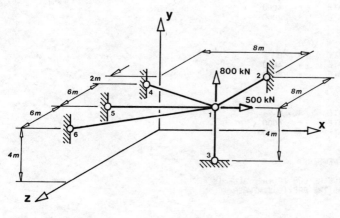

```
-----------------------------------------------------------
PROGRAM   P R E . M A   TO PREPROCESS DATA FOR PROGRAMS
PTRUSS.PT, STRUSS.PT, PFRAME.PF, GRID.GD, AND SFRAME.SF
-----------------------------------------------------------

DO YOU WISH TO
(1) CREATE A NEW DATA FILE
(2) MODIFY AN EXISTING DATA FILE
? 1

NAME OF THE DATA FILE  =  DATA.ST

WAIT - CHAINING PRE.01


SELECT PROBLEM TYPE
(1) PLANE TRUSS
(2) SPACE TRUSS
(3) PLANE FRAME
(4) GRILLAGE
(5) SPACE FRAME
? 2

JOB TITLE  =  SPACE TRUSS EXAMPLE   -   FILE DATA.ST

NUMBER OF NODES  =   6

NODE 1

X-COORDINATE     =    8000
Y-COORDINATE     =    4000
Z-COORDINATE     =    8000

NODE 2

X-COORDINATE     =    8000
Y-COORDINATE     =    4000
Z-COORDINATE     =    0
```

COMPUTER ANALYSIS OF STRUCTURAL FRAMEWORKS

NODE 3

X-COORDINATE = **8000**
Y-COORDINATE = **0**
Z-COORDINATE = **8000**

NODE 4

X-COORDINATE = **8000**
Y-COORDINATE = **4000**
Z-COORDINATE = **2000**

NODE 5

X-COORDINATE = **0**
Y-COORDINATE = **4000**
Z-COORDINATE = **8000**

NODE 6

X-COORDINATE = **0**
Y-COORDINATE = **4000**
Z-COORDINATE = **14000**

NUMBER OF MEMBERS = **5**

HIT CARRIAGE RETURN IF THE VALUE IS THE SAME AS THE PREVIOUS MEMBER

MEMBER 1

LOWER NODE NUMBER = **1**
HIGHER NODE NUMBER = **2**
CROSS-SECTIONAL AREA = **5000**
ELASTIC CONSTANT = **200**

MEMBER 2

LOWER NODE NUMBER = **1**   Note that the values not shown in
HIGHER NODE NUMBER = **3**   bold typeface have been entered
CROSS-SECTIONAL AREA = 5000   simply by hitting RETURN (the computer
ELASTIC CONSTANT = 200   then echoes the previous value).

MEMBER 3

LOWER NODE NUMBER = **1**
HIGHER NODE NUMBER = **4**
CROSS-SECTIONAL AREA = 5000
ELASTIC CONSTANT = 200

MEMBER 4

LOWER NODE NUMBER = **1**
HIGHER NODE NUMBER = **5**
CROSS-SECTIONAL AREA = 5000
ELASTIC CONSTANT = 200

MEMBER 5

LOWER NODE NUMBER = **1**
HIGHER NODE NUMBER = **6**
CROSS-SECTIONAL AREA = 5000
ELASTIC CONSTANT = 200

NUMBER OF RESTRAINED NODES = **5**

RESTRAINT DIRECTIONS ARE
X-TRANSLATION = 1
Y-TRANSLATION = 2
Z-TRANSLATION = 3

SPACE TRUSSES

ENTER RESTRAINED DIRECTIONS AS A COMPOSITE NUMBER
E.G. IF NODE IS RESTRAINED IN DIRECTIONS 1 AND 2 ENTER 12

HIT CARRIAGE RETURN IF THE VALUE IS
THE SAME AS THE PREVIOUS RESTRAINT

RESTRAINED NODE NUMBER 1

NODE TO BE RESTRAINED   = 2
DIRECTION(S)   = **123**

RESTRAINED NODE NUMBER 2

NODE TO BE RESTRAINED   = **5**
DIRECTION(S)   = 123

RESTRAINED NODE NUMBER 3

NODE TO BE RESTRAINED   = **3**
DIRECTION(S)   = 123

RESTRAINED NODE NUMBER 4

NODE TO BE RESTRAINED   = **6**
DIRECTION(S)   = 123

RESTRAINED NODE NUMBER 5

NODE TO BE RESTRAINED   = **4**
DIRECTION(S)   = 123

NUMBER OF NODAL LOADS   = 2

LOAD DIRECTIONS ARE
FORCE IN THE X-DIRECTION = 1
FORCE IN THE Y-DIRECTION = 2
FORCE IN THE Z-DIRECTION = 3

HIT CARRIAGE RETURN IF THE VALUE IS
THE SAME AS THE PREVIOUS LOAD

NODAL LOAD NUMBER 1

NODE CARRYING LOAD   = 1
DIRECTION   = 2
MAGNITUDE OF LOAD   = **800**

NODAL LOAD NUMBER 2

NODE CARRYING LOAD   = 1
DIRECTION   = **1**
MAGNITUDE OF LOAD   = **500**

WAIT - CHAINING PRE.02


```
+ + + + + + + +
+   JOB TITLE   +  -  SPACE TRUSS EXAMPLE  -  FILE DATA.ST
+ + + + + + + +


+ + + + + + + + + + + + +
+   NODAL COORDINATES    +
+ + + + + + + + + + + + +

NODE          X             Y             Z
NUMBER
  1         8000          4000          8000
  2         8000          4000             0
  3         8000             0          8000
  4         8000          4000          2000
```

```
         5            0            4000         8000
         6            0            4000        14000

    + + + + + + + + + + + +
    +   MEMBER PROPERTIES  +
    + + + + + + + + + + + +

    MEMBER         A         E
    NUMBER
     1  2    0.500E+04  0.200E+03
     1  3    0.500E+04  0.200E+03
     1  4    0.500E+04  0.200E+03
     1  5    0.500E+04  0.200E+03
     1  6    0.500E+04  0.200E+03


    + + + + + + + + +
    +  RESTRAINTS   +
    + + + + + + + + +

    NODE     DIRECTION(S)
    NUMBER
      2         123
      3         123
      4         123
      5         123
      6         123


    + + + + + + + + + +
    +   NODAL LOADS   +
    + + + + + + + + + +

    NODE    DIRECTION   VALUE
    NUMBER
      1         1      0.500E+03
      1         2      0.800E+03

    DO YOU WISH TO  (CR = CARRIAGE RETURN)
    (1)  MODIFY DATA
    (2)  LIST ALL THE CURRENT DATA
    (3)  RUN DATA FILE WITH APPROPRIATE ANALYSIS PROGRAM
    (CR) STOP
    ? 1

    DO YOU WISH TO  (CR = CARRIAGE RETURN)
    (1)  CHANGE JOB TITLE
    (2)  MODIFY NODAL DATA
    (3)  MODIFY MEMBER DATA
    (4)  MODIFY RESTRAINTS
    (5)  MODIFY NODAL LOADS
    (CR) STOP MODIFYING DATA
    ? 2

    + + + + + + + + + + + + +
    +   NODAL COORDINATES   +
    + + + + + + + + + + + + +

    NODE           X            Y            Z
    NUMBER
      1          8000         4000         8000
      2          8000         4000            0
      3          8000            0         8000
      4          8000         4000         2000
      5             0         4000         8000
      6             0         4000        14000

    NODE TO BE REVISED/ADDED  =  4
    X-COORDINATE              =  0
    Y-COORDINATE              =  4000
    Z-COORDINATE              =  2000
```

SPACE TRUSSES

```
+ + + + + + + + + + + +
+   NODAL COORDINATES  +
+ + + + + + + + + + + +

NODE         X              Y              Z
NUMBER
  1         8000           4000           8000
  2         8000           4000              0
  3         8000              0           8000
  4            0           4000           2000
  5            0           4000           8000
  6            0           4000          14000

AGAIN  -  (Y)ES OR (CR) ?              Carriage return entered

DO YOU WISH TO  (CR = CARRIAGE RETURN)
(1)  CHANGE JOB TITLE
(2)  MODIFY NODAL DATA
(3)  MODIFY MEMBER DATA
(4)  MODIFY RESTRAINTS
(5)  MODIFY NODAL LOADS
(CR) STOP MODIFYING DATA
?                                      Carriage return entered

DO YOU WISH TO  (CR = CARRIAGE RETURN)
(1)  MODIFY DATA
(2)  LIST ALL THE CURRENT DATA
(3)  RUN DATA FILE WITH APPROPRIATE ANALYSIS PROGRAM
(CR) STOP
?                                      Carriage return entered
```

Note that the program STRUSS.ST could have been chained directly from the preprocessor program, in which case STRUSS.ST would be passed the name of the data file via COMMON.

```
------------------------------------------------
PROGRAM   S T R U S S . S T   FOR THE AUTOMATIC
ANALYSIS OF SPACE TRUSSES
------------------------------------------------

NAME OF THE DATA FILE  =  DATA.ST

READING DATA

+ + + + + + + + +
+   JOB TITLE   +    -    SPACE TRUSS EXAMPLE  -  FILE DATA.ST
+ + + + + + + + +

+ + + + + + + + + + + +
+   NODAL COORDINATES  +
+ + + + + + + + + + + +

NODE         X              Y              Z
NUMBER
  1         8000           4000           8000
  2         8000           4000              0
  3         8000              0           8000
  4            0           4000           2000
  5            0           4000           8000
  6            0           4000          14000
```

COMPUTER ANALYSIS OF STRUCTURAL FRAMEWORKS

```
+ + + + + + + + + + + +
+   MEMBER PROPERTIES  +
+ + + + + + + + + + + +

MEMBER            A            E
NUMBER
  1  2       0.500E+04    0.200E+03
  1  3       0.500E+04    0.200E+03
  1  4       0.500E+04    0.200E+03
  1  5       0.500E+04    0.200E+03
  1  6       0.500E+04    0.200E+03


+ + + + + + + + + +
+   RESTRAINTS    +
+ + + + + + + + + +

NODE      DIRECTION(S)
NUMBER
  2          123
  3          123
  4          123
  5          123
  6          123


+ + + + + + + + + +
+   NODAL LOADS   +
+ + + + + + + + + +

NODE    DIRECTION    VALUE
NUMBER
  1         1      0.500E+03
  1         2      0.800E+03

GENERATING THE FREEDOM VECTOR

GENERATING THE STRUCTURE STIFFNESS MATRIX

ADDING ELEMENT :-  1  2  3  4  5

GENERATING THE LOADING VECTOR

SOLVING EQUATIONS  -  NO. OF EQUATIONS  =  3
USING PIVOT        :-  1  2  3

* * * * * * * * * * *
*   DISPLACEMENTS   *            * :- RESTRAINT
* * * * * * * * * * *

NODE    X-DISPLACEMENT   Y-DISPLACEMENT   Z-DISPLACEMENT

  1        0.1976E+01       0.3200E+01       0.0000E+00
  2             *                *                *
  3             *                *                *
  4             *                *                *
  5             *                *                *
  6             *                *                *


* * * * * * * * * * *
*   MEMBER FORCES   *           TENSION POSITIVE
* * * * * * * * * * *

MEMBER           FORCE

  1  2        0.0000E+00
  1  3        0.8000E+03
  1  4        0.1581E+03
  1  5        0.2470E+03
  1  6        0.1581E+03
```

SPACE TRUSSES

```
* * * * * * * * *
*   REACTIONS   *
* * * * * * * * *
              X-FORCE        Y-FORCE        Z-FORCE
NODE  2     0.0000E+00     0.0000E+00     0.0000E+00
NODE  3     0.0000E+00    -.8000E+03      0.0000E+00
NODE  4    -.1265E+03     0.0000E+00     -.9486E+02
NODE  5    -.2470E+03     0.0000E+00      0.0000E+00
NODE  6    -.1265E+03     0.0000E+00      0.9486E+02
```

7 Plane Frames

7.1 INTRODUCTION

Rigidly jointed frames are extensively used in civil engineering. Examples include roof trusses, electricity pylons, steel offshore platforms and all types of building frameworks. As the members are assumed to be rigidly connected at the joints the angles between members meeting at a joint remain unchanged as the structure deforms under load. Consequently the members of rigidly jointed frames transmit load, not only axially, but also by bending and shear. This contrasts with pin jointed structures where the inability of the joints to transmit bending ensures that all loads are transmitted by axial force only. Further, rigidly jointed frames are often designed to carry loads both at joints and along the lengths of the members, whereas pin jointed frames are usually designed to carry loads that are applied only at the joints. This chapter shows that the equations governing the behaviour of trusses also govern the behaviour of rigidly jointed plane frameworks, and that the techniques developed for the analysis of trusses can easily be adapted for the analysis of rigidly jointed frames.

7.2 THE ELEMENT STIFFNESS MATRIX IN THE MEMBER AXES SYSTEM

The forces at the ends of a plane frame member are related to the displacements at the ends by the element stiffness matrix as follows:

$$\begin{bmatrix} f_{ijx} \\ f_{ijy} \\ m_{ij} \\ f_{jix} \\ f_{jiy} \\ m_{ji} \end{bmatrix} = \begin{bmatrix} k_{11} & k_{12} & k_{13} & k_{14} & k_{15} & k_{16} \\ k_{21} & k_{22} & k_{23} & k_{24} & k_{25} & k_{26} \\ k_{31} & k_{32} & k_{33} & k_{34} & k_{35} & k_{36} \\ k_{41} & k_{42} & k_{43} & k_{44} & k_{45} & k_{46} \\ k_{51} & k_{52} & k_{53} & k_{54} & k_{55} & k_{56} \\ k_{61} & k_{62} & k_{63} & k_{64} & k_{65} & k_{66} \end{bmatrix} \begin{bmatrix} \delta_{ijx} \\ \delta_{ijy} \\ \theta_{ij} \\ \delta_{jix} \\ \delta_{jiy} \\ \theta_{ji} \end{bmatrix}$$

PLANE FRAMES

Figure 7.1 shows a typical plane frame element "i,j", and the member axes and freedom numbers that will be adopted in this text. The reader should note the introduction of rotational components of force and displacement at "i" and "j". The element stiffness matrix can be found by treating the element as an unrestrained structure with six freedoms.

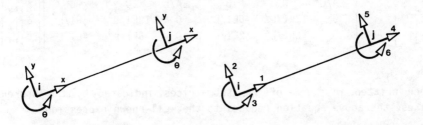

FIGURE 7.1   MEMBER AXES AND FREEDOM NUMBERS FOR A PLANE FRAME ELEMENT

In section 3.4 the column "j" of the structure stiffness matrix was shown to be the force system required to maintain unit displacement in the direction of freedom "j" (displacements in the directions of the other freedoms being held at zero). The relationship between the end displacements and the end forces for a line element was investigated in section 2.12. Consider, for example, the force system associated with rotation at end "i" of member "i,j". That force system is shown in fig 2.32, and it follows that the force system associated with unit rotation is as shown in fig 7.2.

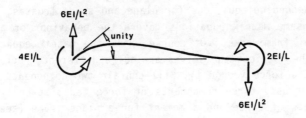

FIGURE 7.2   FORCE SYSTEM ASSOCIATED WITH FREEDOM 3

By inspection of fig 7.2 the stiffness coefficients associated with freedom 3 are

$$k_{13} = 0 \quad k_{23} = 6EI/L^2 \quad k_{33} = 4EI/L$$
$$k_{43} = 0 \quad k_{53} = -6EI/L^2 \quad k_{63} = 2EI/L$$

Treating the other freedoms similarly, and using the stiffness properties developed in section 2.12, the element stiffness relationship can be shown to be

$$\begin{bmatrix} f_{ijx} \\ f_{ijy} \\ m_{ij} \\ \hline f_{jix} \\ f_{jiy} \\ m_{ji} \end{bmatrix} = \begin{bmatrix} EA/L & 0 & 0 & | & -EA/L & 0 & 0 \\ 0 & 12EI/L^3 & 6EI/L^2 & | & 0 & -12EI/L^3 & 6EI/L^2 \\ 0 & 6EI/L^2 & 4EI/L & | & 0 & -6EI/L^2 & 2EI/L \\ \hline -EA/L & 0 & 0 & | & EA/L & 0 & 0 \\ 0 & -12EI/L^3 & -6EI/L^2 & | & 0 & 12EI/L^3 & -6EI/L^2 \\ 0 & 6EI/L^2 & 2EI/L & | & 0 & -6EI/L^2 & 4EI/L \end{bmatrix} \begin{bmatrix} \delta_{ijx} \\ \delta_{ijy} \\ \theta_{ij} \\ \hline \delta_{jix} \\ \delta_{jiy} \\ \theta_{ji} \end{bmatrix}$$

(7.1)

When written in terms of the submatrices indicated by the broken lines, the above equation reduces to the well-known expression

$$\begin{bmatrix} f_{ij} \\ f_{ji} \end{bmatrix} = \begin{bmatrix} k_{ii}^j & k_{ij} \\ k_{ji} & k_{jj}^i \end{bmatrix} \begin{bmatrix} \delta_{ij} \\ \delta_{ji} \end{bmatrix}$$

hence

$$f_{ij} = k_{ii}^j \delta_{ij} + k_{ij} \delta_{ji} \qquad (7.2)$$

$$f_{ji} = k_{ji} \delta_{ij} + k_{jj}^i \delta_{ji} \qquad (7.3)$$

where the matrix product $k_{ii}^j \delta_i$ yields the vector of forces at end "i" due to displacements at end "i", and $k_{ij}\delta_j$ yields the vector of forces at end "i" due to displacements at end "j". The resultant force vector is, of course, the matrix sum of these two force vectors.

It should be noted that equations (7.2) and (7.3) are identical to the corresponding equations for plane and space trusses. This is because the same matrix equations govern the behaviour of all types of structural frameworks. Of course while the matrix equations are identical, the matrices themselves are quite different. For example, the member end force vector f_{ij} will contain two components of force for a plane truss, three components of force for a space truss, and two components of force and a moment for a plane frame (see section 3.2).

7.3 TRANSFORMATION OF FORCE AND DISPLACEMENT

Equations (7.2) and (7.3) show how the member end forces (in the member axes system) are related to the member end displacements (also in the member axes system). Previous chapters have shown that nodal

PLANE FRAMES

equilibrium lies at the heart of the stiffness method, and while member axes are useful when considering member forces they cannot be used when considering nodal equilibrium. This is because members meeting at a node will, in general, lie at different angles. Consequently the member axes will be inconsistently orientated, which precludes their use when summing nodal forces. Before the forces that a member exerts on a node can be included in the equation of nodal equilibrium they must be transformed into the nodal axes system.

When considering plane trusses in section 5.3 the matrix T_{ij} shown below was used to effect the transformation of displacements from global to member axes:

$$\begin{bmatrix} \delta_{ijx} \\ \delta_{ijy} \end{bmatrix} = \begin{bmatrix} \cos\alpha & \sin\alpha \\ -\sin\alpha & \cos\alpha \end{bmatrix} \begin{bmatrix} \Delta_{ix} \\ \Delta_{iy} \end{bmatrix}$$

or

$$\delta_{ij} = T_{ij} \Delta_i$$

where α is *the clockwise rotation of the member about "i" that will make the member axes coincide with the global axes.* Δ_i is the vector of displacement at node "i" described in the nodal axes system and δ_{ij} is the same vector expressed in the member axes system.

The only difference between the nodal displacement vector for a plane truss and the nodal displacement vector for a plane frame is that the plane frame vector has a rotational component of displacement. As the member "x,y" plane lies in the global "x,y" plane rotation will have the same magnitude whether expressed in the global or in the member axes system. Further, if the sense of the rotation (in this text a right hand system is used throughout) is consistent between the axes systems then the rotational component of displacement will have the same sign in both axes systems, and the required transformation becomes

$$\begin{bmatrix} \delta_{ijx} \\ \delta_{ijy} \\ \theta_{ij} \end{bmatrix} = \begin{bmatrix} \cos\alpha & \sin\alpha & 0 \\ -\sin\alpha & \cos\alpha & 0 \\ 0 & 0 & 1 \end{bmatrix} \begin{bmatrix} \Delta_{ix} \\ \Delta_{iy} \\ \theta_i \end{bmatrix}$$

i.e.

$$\delta_{ij} = T_{ij} \Delta_i \tag{7.4}$$

and similarly at end "j" of the member

$$\begin{bmatrix} \delta_{jix} \\ \delta_{jiy} \\ \theta_{ji} \end{bmatrix} = \begin{bmatrix} \cos\alpha & \sin\alpha & 0 \\ -\sin\alpha & \cos\alpha & 0 \\ 0 & 0 & 1 \end{bmatrix} \begin{bmatrix} \Delta_{jx} \\ \Delta_{jy} \\ \theta_j \end{bmatrix}$$

i.e.
$$\delta_{ji} = T_{ij}\, \Delta_j \qquad (7.5)$$

where
- T_{ij} is the transformation matrix for member "i,j".
- Δ_i is the nodal displacement vector at node "i" in the global axes system.
- Δ_j is the nodal displacement vector at node "j" in the global axes system.
- δ_{ij} is the nodal displacement vector at node "i" in the member axes system for member "i,j".
- δ_{ji} is the nodal displacement vector at node "j" in the member axes system for member "i,j".

The transformation matrix T_{ij} can also be used to transform forces from the global to the member axes system:

$$\begin{bmatrix} f_{ijx} \\ f_{ijy} \\ m_{ij} \end{bmatrix} = \begin{bmatrix} \cos\alpha & \sin\alpha & 0 \\ -\sin\alpha & \cos\alpha & 0 \\ 0 & 0 & 1 \end{bmatrix} \begin{bmatrix} F_{ijx} \\ F_{ijy} \\ M_{ij} \end{bmatrix}$$

i.e.
$$f_{ij} = T_{ij}\, F_{ij} \qquad (7.6)$$

and similarly at end "j" of the member

$$\begin{bmatrix} f_{jix} \\ f_{jiy} \\ m_{ji} \end{bmatrix} = \begin{bmatrix} \cos\alpha & \sin\alpha & 0 \\ -\sin\alpha & \cos\alpha & 0 \\ 0 & 0 & 1 \end{bmatrix} \begin{bmatrix} F_{jix} \\ F_{jiy} \\ M_{ji} \end{bmatrix}$$

i.e.
$$f_{ji} = T_{ij}\, F_{ji} \qquad (7.7)$$

where
- T_{ij} is the transformation matrix for member "i,j".
- F_{ij} is the member end force vector at end "i" of member "i,j" in the global axes system.
- F_{ji} is the member end force vector at end "j" of member "i,j" in the global axes system.
- f_{ij} is the member end force vector at end "i" of member "i,j" in the member axes system.

PLANE FRAMES

f_{ji} is the member end force vector at end "j" of member "i,j" in the member axes system.

Section 5.3 showed how the reverse transformation (from member to global axes) is effected by the inverse of the transformation matrix,

i.e.
$$\Delta_i = T_{ij}^{-1} \delta_{ij}$$
$$\Delta_j = T_{ij}^{-1} \delta_{ji}$$
$$F_{ij} = T_{ij}^{-1} f_{ij}$$
$$F_{ji} = T_{ij}^{-1} f_{ji}$$

Summary

The transformation matrix T_{ij} for plane frame member "i,j" transforms vectors of force and displacement from the global to the member axes system, and hence its inverse transforms vectors from the member to the global axes system,

where
$$T_{ij} = \begin{bmatrix} \cos\alpha & \sin\alpha & 0 \\ -\sin\alpha & \cos\alpha & 0 \\ 0 & 0 & 1 \end{bmatrix}$$

and
$$T_{ij}^{-1} = \begin{bmatrix} \cos\alpha & -\sin\alpha & 0 \\ \sin\alpha & \cos\alpha & 0 \\ 0 & 0 & 1 \end{bmatrix}$$

7.4 THE ELEMENT STIFFNESS MATRIX IN THE GLOBAL AXES SYSTEM

In section 7.2 the relationship between the forces at end "i" of member "i,j" and the displacements at ends "i" and "j" was shown to be

$$f_{ij} = k_{ii}^j \delta_{ij} + k_{ij} \delta_{ji} \qquad (7.8)$$

Section 7.3 showed that the inverse of the transformation matrix T_{ij} could transform member end forces from the member axes system used in the above equation to the global axes system. Hence the member end forces can be expressed in terms of the global axes system as follows:

$$F_{ij} = T_{ij}^{-1} f_{ij}$$

$$= T_{ij}^{-1} k_{ii}^j \delta_{ij} + T_{ij}^{-1} k_{ij} \delta_{ji} \qquad (7.9)$$

Now, in order to generate and solve the stiffness equation for the complete structure, the force and displacements at each node must be expressed in a common axes system (the global axes system). Consequently the member end displacements δ_{ij} and δ_{ji} must be transformed from the member axes system into the common axes system, and equation (7.9) becomes

$$F_{ij} = T_{ij}^{-1} k_{ii}^j T_{ij} \Delta_i + T_{ij}^{-1} k_{ij} T_{ij} \Delta_j$$

The above equation can be rewritten as

$$F_{ij} = K_{ii}^j \Delta_i + K_{ij} \Delta_j \qquad (7.10)$$

where

$$K_{ii}^j = T_{ij}^{-1} k_{ii}^j T_{ij}$$

and

$$K_{ij} = T_{ij}^{-1} k_{ij} T_{ij}$$

The matrix product $K_{ii}^j \Delta_{ij}$ gives the forces (in the global axes system) at end "i" of member "i,j" caused by displacements at end "i" (also in the global axes system). Similarly the matrix product $K_{ij} \Delta_j$ gives the member end forces at end "i" of member "i,j" caused by displacements at end "j". Explicit evaluation of the global element stiffness submatrices K_{ii}^j, K_{ij}, K_{ji}, and K_{jj}^i is computationally advantageous:

$$K_{ii}^j = T_{ij}^{-1} k_{ii}^j T_{ij}$$

$$= \begin{bmatrix} \cos\alpha & -\sin\alpha & 0 \\ \sin\alpha & \cos\alpha & 0 \\ 0 & 0 & 1 \end{bmatrix} \begin{bmatrix} EA/L & 0 & 0 \\ 0 & 12EI/L^3 & 6EI/L^2 \\ 0 & 6EI/L^2 & 4EI/L \end{bmatrix} \begin{bmatrix} \cos\alpha & \sin\alpha & 0 \\ -\sin\alpha & \cos\alpha & 0 \\ 0 & 0 & 1 \end{bmatrix}$$

$$= \begin{bmatrix} (EA\cos^2\alpha/L + 12EI\sin^2\alpha/L^3) & (EA/L - 12EI/L^3)\sin\alpha\cos\alpha & -6EI\sin\alpha/L^2 \\ (EA/L - 12EI/L^3)\sin\alpha\cos\alpha & (EA\sin^2\alpha/L + 12EI\cos^2\alpha/L^3) & 6EI\cos\alpha/L^2 \\ -6EI\sin\alpha/L^2 & 6EI\cos\alpha/L^2 & 4EI/L \end{bmatrix}$$

$$(7.11)$$

and

$$K_{ij} = T_{ij}^{-1} k_{ij} T_{ij}$$

$$= \begin{bmatrix} \cos\alpha & -\sin\alpha & 0 \\ \sin\alpha & \cos\alpha & 0 \\ 0 & 0 & 1 \end{bmatrix} \begin{bmatrix} -EA/L & 0 & 0 \\ 0 & -12EI/L^3 & 6EI/L^2 \\ 0 & -6EI/L^2 & 2EI/L \end{bmatrix} \begin{bmatrix} \cos\alpha & \sin\alpha & 0 \\ -\sin\alpha & \cos\alpha & 0 \\ 0 & 0 & 1 \end{bmatrix}$$

$$= \begin{bmatrix} -(EA\cos^2\alpha/L + 12EI\sin^2\alpha/L^3) & -(EA/L - 12EI/L^3)\sin\alpha\cos\alpha & -6EI\sin\alpha/L^2 \\ -(EA/L - 12EI/L^3)\sin\alpha\cos\alpha & -(EA\sin^2\alpha/L + 12EI\cos^2\alpha/L^3) & 6EI\cos\alpha/L^2 \\ 6EI\sin\alpha/L^2 & -6EI\cos\alpha/L^2 & 2EI/L \end{bmatrix}$$

The K_{ji} and K_{jj}^{i} submatrices can be found by similar expansions. In practice only the K_{ii}^{j} submatrix need be evaluated for each element as the other submatrices can found by multiplying the elements of the K_{ii}^{j} submatrix as follows:

Find K_{ij} by multiplying corresponding elements of K_{ii}^{j} by
$$\begin{bmatrix} -1 & -1 & 1 \\ -1 & -1 & 1 \\ -1 & -1 & 0.5 \end{bmatrix}$$

Find K_{ji} by multiplying corresponding elements of K_{ii}^{j} by
$$\begin{bmatrix} -1 & -1 & -1 \\ -1 & -1 & -1 \\ 1 & 1 & 0.5 \end{bmatrix}$$

Find K_{jj}^{i} by multiplying corresponding elements of K_{ii}^{j} by
$$\begin{bmatrix} 1 & 1 & -1 \\ 1 & 1 & -1 \\ -1 & -1 & 1 \end{bmatrix}$$

(7.12)

Note that this is not a matrix multiplication, but simply a multiplication of corresponding elements.

7.5 NODAL EQUILIBRIUM

Figure 7.3(a) shows a pitched roof portal, and fig 7.3(b) shows how it might be idealised.

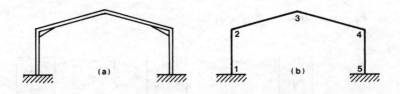

FIGURE 7.3   A PITCHED ROOF PORTAL

COMPUTER ANALYSIS OF STRUCTURAL FRAMEWORKS

In general, each member of a plane frame carries axial force, shear, and bending. Hence the freebody diagram for node 2 will be as shown in fig 7.4(a). As the node is part of a structure in equilibrium then it too must be in equilibrium under the action of the internal member forces (f_{21x}, f_{21y}, m_{21}, f_{23x}, etc.) and the external applied loads (P_{2x}, P_{2y}, M_2). Hence if the member end forces are transformed into the nodal axes system as indicated in fig 7.4(b), then the horizontal forces, the vertical forces, and the moments must sum to zero, i.e.

$$P_{2x} + F'_{21x} + F'_{23x} = 0$$

$$P_{2y} + F'_{21y} + F'_{23y} = 0$$

$$M_2 + M'_{21} + M'_{23} = 0$$

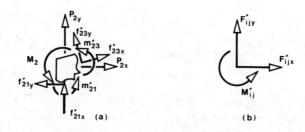

FIGURE 7.4 FREEBODY DIAGRAM FOR NODE 2

In section 5.5 the forces exerted on a node by a member were seen to be equal and opposite to the member end forces. Hence

$$P_{2x} - F_{21x} - F_{23x} = 0$$
$$P_{2y} - F_{21y} - F_{23y} = 0$$
$$M_2 - M_{21} - M_{23} = 0 \quad \text{(note } m_{21} = M_{21}, \; m_{23} = M_{23}\text{)}$$

or, in matrix notation

$$\mathbf{P}_2 = \mathbf{F}_{21} + \mathbf{F}_{22}$$

where

$$\mathbf{P}_2 = \begin{bmatrix} P_{2x} \\ P_{2y} \\ M_2 \end{bmatrix} \quad \mathbf{F}_{21} = \begin{bmatrix} F_{21x} \\ F_{21y} \\ M_{21} \end{bmatrix} \quad \mathbf{F}_{23} = \begin{bmatrix} F_{23x} \\ F_{23y} \\ M_{23} \end{bmatrix}$$

PLANE FRAMES

The equilibrium equation for a general node "i" which has members "i,a", "i,b", "i,c", ... "i,n" framing into it is

$$P_i = F_{ia} + F_{ib} + F_{ic} \ldots + F_{in}$$

This equation simply states that the external applied load vector P_i at node "i" must be balanced by the vectors of internal member end force F_{ia}, F_{ib}, F_{ic}, ... F_{in}. Obviously for the summation to be meaningful the force components must be described in a consistent axes system (see section 3.6).

Equation (7.10) gives the relationship between member end forces and the nodal displacements in the gobal axes system:

$$F_{ij} = K_{ii}^j \Delta_i + K_{ij} \Delta_j$$

where

$$K_{ii}^j = T_{ij}^{-1} k_{ii}^j T_{ij}$$

and

$$K_{ij} = T_{ij}^{-1} k_{ij} T_{ij}$$

Substituting for the member end forces in the equation of nodal equilibrium:

$$P_i = K_{ii}^a \Delta_i + K_{ia} \Delta_a + K_{ii}^b \Delta_i + K_{ib} \Delta_b +$$
$$K_{ii}^c \Delta_i + K_{ic} \Delta_c + \ldots + K_{ii}^n \Delta_i + K_{in} \Delta_n$$

$$= K_{ii} \Delta_i + K_{ia} \Delta_a + K_{ib} \Delta_b \ldots + K_{in} \Delta_n$$

where

$$K_{ii} = \sum K_{ii}^j$$

The summation is for all members connected to node "i".

7.6 THE INITIAL STIFFNESS MATRIX - Programs ESTIFF.PF and ISTIFF.PF

Chapter 5 (section 5.7) showed that, prior to the application of the boundary conditions, the equations of nodal equilibrium form a mixed set of simultaneous equations. The coefficient matrix of these equations is known as the initial structure stiffness matrix. The procedure to generate the initial structure stiffness matrix for a plane frame is identical to that used for two- and three-dimensional trusses. The global element stiffness submatrices K_{ii}^j, K_{ij}, K_{ji}, and K_{jj}^i for a plane frame are 3 x 3 square matrices. Hence they must be added to the initial structure stiffness matrix such that their (1,1) elements are added to locations (3i-2,3i-2), (3i-2,3j-2),

(3j-2,3i-2), and (3j-2, 3j-2) respectively. Figure 7.5 shows where the global element stiffness submatrices for member 2,5 would be added to the initial structure stiffness matrix. The initial structure matrix is 18 x 18, indicating that the frame has 6 nodes.

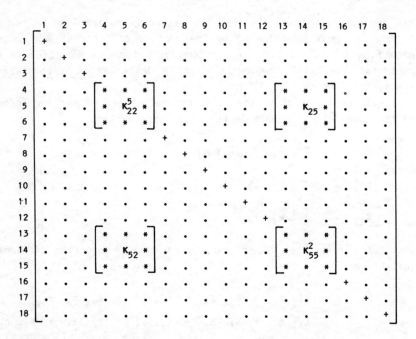

FIGURE 7.5   ADDING ELEMENT 2,5 TO THE INITIAL STRUCTURE STIFFNESS MATRIX

Example 7.1

Generate the initial stiffness equation, in terms of submatrices, for the portal frame shown below.

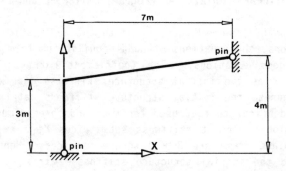

PLANE FRAMES

First remove the boundary conditions, leaving the structure completely unrestrained.

Zero the initial structure stiffness matrix (note that each zero represents a 3 x 3 submatrix of zeros).

$$\begin{bmatrix} 0 & 0 & 0 \\ 0 & 0 & 0 \\ 0 & 0 & 0 \end{bmatrix}$$

Add the global element stiffness submatrices for member 1,2

$$\begin{bmatrix} K_{11} & K_{12} & 0 \\ K_{21} & K_{22} & 0 \\ 0 & 0 & 0 \end{bmatrix} \quad \begin{aligned} K_{11} &= K_{11}^2 \\ K_{22} &= K_{22}^1 \end{aligned}$$

Add the global element stiffness submatrices for member 2,3

$$\begin{bmatrix} K_{11} & K_{12} & 0 \\ K_{21} & K_{22} & K_{23} \\ 0 & K_{32} & K_{33} \end{bmatrix} \quad \begin{aligned} K_{11} &= K_{11}^2 \\ K_{22} &= K_{22}^1 + K_{22}^3 \\ K_{33} &= K_{33}^2 \end{aligned}$$

Hence the initial structure stiffness equation, in terms of submatrices, is

$$\begin{bmatrix} K_{11} & K_{12} & 0 \\ K_{21} & K_{22} & K_{23} \\ 0 & K_{32} & K_{33} \end{bmatrix} \begin{bmatrix} \Delta_1 \\ \Delta_2 \\ \Delta_3 \end{bmatrix} = \begin{bmatrix} P_1 \\ P_2 \\ P_3 \end{bmatrix}$$

Note that the initial stiffness matrix is singular (see section 5.6).

COMPUTER ANALYSIS OF STRUCTURAL FRAMEWORKS

Example 7.2

If the frame considered in example 7.1 is constructed from steel beam sections having the following properties, $I = 0.25 \times 10^9$ mm^4, $A = 9000$ mm^2, $E = 200$ kN/mm^2, evaluate the initial structure stiffness matrix.

The structure has 3 nodes; hence the initial structure stiffness matrix will be 9 x 9. As in example 7.1 the initial structure stiffness matrix is first zeroed, and then the global element stiffness submatrices are added for each element in turn.

Element 1,2 $A = 9000$ mm^2, $I = 0.25 \times 10^9$ mm^4, $L = 3000$ mm,
                 $E = 200$ kN/mm^2, $\alpha = 90°$.

$$K_{11}^2 = T_{12}^{-1} k_{11}^2 T_{12}$$

$$= \begin{bmatrix} (EA\cos^2\alpha/L + 12EI\sin^2\alpha/L^3) & (EA/L - 12EI/L^3)\sin\alpha\cos\alpha & -6EI\sin\alpha/L^2 \\ (EA/L - 12EI/L^3)\sin\alpha\cos\alpha & (EA\sin^2\alpha/L + 12EI\cos^2\alpha/L^3) & 6EI\cos\alpha/L^2 \\ -6EI\sin\alpha/L^2 & 6EI\cos\alpha/L^2 & 4EI/L \end{bmatrix}$$

$$= \begin{bmatrix} 22.2 & 0 & -33333 \\ 0 & 600.0 & 0 \\ -33333 & 0 & 66.7 \times 10^6 \end{bmatrix}$$

Adding K_{11}^2, K_{12}, and K_{22}^1 to the initial structure stiffness matrix gives (note that K_{21} does not have to be added as symmetry will be used later to complete the matrix):

$$\begin{bmatrix} 22.2 & 0 & -33333 & -22.2 & 0 & -33333 & 0 & 0 & 0 \\ & 600.0 & 0 & 0 & -600.0 & 0 & 0 & 0 & 0 \\ & & 66.7 \times 10^6 & 33333 & 0 & 33.3 \times 10^6 & 0 & 0 & 0 \\ & & & 22.2 & 0 & 33333 & 0 & 0 & 0 \\ & & & & 600.0 & 0 & 0 & 0 & 0 \\ & \text{symmetrical} & & & & 66.7 \times 10^6 & 0 & 0 & 0 \\ & & & & & & 0 & 0 & 0 \\ & & & & & & & 0 & 0 \\ & & & & & & & & 0 \end{bmatrix}$$

Element 2,3 $A = 9000$ mm^2, $I = 0.25 \times 10^9$ mm^4, $L = 7071$ mm,
                $E = 200$ kN/mm^2, $\alpha = 8.13°$.

PLANE FRAMES

$$K_{22}^3 = \begin{bmatrix} 249.5 & 35.40 & -848.5 \\ 35.40 & 6.754 & 5940 \\ -848.5 & 5940 & 28.3 \times 10^6 \end{bmatrix}$$

Adding K_{22}^3, K_{23}, and K_{33}^2 to the structure stiffness matrix completes the assembly of the initial structure stiffness matrix.

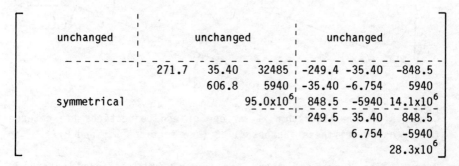

The calculation of element stiffness submatrices is a tedious and error-prone procedure that is best done with the aid of a computer. There follows a description of a computer program that calculates the element stiffness matrix for plane frame elements having a variety of end connections. The theory for plane frame elements containing pins is given in section 8.5.

Program ESTIFF.PF

This program calculates the element stiffness matrix for plane frame elements having a variety of end connections.

Input The following data is input for each element:

- A   — the cross-sectional area of the element.
- I   — the second moment of area of the element.
- L   — the length of the element.
- E   — the elastic constant for the element.
- α   — the angle the element "i,j" makes with the global "x" axis. Defined as shown in the following diagram.
- MTYPE — the element type. Thus far only fixed/fixed elements have been considered. Elements with pins are dealt with in section 8.5.

A, I, L, and E must be in consistent units, the unit of force used in E must correspond to the unit of force used for the applied loading, and α is in degrees.

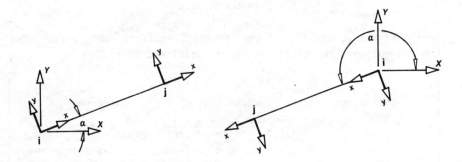

Comments on the Algorithm A variety of end connections are catered for by using stiffness factors K1-K8 (see sections 8.5 and 8.7).

Output The input is echoed and the element stiffness matrix is output - first in the member axes system and then in the global axes system.

Listing

```
1000 REM ************************************************************
1010 REM *                                                            *
1020 REM *    PROGRAM   E S T I F F . P F   TO CALCULATE THE ELEMENT  *
1030 REM *    STIFFNESS MATRIX IN THE MEMBER AND THE GLOBAL AXES      *
1032 REM *    SYSTEMS FOR A PLANE FRAME MEMBER                        *
1040 REM *                                                            *
1050 REM *         J.A.D.BALFOUR                                      *
1060 REM *                                                            *
1070 REM ************************************************************
1080 REM
1090 REM ---- NOTE THAT VARIABLES ARE DEFINED IN APPENDIX A
1100 REM
1110 OPTION BASE 1
1120 DIM ESTIFF(6,6)
1130 PRINT : PRINT
1140 PRINT "-----------------------------------------------------------"
1150 PRINT
1160 PRINT "PROGRAM    E S T I F F . P F    TO CALCULATE THE ELEMENT"
1170 PRINT "STIFFNESS MATRIX IN THE MEMBER AND THE GLOBAL AXES SYSTEMS"
1180 PRINT "FOR A PLANE FRAME ELEMENT"
1190 PRINT
1200 PRINT "-----------------------------------------------------------"
1210 PRINT : PRINT
1220 INPUT "NUMBER OF ELEMENTS   = ", NMEMB
1230 REM
1240 REM --- LOOP FOR ALL ELEMENTS
1250 REM
1260 FOR K = 1 TO NMEMB
1270 REM
1280 REM --- INPUT THE ELEMENT DATA
1290 REM
1300 GOSUB 1450
1310 REM
1320 REM --- DETERMINE K VALUES
1330 REM
1340 GOSUB 1760
```

PLANE FRAMES

```
1350 REM
1360 REM --- CALCULATE THE ELEMENT STIFFNESS MATRIX IN THE MEMBER AXES SYSTEM
1370 REM
1380 GOSUB 2050
1390 REM
1400 REM --- CALCULATE THE ELEMENT STIFFNESS MATRIX IN THE GLOBAL AXES SYSTEM
1410 REM
1420 GOSUB 2550
1430 NEXT K
1440 END
1450 REM *****************************************
1460 REM *                                       *
1470 REM *   SUBROUTINE TO INPUT THE ELEMENT DATA *
1480 REM *                                       *
1490 REM *****************************************
1500 REM
1510 IF K > 1 THEN GOTO 1600
1520 PRINT
1530 PRINT "ELEMENT      LOWER       HIGHER"
1540 PRINT " TYPE        NODE         NODE"
1550 PRINT
1560 PRINT "   1         FIXED       FIXED"
1570 PRINT "   2         FIXED       PINNED"
1580 PRINT "   3         PINNED      FIXED"
1590 PRINT "   4         PINNED      PINNED"
1600 PRINT : PRINT
1610 PRINT "ELEMENT "; K : PRINT
1620 INPUT "LOWER  NODE NUMBER  = ", IN
1630 INPUT "HIGHER NODE NUMBER  = ", JN
1640 PRINT
1650 PRINT        "+ + + + + + + + + +"
1660 PRINT USING "+    ELEMENT ## ##   +"; IN, JN
1670 PRINT        "+ + + + + + + + + +"
1680 PRINT
1690 INPUT "AREA                       = ", A
1700 INPUT "SECOND MOMENT OF AREA      = ", A2
1710 INPUT "LENGTH                     = ", L
1720 INPUT "ELASTIC CONSTANT           = ", E
1730 INPUT "ANGLE WITH THE GLOBAL X-AXIS = ", ALPHA
1740 INPUT "ELEMENT TYPE               = ", MTYPE
1750 RETURN
1760 REM *****************************************
1770 REM *                                       *
1780 REM *   SUBROUTINE TO SET UP ELEMENT STIFFNESS FACTORS FOR *
1790 REM *   DIFFERENT END CONDITIONS            *
1800 REM *                                       *
1810 REM *****************************************
1820 REM
1830 REM
1840 REM --- ZERO STIFFNESS FACTORS (I.E. PINNED/PINNED MEMBER)
1850 REM
1860 K1 =  0 : K2 =  0 : K3 =  0 : K4 =  0
1870 K5 =  0 : K6 =  0 : K7 =  0 : K8 =  0
1880 ON MTYPE GOTO 1920, 1980, 2030, 1940
1890 REM
1900 REM --- FIXED/FIXED MEMBER
1910 REM
1920 K1 = 12 : K2 =  6 : K3 =  6 : K4 =  4
1930 K5 = -6 : K6 =  2 : K7 = -6 : K8 =  4
1940 GOTO 2040
1950 REM
1960 REM --- FIXED/PINNED MEMBER
1970 REM
1980 K1 =  3 : K2 =  3 : K4 =  3 : K5 = -3
1990 GOTO 2040
2000 REM
2010 REM --- PINNED/FIXED MEMBER
2020 REM
2030 K1 =  3 : K3 =  3 : K7 = -3 : K8 =  3
2040 RETURN
```

```
2050 REM ***********************************
2060 REM *                                                           *
2070 REM *     SUBROUTINE TO CALCULATE AND OUTPUT THE ELEMENT STIFFNESS  *
2080 REM *     MATRIX IN THE MEMBER AXES SYSTEM                       *
2090 REM *                                                           *
2100 REM ***********************************
2110 REM
2180 REM --- CALCULATE THE STIFFNESS COEFFICIENTS (UPPER TRIANGLE ONLY)
2190 REM
2200 FOR I = 1 TO 6
2210 FOR J = 1 TO 6
2220 ESTIFF(I,J) = 0
2230 NEXT J
2240 NEXT I
2250 ESTIFF(1,1) =     E * A / L
2260 ESTIFF(1,4) = - ESTIFF(1,1)
2270 ESTIFF(2,2) =    K1 * E * A2 / L^3
2280 ESTIFF(2,3) =    K2 * E * A2 / L^2
2290 ESTIFF(2,5) = - ESTIFF(2,2)
2300 ESTIFF(2,6) =    K3 * E * A2 / L^2
2310 ESTIFF(3,3) =    K4 * E * A2 / L
2320 ESTIFF(3,5) =    K5 * E * A2 / L^2
2330 ESTIFF(3,6) =    K6 * E * A2 / L
2340 ESTIFF(4,4) =    ESTIFF(1,1)
2350 ESTIFF(5,5) =    ESTIFF(2,2)
2360 ESTIFF(5,6) =    K7 * E * A2 / L^2
2370 ESTIFF(6,6) =    K8 * E * A2 / L
2380 REM
2390 REM --- OUTPUT ELEMENT STIFFNESS MATRIX IN THE MEMBER AXES SYSTEM
2400 REM
2410 PRINT : PRINT
2420 PRINT "* * * * * * * * * * * * * * * *"
2430 PRINT "*    ELEMENT STIFFNESS MATRIX IN    *"
2440 PRINT "*    THE MEMBER AXES SYSTEM         *"
2450 PRINT "* * * * * * * * * * * * * * * *"
2460 PRINT
2470 FOR I = 1 TO 6
2480 FOR J = 1 TO 6
2490 IF I <= J THEN PRINT USING "#.####^^^^    "; ESTIFF(I,J);
2500 IF I >  J THEN PRINT USING "#.####^^^^    "; ESTIFF(J,I);
2510 NEXT J
2520 PRINT
2530 NEXT I
2540 RETURN
2550 REM ***********************************
2560 REM *                                                           *
2570 REM *     SUBROUTINE TO CALCULATE AND OUTPUT THE ELEMENT STIFFNESS  *
2580 REM *     MATRIX IN THE GLOBAL AXES SYSTEM                       *
2590 REM *                                                           *
2600 REM ***********************************
2610 REM
2612 REM --- CONVERT THE ANGLE TO RADIANS
2614 REM
2616 AR = 4 * ATN(1) * ALPHA / 180
2617 C  = COS(AR)
2618 S  = SIN(AR)
2620 REM
2622 REM --- CALCULATE THE STIFFNESS COEFFICIENTS (UPPER TRIANGLE ONLY)
2624 REM
2628 ESTIFF(1,1) =     E * A * C * C / L  +  K1 * E * A2 * S * S / L^3
2630 ESTIFF(1,2) =     E * A * S * C / L  -  K1 * E * A2 * S * C / L^3
2640 ESTIFF(1,3) = - K2 * E * A2 * S / L^2
2650 ESTIFF(1,4) = - ESTIFF(1,1)
2660 ESTIFF(1,5) = - ESTIFF(1,2)
2670 ESTIFF(1,6) = - K3 * E * A2 * S / L^2
2680 ESTIFF(2,2) =     E * A * S * S / L  +  K1 * E * A2 * C * C / L^3
2690 ESTIFF(2,3) =    K2 * E * A2 * C / L^2
2700 ESTIFF(2,4) = - ESTIFF(1,2)
2710 ESTIFF(2,5) = - ESTIFF(2,2)
2720 ESTIFF(2,6) =    K3 * E * A2 * C / L^2
2730 ESTIFF(3,3) =    K4 * E * A2 / L
```

PLANE FRAMES

```
2740 ESTIFF(3,4) = - K5 * E * A2 * S / L^2
2750 ESTIFF(3,5) =   K5 * E * A2 * C / L^2
2760 ESTIFF(3,6) =   K6 * E * A2 / L
2770 ESTIFF(4,4) =   ESTIFF(1,1)
2780 ESTIFF(4,5) =   ESTIFF(1,2)
2790 ESTIFF(4,6) = - K7 * E * A2 * S / L^2
2800 ESTIFF(5,5) =   ESTIFF(2,2)
2810 ESTIFF(5,6) =   K7 * E * A2 * C / L^2
2820 ESTIFF(6,6) =   K8 * E * A2 / L
2830 REM
2840 REM  ---  OUTPUT THE ELEMENT STIFFNESS MATRIX IN THE GLOBAL AXES SYSTEM
2850 REM
2860 PRINT : PRINT
2870 PRINT "* * * * * * * * * * * * * * * * *"
2880 PRINT "*    ELEMENT STIFFNESS MATRIX IN  *"
2890 PRINT "*    THE GLOBAL AXES SYSTEM       *"
2900 PRINT "* * * * * * * * * * * * * * * * *"
2910 PRINT
2920 FOR I = 1 TO 6
2930 FOR J = 1 TO 6
2940 IF I <= J THEN PRINT USING "#.####^^^^   "; ESTIFF(I,J);
2950 IF I >  J THEN PRINT USING "#.####^^^^   "; ESTIFF(J,I);
2960 NEXT J
2970 PRINT
2980 NEXT I
2990 RETURN
```

Sample Run The following sample run shows program ESTIFF.PF being used to evaluate the element stiffness matrices for the structure analysed in example 7.2. Input from the keyboard is shown in bold typeface.

```
-----------------------------------------------------------
PROGRAM    E S T I F F . P F    TO CALCULATE THE ELEMENT
STIFFNESS MATRIX IN THE MEMBER AND THE GLOBAL AXES SYSTEMS
FOR A PLANE FRAME ELEMENT
-----------------------------------------------------------

NUMBER OF ELEMENTS     =  2


ELEMENT    LOWER       HIGHER
 TYPE      NODE        NODE

   1       FIXED       FIXED
   2       FIXED       PINNED
   3       PINNED      FIXED
   4       PINNED      PINNED
ELEMENT  1

LOWER  NODE NUMBER  =  1
HIGHER NODE NUMBER  =  2

+ + + + + + + + + + +
+    ELEMENT  1  2   +
+ + + + + + + + + + +
AREA                           =  9000
SECOND MOMENT OF AREA          =  .25E9
LENGTH                         =  3000
ELASTIC CONSTANT               =  200
ANGLE WITH THE GLOBAL X-AXIS   =  90
ELEMENT TYPE                   =  1
```

```
* * * * * * * * * * * * * * * *
*    ELEMENT STIFFNESS MATRIX IN   *
*      THE MEMBER AXES SYSTEM      *
* * * * * * * * * * * * * * * *

 0.6000E+03   0.0000E+00   0.0000E+00  -.6000E+03   0.0000E+00   0.0000E+00
 0.0000E+00   0.2222E+02   0.3333E+05   0.0000E+00  -.2222E+02   0.3333E+05
 0.0000E+00   0.3333E+05   0.6667E+08   0.0000E+00  -.3333E+05   0.3333E+08
-.6000E+03    0.0000E+00   0.0000E+00   0.6000E+03   0.0000E+00   0.0000E+00
 0.0000E+00  -.2222E+02   -.3333E+05   0.0000E+00   0.2222E+02  -.3333E+05
 0.0000E+00   0.3333E+05   0.3333E+08   0.0000E+00  -.3333E+05   0.6667E+08


* * * * * * * * * * * * * * * *
*    ELEMENT STIFFNESS MATRIX IN   *
*      THE GLOBAL AXES SYSTEM      *
* * * * * * * * * * * * * * * *

 0.2222E+02  -.2164E-03  -.3333E+05  -.2222E+02   0.2164E-03  -.3333E+05
-.2164E-03   0.6000E+03  -.1248E-01   0.2164E-03  -.6000E+03  -.1248E-01
-.3333E+05  -.1248E-01   0.6667E+08   0.3333E+05   0.1248E-01   0.3333E+08
-.2222E+02   0.2164E-03   0.3333E+05   0.2222E+02  -.2164E-03   0.3333E+05
 0.2164E-03  -.6000E+03   0.1248E-01  -.2164E-03   0.6000E+03   0.1248E-01
-.3333E+05  -.1248E-01   0.3333E+08   0.3333E+05   0.1248E-01   0.6667E+08
```

ELEMENT 2

LOWER NODE NUMBER = 2
HIGHER NODE NUMBER = 3

```
+ + + + + + + + + +
+    ELEMENT  2  3   +
+ + + + + + + + + +
```

AREA   = 9000
SECOND MOMENT OF AREA   = .25E9
LENGTH   = 7071
ELASTIC CONSTANT   = 200
ANGLE WITH THE GLOBAL X-AXIS = 8.13
ELEMENT TYPE   = 1

```
* * * * * * * * * * * * * * * *
*    ELEMENT STIFFNESS MATRIX IN   *
*      THE MEMBER AXES SYSTEM      *
* * * * * * * * * * * * * * * *

 0.2546E+03   0.0000E+00   0.0000E+00  -.2546E+03   0.0000E+00   0.0000E+00
 0.0000E+00   0.1697E+01   0.6000E+04   0.0000E+00  -.1697E+01   0.6000E+04
 0.0000E+00   0.6000E+04   0.2828E+08   0.0000E+00  -.6000E+04   0.1414E+08
-.2546E+03    0.0000E+00   0.0000E+00   0.2546E+03   0.0000E+00   0.0000E+00
 0.0000E+00  -.1697E+01  -.6000E+04    0.0000E+00   0.1697E+01  -.6000E+04
 0.0000E+00   0.6000E+04   0.1414E+08   0.0000E+00  -.6000E+04   0.2828E+08


* * * * * * * * * * * * * * * *
*    ELEMENT STIFFNESS MATRIX IN   *
*      THE GLOBAL AXES SYSTEM      *
* * * * * * * * * * * * * * * *

 0.2495E+03   0.3540E+02  -.8485E+03  -.2495E+03  -.3540E+02  -.8485E+03
 0.3540E+02   0.6754E+01   0.5940E+04  -.3540E+02  -.6754E+01   0.5940E+04
-.8485E+03    0.5940E+04   0.2828E+08   0.8485E+03  -.5940E+04   0.1414E+08
-.2495E+03   -.3540E+02   0.8485E+03   0.2495E+03   0.3540E+02   0.8485E+03
-.3540E+02   -.6754E+01  -.5940E+04    0.3540E+02   0.6754E+01  -.5940E+04
-.8485E+03    0.5940E+04   0.1414E+08   0.8485E+03  -.5940E+04   0.2828E+08
```

PLANE FRAMES

Program ISTIFF.PF

This program illustrates how ESTIFF.PF can be developed to automatically generate the initial structure stiffness matrix for a plane frame structure.

Input The number of nodes and the number of elements are input, the program then loops for each element in turn requesting the member data. The element data is used to calculate the element stiffness which is then added to the initial structure stiffness matrix.

Comments on the Algorithm Only the upper triangle of the initial structure stiffness matrix is generated.

Output The initial structure stiffness matrix is output, with advantage being taken of the symmetry of the matrix to output the sub-diagonal elements.

Listing

```
1000 REM  ***********************************************************************
1010 REM  *                                                                     *
1020 REM  *     PROGRAM    I S T I F F . P F    TO GENERATE THE INITIAL         *
1030 REM  *     STIFFNESS MATRIX FOR A PLANE FRAME STRUCTURE                    *
1040 REM  *                                                                     *
1050 REM  *     J.A.D.BALFOUR                                                   *
1060 REM  *                                                                     *
1070 REM  ***********************************************************************
1080 REM
1090 REM  --- NOTE THAT VARIABLES ARE DEFINED IN APPENDIX A
1100 REM
1110 OPTION BASE 1
1120 REM
1130 REM  --- MAX PROBLEM SIZE SET BY THE FOLLOWING DIMENSION STATEMENT
1140 REM
1150 DIM ISTIFF(21,21), ESTIFF(6,6), EFREE(6)
1160 PRINT : PRINT
1170 PRINT "-----------------------------------------------"
1180 PRINT
1190 PRINT "PROGRAM       I S T I F F . P F      TO GENERATE"
1200 PRINT "THE INITIAL STIFFNESS MATRIX FOR A PLANE FRAME"
1210 PRINT
1220 PRINT "-----------------------------------------------"
1230 PRINT : PRINT
1240 INPUT "NUMBER OF NODES     = ", NNODE
1250 INPUT "NUMBER OF ELEMENTS  = ", NMEMB
1260 REM
1270 REM  --- ZERO THE INITIAL STIFFNESS MATRIX
1280 REM
1290 NDOF = 3 * NNODE
1300 FOR I = 1 TO NDOF
1310 FOR J = 1 TO NDOF
1320 ISTIFF(I,J) = 0
1330 NEXT J
1340 NEXT I
1350 REM
1360 REM  --- LOOP FOR ALL ELEMENTS
1370 REM
1380 FOR K = 1 TO NMEMB
1390 REM
```

```
1400 REM  ---  INPUT THE ELEMENT DATA
1410 REM
1420 GOSUB 1530
1430 REM
1440 REM  ---  ADD THE ELEMENT STIFFNESS TO THE INITIAL STIFFNESS MATRIX
1450 REM
1460 GOSUB 1740
1470 NEXT K
1480 REM
1490 REM  ---  OUTPUT THE INITIAL STIFFNESS MATRIX
1500 REM
1510 GOSUB 2310
1520 END
1530 REM  * * * * * * * * * * * * * * * * * * * * * * * * * * * * * * * *
1540 REM  *                                                              *
1550 REM  *      SUBROUTINE TO INPUT THE ELEMENT DATA                    *
1560 REM  *                                                              *
1570 REM  * * * * * * * * * * * * * * * * * * * * * * * * * * * * * * * *
1580 REM
1590 PRINT : PRINT
1600 PRINT "ELEMENT "; K : PRINT
1610 INPUT "LOWER  NODE NUMBER  =  ", IN
1620 INPUT "HIGHER NODE NUMBER  =  ", JN
1630 PRINT
1640 PRINT           "+ + + + + + + + + +"
1650 PRINT USING "+    ELEMENT ## ##   +"; IN, JN
1660 PRINT           "+ + + + + + + + + +"
1670 PRINT
1680 INPUT "AREA                         =  ", A
1690 INPUT "SECOND MOMENT OF AREA        =  ", A2
1700 INPUT "LENGTH                       =  ", L
1710 INPUT "ELASTIC CONSTANT             =  ", E
1720 INPUT "ANGLE WITH THE GLOBAL X-AXIS =  ", ALPHA
1730 RETURN
1740 REM  * * * * * * * * * * * * * * * * * * * * * * * * * * * * * * * *
1750 REM  *                                                              *
1760 REM  *     SUBROUTINE TO ADD THE ELEMENT STIFFNESS TO THE INITIAL   *
1770 REM  *     STRUCTURE STIFFNESS MATRIX                               *
1780 REM  *                                                              *
1790 REM  * * * * * * * * * * * * * * * * * * * * * * * * * * * * * * * *
1800 REM
1810 REM  ---  CONVERT THE ANGLE TO RADIANS
1820 REM
1830 AR = 4 * ATN(1) * ALPHA / 180
1840 C  = COS(AR)
1850 S  = SIN(AR)
1860 REM
1870 REM  ---  GENERATE THE UPPER TRIANGLE OF THE ELEMENT STIFFNESS MATRIX
1880 REM
1890 ESTIFF(1,1) =    E * A * C * C / L  +  12 * E * A2 * S * S / L^3
1900 ESTIFF(1,2) =    E * A * S * C / L  -  12 * E * A2 * S * C / L^3
1910 ESTIFF(1,3) = - 6 * E * A2 * S / L^2
1920 ESTIFF(1,4) = - ESTIFF(1,1)
1930 ESTIFF(1,5) = - ESTIFF(1,2)
1940 ESTIFF(1,6) =   ESTIFF(1,3)
1950 ESTIFF(2,2) =    E * A * S * S / L  +  12 * E * A2 * C * C / L^3
1960 ESTIFF(2,3) =    6 * E * A2 * C / L^2
1970 ESTIFF(2,4) = - ESTIFF(1,2)
1980 ESTIFF(2,5) = - ESTIFF(2,2)
1990 ESTIFF(2,6) =   ESTIFF(2,3)
2000 ESTIFF(3,3) =    4 * E * A2 / L
2010 ESTIFF(3,4) = - ESTIFF(1,3)
2020 ESTIFF(3,5) = - ESTIFF(2,3)
2030 ESTIFF(3,6) =   .5 * ESTIFF(3,3)
2040 ESTIFF(4,4) =   ESTIFF(1,1)
2050 ESTIFF(4,5) =   ESTIFF(1,2)
2060 ESTIFF(4,6) = - ESTIFF(1,3)
2070 ESTIFF(5,5) =   ESTIFF(2,2)
2080 ESTIFF(5,6) = - ESTIFF(2,3)
2090 ESTIFF(6,6) =   ESTIFF(3,3)
2100 REM
2110 REM  ---  SET UP THE ELEMENT CODE NUMBER (INITIAL)
2120 REM
```

PLANE FRAMES

```
2130 EFREE(1) = 3*IN-2
2140 EFREE(2) = 3*IN-1
2150 EFREE(3) = 3*IN
2160 EFREE(4) = 3*JN-2
2170 EFREE(5) = 3*JN-1
2180 EFREE(6) = 3*JN
2190 REM
2200 REM  ---  ADD THE ELEMENT STIFFNESS TERMS TO UPPER TRIANGLE
2210 REM  ---  OF THE INITIAL STRUCTURE STIFFNESS MATRIX
2220 REM
2230 FOR I = 1 TO 6
2240 FOR J = 1 TO 6
2250 L = EFREE(I)
2260 M = EFREE(J)
2270 ISTIFF(L,M) = ISTIFF(L,M) + ESTIFF(I,J)
2280 NEXT J
2290 NEXT I
2300 RETURN
2310 REM * * * * * * * * * * * * * * * * * * * * * * * * * * * * * * *
2320 REM *                                                           *
2330 REM *   SUBROUTINE TO OUTPUT THE INITIAL STRUCTURE STIFFNESS MATRIX  *
2340 REM *                                                           *
2350 REM * * * * * * * * * * * * * * * * * * * * * * * * * * * * * * *
2360 REM
2370 PRINT : PRINT
2380 PRINT "* * * * * * * * * * * * *"
2390 PRINT "*    INITIAL STRUCTURE  *"
2400 PRINT "*    STIFFNESS MATRIX   *"
2410 PRINT "* * * * * * * * * * * * *"
2420 PRINT
2430 FOR I = 1 TO NDOF
2440 FOR J = 1 TO NDOF
2450 IF I <= J THEN PRINT ISTIFF(I,J), ELSE PRINT ISTIFF(J,I),
2460 NEXT J
2470 PRINT : PRINT
2480 NEXT I
2490 END
```

Sample Run The following sample run shows program ISTIFF.PF being used to evaluate the initial stiffness matrices for the structure analysed in example 7.2. Input from the keyboard is shown in bold typeface.

```
-----------------------------------------------
PROGRAM      I S T I F F . P F      TO GENERATE
THE INITIAL STIFFNESS MATRIX FOR A PLANE FRAME
-----------------------------------------------


NUMBER OF NODES       =  3
NUMBER OF ELEMENTS    =  2

ELEMENT 1

LOWER  NODE NUMBER  =  1
HIGHER NODE NUMBER  =  2

+ + + + + + + + + + +
+    ELEMENT 1  2   +
+ + + + + + + + + + +

AREA                          =  9000
SECOND MOMENT OF AREA         =  .25E9
LENGTH                        =  3000
ELASTIC CONSTANT              =  200
ANGLE WITH THE GLOBAL X-AXIS  =  90
```

213

ELEMENT 2

LOWER NODE NUMBER = 2
HIGHER NODE NUMBER = 3

+ + + + + + + + + + +
+   ELEMENT 2 3   +
+ + + + + + + + + + +

AREA   = 9000
SECOND MOMENT OF AREA   = .25E9
LENGTH   = 7071
ELASTIC CONSTANT   = 200
ANGLE WITH THE GLOBAL X-AXIS = 8.13

* * * * * * * * * * * *
*   INITIAL STRUCTURE   *
*   STIFFNESS MATRIX   *
* * * * * * * * * * * *

   22.2222   -2.16382E-04   -33333.3   -22.2222   2.16382E-04
  -33333.3   0   0   0

   -2.16382E-04   600   -.0124836   2.16382E-04   -600
   -.0124836   0   0   0

  -33333.3   -.0124836   6.66667E+07   33333.3   .0124836
   3.33333E+07   0   0   0

  -22.2222   2.16382E-04   33333.3   271.726   35.4003
   32484.8   -249.504   -35.4005   -848.534

    2.16382E-04 -600   .0124836   35.4003   606.754
    5939.82   -35.4005   -6.75425   5939.81

  -33333.3   -.0124836   3.33333E+07   32484.8   5939.82
   9.49512E+07   848.534   -5939.81   1.41423E+07

    0   0   0   -249.504   -35.4005
    848.534   249.504   35.4005   848.534

    0   0   0   -35.4005   -6.75425
   -5939.81   35.4005   6.75425   -5939.81

    0   0   0   -848.534   5939.81
    1.41423E+07   848.534   -5939.81   2.82845E+07

7.7 APPLICATION OF THE BOUNDARY CONDITIONS

In section 5.7 the initial structure stiffness equation was shown to be a mixed set of simultaneous equations containing known displacements (boundary conditions) and unknown displacements in the displacement vector, and known forces (applied loads) and unknown forces (reactions) in the force vector. Section 5.7 also showed that the final structure stiffness equation can be obtained from the initial structure stiffness equation by deleting the rows and columns associated with the boundary conditions. If freedom "m" is restrained (i.e. has zero displacement) then row "m" and column "m" are deleted from the initial stiffness equation.

PLANE FRAMES

Example 7.3 Find the displacements for the frame shown below.

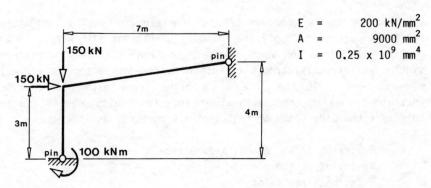

$E = 200$ kN/mm^2
$A = 9000$ mm^2
$I = 0.25 \times 10^9$ mm^4

Nodes 1 and 3 are restrained against translation in the global "x" and "y" directions, therefore $\Delta_{1x} = \Delta_{1y} = \Delta_{3x} = \Delta_{3y} = 0$. Hence the final stiffness equation is obtained by deleting rows and columns centred on diagonal elements 1, 2, 7, and 8 of the initial stiffness equation. The frame in this example is, in fact, identical to the frame analysed in example 7.2, making unnecessary the evaluation of the initial stiffness matrix, and allowing the initial structure stiffness equation to be written directly as follows:

$$
\begin{bmatrix}
22.2 & 0 & -33333 & -22.2 & 0 & -33333 & 0 & 0 & 0 \\
 & 600.0 & 0 & 0 & -600.0 & 0 & 0 & 0 & 0 \\
 & & 66.7E6 & 33333 & 0 & 33.3E6 & 0 & 0 & 0 \\
 & & & 271.7 & 35.40 & 32485 & -249.4 & -35.4 & -848.5 \\
 & & & & 606.8 & 5940 & -35.40 & -6.75 & 5940 \\
 & & \text{symmetrical} & & & 95.0E6 & 848.5 & -5940 & 14.1E6 \\
 & & & & & & 249.5 & 35.40 & 848.5 \\
 & & & & & & & 6.754 & -5940 \\
 & & & & & & & & 28.3E6
\end{bmatrix}
\begin{bmatrix}
0 \\ 0 \\ \theta_1 \\ \Delta_{2x} \\ \Delta_{2y} \\ \theta_2 \\ 0 \\ 0 \\ \theta_3
\end{bmatrix}
=
\begin{bmatrix}
-R_{1x} \\ -R_{1y} \\ -1E5 \\ 150 \\ -150 \\ 0 \\ -R_{3x} \\ -R_{3y} \\ 0
\end{bmatrix}
$$

The reader should note the mixed nature of the equations with unknown and known quantities appearing on either side of the equality. The boundary conditions are applied by striking out rows and columns as shown. Solution of the remaining equations (using any of the equation solving programs in Chapter 4) yields

$$
\begin{bmatrix} \theta_1 \\ \Delta_{2x} \\ \Delta_{2y} \\ \theta_2 \\ \theta_3 \end{bmatrix}
=
\begin{bmatrix}
-2.16 \times 10^{-3} \text{ rads} \\
0.792 \text{ mm} \\
-0.297 \text{ mm} \\
0.532 \times 10^{-3} \text{ rads} \\
-0.179 \times 10^{-3} \text{ rads}
\end{bmatrix}
$$

7.8 THE FINAL STIFFNESS MATRIX - Program FSTIFF.PF

In practice it is uneconomical to generate the initial structure stiffness matrix and apply the boundary conditions afterwards. When dealing with plane and space trusses the final stiffness matrix was synthesised directly through the use of a freedom vector and element code numbers. In the case of a plane frame each node can be restrained against translational or rotational displacement and in this text the directions of restraint are numbered as follows:

    Restraint in the global x-direction   -   1
    Restraint in the global y-direction   -   2
    Rotational restraint   -   3

Consider the frame shown below which was analysed in example 7.3.

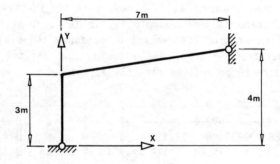

FIGURE 7.6   THE FRAME ANALYSED IN EXAMPLE 7.3

Using the adopted restraint numbering convention the boundary conditions would be input as follows:

NODE   DIRECTION(S)
 1   12   (node 1 restrained in the x- and y-directions)
 3   12   (node 3 restrained in the x- and y-directions)

Note that the restraint directions have been described using composite numbers. If a node is restrained in all directions then the composite restraint number would be 123 or 213 or 231, etc. From the boundary conditions a vector of the freedom numbers can be produced. The procedure used is exactly the same as was employed in the analysis of plane and space trusses; the freedom vector for the

PLANE FRAMES

frame in example 7.3 being found as follows:

Zero the freedom vector $\begin{bmatrix} 0\\0\\0\\0\\0\\0\\0\\0\\0 \end{bmatrix}$ Apply the boundary conditions $\begin{bmatrix} 1\\1\\0\\0\\0\\0\\1\\1\\0 \end{bmatrix}$ Number the freedoms $\begin{bmatrix} 0\\0\\1\\2\\3\\4\\0\\0\\5 \end{bmatrix}$

To add the stiffness of element "i,j" to the final structure stiffness matrix the element code number is constructed from elements (3i-2), (3i-1), (3i), (3j-2), (3j-1), and (3j) of the freedom vector. The code number, when applied to the global element stiffness matrix, gives the destinations of the element stiffness coefficients in the final structure stiffness matrix.

$$\begin{array}{c} \\ 2\\3\\4\\0\\0\\5 \end{array} \begin{array}{c} 234005 \\ \begin{bmatrix} K_{11} & K_{12} & K_{13} & K_{14} & K_{15} & K_{16} \\ K_{21} & K_{22} & K_{23} & K_{24} & K_{25} & K_{26} \\ K_{31} & K_{32} & K_{33} & K_{34} & K_{35} & K_{36} \\ K_{41} & K_{42} & K_{43} & K_{44} & K_{45} & K_{46} \\ K_{51} & K_{52} & K_{53} & K_{54} & K_{55} & K_{56} \\ K_{61} & K_{62} & K_{63} & K_{64} & K_{65} & K_{66} \end{bmatrix} \end{array}$$

For example, the code number of element 2,3 in example 7.3 is [2 3 4 0 0 5] which, when applied to the global element stiffness matrix as shown, indicates that

      element K_{11} should be added to location (2,2) in the final structure stiffness matrix.

      element K_{12} should be added to location (2,3) in the final structure stiffness matrix.

                     .
                     .

      element K_{33} should be added to location (4,4) in the final structure stiffness matrix.

                     .
                     .

      element K_{66} should be added to location (5,5) in the final structure stiffness matrix.

COMPUTER ANALYSIS OF STRUCTURAL FRAMEWORKS

Note that if a zero element appears in either the "horizontal" or the "vertical" code number then that stiffness term is ignored. The final structure stiffness matrix is complete when the stiffness contributions of all elements have been added using this technique.

Example 7.4

Assemble the final stiffness matrix for the frame shown below using the structure freedom vector and element code numbers. Note that the frame is identical to that analysed in example 7.2, and the freedom vector has already been found to be $\{0\ 0\ 1\ 2\ 3\ 4\ 0\ 0\ 5\}$.

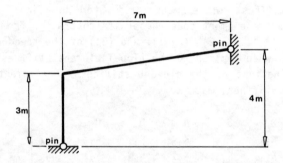

Member 1,2 - code number [0 0 1 2 3 4]

Apply the code number to the global element stiffness matrix (from example 7.2).

$$\begin{array}{c} \\ 0 \\ 0 \\ 1 \\ 2 \\ 3 \\ 4 \end{array} \begin{array}{c} 0 0 1 2 3 4 \end{array} \\ \begin{bmatrix} 22.2 & 0 & -33333 & -22.2 & 0 & -33333 \\ 0 & 600.0 & 0 & 0 & -600.0 & 0 \\ -33333 & 0 & 66.7E6 & 33333 & 0 & 33.3E6 \\ -22.2 & 0 & 33333 & 22.2 & 0 & 33333 \\ 0 & -600.0 & 0 & 0 & 600.0 & 0 \\ -33333 & 0 & 33.3E6 & 33333 & 0 & 66.6E6 \end{bmatrix}$$

Adding the stiffness terms to the structure stiffness matrix gives

$$\begin{bmatrix} 66.7E6 & -33333 & 0 & 33.3E6 & 0 \\ -33333 & 22.2 & 0 & -33333 & 0 \\ 0 & 0 & 600.0 & 0 & 0 \\ 33.3E6 & -33333 & 0 & 66.7E6 & 0 \\ 0 & 0 & 0 & 0 & 0 \end{bmatrix}$$

PLANE FRAMES

Member 2,3   -   code number   [2 3 4 0 0 5]

Apply the code number to the element stiffness matrix.

$$\begin{array}{c} \\ 2 \\ 3 \\ 4 \\ 0 \\ 0 \\ 5 \end{array} \begin{bmatrix} 2 & 3 & 4 & 0 & 0 & 5 \\ 249.5 & 35.40 & -848.5 & -249.5 & -35.40 & -848.5 \\ 35.40 & 6.754 & 5940 & -35.40 & -6.754 & 5940 \\ -848.5 & 5940 & 28.3E6 & 848.5 & -5940 & 14.1E6 \\ -249.5 & -35.40 & 848.5 & 249.5 & 35.40 & 848.5 \\ -35.40 & -6.754 & -5940 & 35.40 & 6.754 & -5940 \\ -848.5 & 5940 & 14.1E6 & 848.5 & -5940 & 28.3E6 \end{bmatrix}$$

Adding the stiffness terms completes the final structure stiffness matrix.

$$\begin{bmatrix} 66.7E6 & 33333 & 0 & 33.3E6 & 0 \\ 33333 & 271.7 & 35.40 & 32484 & -848.5 \\ 0 & 35.40 & 606.8 & 5940 & 5940 \\ 33.3E6 & 32484 & 5940 & 95.0E6 & 14.1E6 \\ 0 & -848.5 & 5940 & 14.1E6 & 28.3E6 \end{bmatrix}$$

Program FSTIFF.PF

This program shows how the freedom vector is set up and then used to assemble the final structure stiffness matrix from the global element stiffness matrices for a plane frame. Note that advantage has been taken of the symmetry and the bandwidth of the structure stiffness matrix to save space by generating only the upper semi-band (i.e. elements that are on or above the diagonal and within the bandwidth).

Input   The program asks for the number of nodes, the number of elements, and the number of restrained nodes (note that restraint directions are entered as composite numbers). It then requests details of the boundary conditions, and then the data for each element in turn.

Comments on the Algorithm   Once the boundary conditions have been established the freedom vector is developed. The program then loops for each element in turn; first generating the global element stiffness matrix and then the element code number which is used to add the element stiffness to the structure stiffness matrix.

Output   The upper semi-band of the final structure stiffness matrix is output. The reader should observe that the matrix is in the

COMPUTER ANALYSIS OF STRUCTURAL FRAMEWORKS

correct form to be used directly by the equation solving programs GAUSS.EQ and UDU.EQ described in Chapter 4.

Listing

```
1000 REM ************************************************************
1010 REM *                                                          *
1020 REM *      PROGRAM    F S T I F F . P F    TO GENERATE THE FINAL *
1030 REM *      STIFFNESS MATRIX FOR A PLANE FRAME STRUCTURE         *
1040 REM *                                                          *
1050 REM *             J.A.D.BALFOUR                                *
1060 REM *                                                          *
1070 REM ************************************************************
1080 REM
1090 REM  ---  NOTE THAT VARIABLES ARE DEFINED IN APPENDIX A
1100 REM
1110 OPTION BASE 1
1120 REM
1130 REM  ---  PROBLEM SIZE SET BY THE FOLLOWING DIMENSION STATEMENT
1140 REM
1150 DIM ESTIFF(6,6), FSTIFF(20,20), EFREE(6), FREE(20)
1160 PRINT : PRINT
1170 PRINT "--------------------------------------------"
1180 PRINT
1190 PRINT "PROGRAM    F S T I F F . P F    TO GENERATE"
1200 PRINT "THE FINAL STIFFNESS MATRIX FOR A PLANE FRAME"
1210 PRINT
1220 PRINT "--------------------------------------------"
1230 PRINT : PRINT
1240 INPUT "NUMBER OF NODES            = ", NNODE
1250 INPUT "NUMBER OF ELEMENTS         = ", NMEMB
1260 INPUT "NUMBER OF RESTRAINED NODES = ", NREST
1270 REM
1280 REM  ---  INPUT THE BOUNDARY CONDITIONS AND GENERATE THE FREEDOM VECTOR
1290 REM
1300 GOSUB 1540
1310 REM
1320 REM  ---  ZERO THE STRUCTURE STIFFNESS MATRIX
1330 REM
1340 FOR I       = 1 TO NDOF
1350 FOR J       = 1 TO NDOF
1360 FSTIFF(I,J) = 0
1370 NEXT J
1380 NEXT I
1390 REM
1400 REM  ---  LOOP FOR ALL ELEMENTS
1410 REM
1420 BAND = 0
1430 FOR K = 1 TO NMEMB
1440 REM
1450 REM  ---  INPUT ELEMENT DATA AND ADD THE ELEMENT STIFFNESS
1460 REM
1470 GOSUB 2010
1480 NEXT K
1490 REM
1500 REM  ---  OUTPUT THE FINAL STRUCTURE STIFFNESS MATRIX
1510 REM
1520 GOSUB 2750
1530 END
1540 REM * * * * * * * * * * * * * * * * * * * * * * * * * * * * * *
1550 REM *                                                          *
1560 REM *      SUBROUTINE TO INPUT THE BOUNDARY CONDITIONS AND GENERATE *
1570 REM *      THE FREEDOM VECTOR                                  *
1580 REM *                                                          *
1590 REM * * * * * * * * * * * * * * * * * * * * * * * * * * * * * *
1600 REM
1610 REM  ---  ZERO THE FREEDOM VECTOR
1620 REM
1630 FOR I   = 1 TO 3*NNODE
1640 FREE(I) = 0
```

PLANE FRAMES

```
1650 NEXT I
1660 PRINT : PRINT
1670 PRINT "+ + + + + + + + + + + + + +"
1680 PRINT "+     BOUNDARY CONDITIONS    +"
1690 PRINT "+ + + + + + + + + + + + + +"
1700 PRINT
1710 PRINT "INPUT BOUNDARY CONDITIONS"
1720 PRINT "RESTRAINT IN THE X-DIRECTION   =   1"
1730 PRINT "RESTRAINT IN THE Y-DIRECTION   =   2"
1740 PRINT "ROTATIONAL RESTRAINT           =   3"
1750 PRINT "ENTER RESTRAINTS AS A COMPOSITE NUMBER"
1760 PRINT "(E.G. IF RESTRAINED IN THE X- AND Y-DIRECTIONS ENTER 12)"
1770 FOR I = 1 TO NREST
1780 PRINT
1790 INPUT "NODE NUMBER    = ", IN
1800 INPUT "DIRECTION(S)   = ", DIRN
1810 REM
1820 REM  ---  EVALUATE THE FREEDOM TO BE RESTRAINED FROM THE RIGHT-MOST DIGIT
1830 REM
1840 J       = 3*IN - 3 + (DIRN MOD 10)
1850 FREE(J) = 1
1860 DIRN    = INT(DIRN/10)
1870 IF DIRN > 0 THEN GOTO 1840
1880 NEXT I
1890 REM
1900 REM  ---  NUMBER THE FREEDOMS (RESTRAINTS SET TO ZERO)
1910 REM
1920 NDOF        = 0
1930 FOR I       = 1 TO 3*NNODE
1940 IF FREE(I)  = 1 THEN GOTO 1980
1950 NDOF        = NDOF + 1
1960 FREE(I)     = NDOF
1970 GOTO 1990
1980 FREE(I)     = 0
1990 NEXT I
2000 RETURN
2010 REM * * * * * * * * * * * * * * * * * * * * * * * * * * * * * *
2020 REM *                                                          *
2030 REM *    SUBROUTINE TO INPUT THE ELEMENT DATA AND ADD THE ELEMENT *
2040 REM *    STIFFNESS TO THE FINAL STRUCTURE STIFFNESS MATRIX     *
2050 REM *                                                          *
2060 REM * * * * * * * * * * * * * * * * * * * * * * * * * * * * * *
2070 REM
2080 PRINT : PRINT
2090 PRINT "ELEMENT "; K : PRINT
2100 INPUT "LOWER  NODE NUMBER  = ", IN
2110 INPUT "HIGHER NODE NUMBER  = ", JN
2120 PRINT
2130 PRINT         "+ + + + + + + + + +"
2140 PRINT USING "+   ELEMENT ## ##   +"; IN, JN
2150 PRINT         "+ + + + + + + + + +"
2160 PRINT
2170 INPUT "AREA                          = ", A
2180 INPUT "SECOND MOMENT OF AREA         = ", A2
2190 INPUT "LENGTH                        = ", L
2200 INPUT "ELASTIC CONSTANT              = ", E
2210 INPUT "ANGLE WITH THE GLOBAL X-AXIS  = ", ALPHA
2220 REM
2230 REM  ---  CONVERT THE ANGLE TO RADIANS
2240 REM
2250 AR = 4 * ATN(1) * ALPHA / 180
2260 C  = COS(AR)
2270 S  = SIN(AR)
2280 REM
2290 REM  ---  GENERATE THE UPPER TRIANGLE OF THE ELEMENT STIFFNESS MATRIX
2300 REM
2310 ESTIFF(1,1) =   E * A * C * C / L  +  12 * E * A2 * S * S / L^3
2320 ESTIFF(1,2) =   E * A * S * C / L  -  12 * E * A2 * S * C / L^3
2330 ESTIFF(1,3) = - 6 * E * A2 * S / L^2
2340 ESTIFF(1,4) = - ESTIFF(1,1)
2350 ESTIFF(1,5) = - ESTIFF(1,2)
2360 ESTIFF(1,6) =   ESTIFF(1,3)
2370 ESTIFF(2,2) =   E * A * S * S / L  +  12 * E * A2 * C * C / L^3
```

```
2380 ESTIFF(2,3) =   6 * E * A2 * C / L^2
2390 ESTIFF(2,4) = - ESTIFF(1,2)
2400 ESTIFF(2,5) = - ESTIFF(2,2)
2410 ESTIFF(2,6) =   ESTIFF(2,3)
2420 ESTIFF(3,3) =   4 * E * A2 / L
2430 ESTIFF(3,4) = - ESTIFF(1,3)
2440 ESTIFF(3,5) = - ESTIFF(2,3)
2450 ESTIFF(3,6) =   .5 * ESTIFF(3,3)
2460 ESTIFF(4,4) =   ESTIFF(1,1)
2470 ESTIFF(4,5) =   ESTIFF(1,2)
2480 ESTIFF(4,6) = - ESTIFF(1,3)
2490 ESTIFF(5,5) =   ESTIFF(2,2)
2500 ESTIFF(5,6) = - ESTIFF(2,3)
2510 ESTIFF(6,6) =   ESTIFF(3,3)
2520 REM
2530 REM  ---   SET UP THE ELEMENT CODE NUMBER
2540 REM
2550 EFREE(1) = FREE(3*IN-2)
2560 EFREE(2) = FREE(3*IN-1)
2570 EFREE(3) = FREE(3*IN)
2580 EFREE(4) = FREE(3*JN-2)
2590 EFREE(5) = FREE(3*JN-1)
2600 EFREE(6) = FREE(3*JN)
2610 REM
2620 REM  ---   ADD THE ELEMENT STIFFNESS TERMS
2630 REM
2640 FOR I      = 1 TO 6
2650 IF EFREE(I) = 0 THEN GOTO 2730
2660 FOR J      = 1 TO 6
2670 IF EFREE(J) = 0 THEN GOTO 2720
2680 L          = EFREE(I)
2690 M          = EFREE(J) - L + 1
2700 IF M       > BAND THEN BAND = M
2710 FSTIFF(L,M) = FSTIFF(L,M) + ESTIFF(I,J)
2720 NEXT J
2730 NEXT I
2740 RETURN
2750 REM * * * * * * * * * * * * * * * * * * * * * * * * * * * * * *
2760 REM *                                                          *
2770 REM *     SUBROUTINE TO OUTPUT THE FINAL STRUCTURE STIFFNESS MATRIX    *
2780 REM *                                                          *
2790 REM * * * * * * * * * * * * * * * * * * * * * * * * * * * * * *
2800 REM
2810 PRINT : PRINT
2820 PRINT "* * * * * * * * * * * * * *"
2830 PRINT "*    UPPER SEMI-BAND OF THE    *"
2840 PRINT "*     FINAL STIFFNESS MATRIX    *"
2850 PRINT "* * * * * * * * * * * * * *"
2860 PRINT
2870 FOR I = 1 TO NDOF
2880 FOR J = 1 TO BAND
2890 PRINT FSTIFF(I,J),
2900 NEXT J
2910 PRINT : PRINT
2920 NEXT I
2930 RETURN
```

Sample Run The following sample run shows program FSTIFF.PF being used to evaluate the final stiffness matrix for the structure analysed in example 7.2. Input from the keyboard is shown in bold typeface.

```
---------------------------------------------
PROGRAM    F S T I F F . P F      TO GENERATE
THE FINAL STIFFNESS MATRIX FOR A PLANE FRAME
---------------------------------------------
```

PLANE FRAMES

```
NUMBER OF NODES            =  3
NUMBER OF ELEMENTS         =  2
NUMBER OF RESTRAINED NODES =  2


+ + + + + + + + + + + + +
+    BOUNDARY CONDITIONS  +
+ + + + + + + + + + + + +

INPUT BOUNDARY CONDITIONS
RESTRAINT IN THE X-DIRECTION  =  1
RESTRAINT IN THE Y-DIRECTION  =  2
ROTATIONAL RESTRAINT          =  3
ENTER RESTRAINTS AS A COMPOSITE NUMBER
(E.G. IF RESTRAINED IN THE X- AND Y-DIRECTIONS ENTER 12)

NODE NUMBER   =  1
DIRECTION(S)  =  12

NODE NUMBER   =  3
DIRECTION(S)  =  12


ELEMENT  1

LOWER  NODE NUMBER  =  1
HIGHER NODE NUMBER  =  2

+ + + + + + + + + + +
+    ELEMENT  1  2   +
+ + + + + + + + + + +

AREA                       =  9000
SECOND MOMENT OF AREA      =  .25E9
LENGTH                     =  3000
ELASTIC CONSTANT           =  200
ANGLE WITH THE GLOBAL X-AXIS  =  90


ELEMENT  2

LOWER  NODE NUMBER  =  2
HIGHER NODE NUMBER  =  3

+ + + + + + + + + + +
+    ELEMENT  2  3   +
+ + + + + + + + + + +

AREA                       =  9000
SECOND MOMENT OF AREA      =  .25E9
LENGTH                     =  7071
ELASTIC CONSTANT           =  200
ANGLE WITH THE GLOBAL X-AXIS  =  8.13

* * * * * * * * * * * * * *
*   UPPER SEMI-BAND OF THE   *
*    FINAL STIFFNESS MATRIX  *
* * * * * * * * * * * * * *
```

6.66667E+07	33333.3	.0124836	3.33333E+07
271.726	35.4003	32484.8	-848.534
606.754	5939.82	5939.81	0
9.49512E+07	1.41423E+07	0	0
2.82845E+07	0	0	0

7.9 MEMBER END FORCES AND REACTIONS - Program MFORCE.PF

The element stiffness submatrices in the member axes system are used to calculate the member end forces from the member end displacements using the following equation (see section 7.2):

$$f_{ij} = k_{ii}^j \delta_{ij} + k_{ij} \delta_{ji}$$

Solution of the final stiffness equation yields the nodal displacements in terms of the global axes systems. Before the above equation can be used to find the member end forces the nodal displacements must first be transformed into the member axes system:

$$f_{ij} = k_{ii}^j T_{ij} \Delta_i + k_{ij} T_{ij} \Delta_j$$

hence

$$\begin{bmatrix} f_{ijx} \\ f_{ijy} \\ m_{ij} \end{bmatrix} = \begin{bmatrix} EA/L & 0 & 0 \\ 0 & 12EI/L^3 & 6EI/L^2 \\ 0 & 6EI/L^2 & 4EI/L \end{bmatrix} \begin{bmatrix} \cos\alpha & \sin\alpha & 0 \\ -\sin\alpha & \cos\alpha & 0 \\ 0 & 0 & 1 \end{bmatrix} \begin{bmatrix} \Delta_{ix} \\ \Delta_{iy} \\ \theta_i \end{bmatrix} +$$

$$\begin{bmatrix} -EA/L & 0 & 0 \\ 0 & -12EI/L^3 & 6EI/L^2 \\ 0 & -6EI/L^2 & 2EI/L \end{bmatrix} \begin{bmatrix} \cos\alpha & \sin\alpha & 0 \\ -\sin\alpha & \cos\alpha & 0 \\ 0 & 0 & 1 \end{bmatrix} \begin{bmatrix} \Delta_{jx} \\ \Delta_{jy} \\ \theta_j \end{bmatrix}$$

The forces at end "j" can be found in a similar manner by evaluation of the following matrix equation:

$$f_{ji} = k_{ji} T_{ij} \Delta_i + k_{jj}^i T_{ij} \Delta_j$$

Example 7.5

Find the member end forces for member 2,3 of the structure shown in the following diagram.

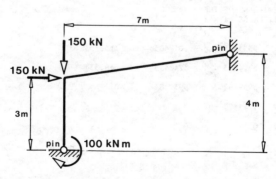

PLANE FRAMES

Note that the final stiffness matrix for this structure was found in example 7.4. Hence the final stiffness equation is

$$\begin{bmatrix} 66.7E6 & 33333 & 0 & 33.3E6 & 0 \\ 33333 & 271.7 & 35.40 & 32484 & -848.5 \\ 0 & 35.40 & 606.8 & 5940 & 5940 \\ 33.3E6 & 32484 & 5940 & 95.0E6 & 14.1E6 \\ 0 & -848.5 & 5940 & 14.1E6 & 28.3E6 \end{bmatrix} \begin{bmatrix} \theta_1 \\ \Delta_{2x} \\ \Delta_{2y} \\ \theta_2 \\ \theta_3 \end{bmatrix} = \begin{bmatrix} -100000 \\ 150 \\ -150 \\ 0 \\ 0 \end{bmatrix}$$

Solving the above equation using any of the equation solvers listed in Chapter 4 yields

$$\begin{bmatrix} \theta_1 \\ \Delta_{2x} \\ \Delta_{2y} \\ \theta_2 \\ \theta_3 \end{bmatrix} = \begin{bmatrix} -2.16 \times 10^{-3} \text{ rads} \\ 0.792 \text{ mm} \\ -0.297 \text{ mm} \\ 0.532 \times 10^{-3} \text{ rads} \\ -0.179 \times 10^{-3} \text{ rads} \end{bmatrix}$$

Member forces for member 2,3 are found using the following equation:

$$f_{23} = k_{22}^3 T_{23} \Delta_2 + k_{23} T_{23} \Delta_3$$

$$\begin{bmatrix} f_{23x} \\ f_{23y} \\ m_{23} \end{bmatrix} = \begin{bmatrix} 255 & 0 & 0 \\ 0 & 1.7 & 6000 \\ 0 & 6000 & 28.3 \times 10^6 \end{bmatrix} \begin{bmatrix} 0.990 & 0.141 & 0 \\ -0.141 & 0.990 & 0 \\ 0 & 0 & 1 \end{bmatrix} \begin{bmatrix} 0.792 \\ -0.297 \\ 0.532 \times 10^{-3} \end{bmatrix} +$$

$$\begin{bmatrix} -255 & 0 & 0 \\ 0 & -1.7 & 6000 \\ 0 & -6000 & 14.1 \times 10^6 \end{bmatrix} \begin{bmatrix} 0.990 & 0.141 & 0 \\ -0.141 & 0.990 & 0 \\ 0 & 0 & 1 \end{bmatrix} \begin{bmatrix} 0 \\ 0 \\ -0.179 \times 10^{-3} \end{bmatrix}$$

$$= \begin{bmatrix} 189 \\ 2.50 \\ 12621 \end{bmatrix} + \begin{bmatrix} 0 \\ -1.07 \\ -2524 \end{bmatrix} = \begin{bmatrix} 189 \text{ kN} \\ 1.43 \text{ kN} \\ 10097 \text{ kN mm} \end{bmatrix}$$

Similarly

$$\begin{bmatrix} f_{32x} \\ f_{32y} \\ m_{32} \end{bmatrix} = \begin{bmatrix} -189 \\ -2.05 \\ 5067 \end{bmatrix} + \begin{bmatrix} 0 \\ 1.07 \\ -5066 \end{bmatrix} = \begin{bmatrix} -189 \text{ kN} \\ -1.43 \text{ kN} \\ 1.00 \text{ kN mm} \end{bmatrix}$$

Note that at end 3, where the moment should be zero, there is a small moment of 1.0 kN mm. This error arises from the limited number of significant figures used in the calculations.

It is computationally advantageous to expand the expressions for the member end forces at end "i", i.e.

$$f_{ijx} = EA/L\,((\Delta_{ix} - \Delta_{jx})\cos\alpha + (\Delta_{iy} - \Delta_{jy})\sin\alpha)$$

$$f_{ijy} = 12EI/L^3((-\Delta_{ix} + \Delta_{jx})\sin\alpha + (\Delta_{iy} - \Delta_{jy})\cos\alpha) + 6EI/L^2(\theta_i + \theta_j)$$

$$m_{ij} = 6EI/L^2((-\Delta_{ix} + \Delta_{jx})\sin\alpha + (\Delta_{iy} - \Delta_{jy})\cos\alpha) + 2EI/L\,(2\theta_i + \theta_j)$$

By considering the equilibrium of the element it is a simple matter to show that the forces at end "j" of the element are as follows:

$$f_{jix} = -f_{ijx}$$
$$f_{jiy} = -f_{ijy}$$
$$m_{ji} = -m_{ij} + L\,f_{ijy}$$

Reactions

If node "i" is restrained then the reactive forces (in the global axes system) can be found using equation (5.23) which, although developed for plane trusses, is generally applicable.

i.e.
$$R_i = \sum T_{ij}^{-1} f_{ij} - P_i$$

$$\begin{bmatrix} R_{ix} \\ R_{iy} \\ RM_i \end{bmatrix} = \sum \begin{bmatrix} \cos\alpha & -\sin\alpha & 0 \\ \sin\alpha & \cos\alpha & 0 \\ 0 & 0 & 1 \end{bmatrix} \begin{bmatrix} f_{ijx} \\ f_{ijy} \\ m_{ij} \end{bmatrix} - \begin{bmatrix} P_{ix} \\ P_{iy} \\ M_i \end{bmatrix}$$

where RM_i is the reactive moment at node "i".

Program MFORCE.PF

This program calculates the axial force, shear force, and bending moment at the ends of a plane frame member from a knowledge of the end displacements.

Input The number of members for which the forces are to be found is input to allow the program to loop for all members. Then the member properties plus end displacements are input for each member in turn thus allowing the axial force, shear force, and bending moment at each end of the member to be calculated.

PLANE FRAMES

Comments on the Algorithm A variety of end connections are catered for by using stiffness factors K1-K8. The theory for plane frame members having pins at one or both ends is dealt with in the next chapter (see sections 8.5 and 8.7).

Output Axial force, shear force and bending moment at each end of the member are output.

Listing

```
1000 REM *********************************************************************
1010 REM *                                                                   *
1020 REM *     PROGRAM     M F O R C E . P F     TO CALCULATE THE MEMBER     *
1030 REM *     END FORCES FROM END DISPLACEMENTS FOR A PLANE FRAME MEMBER    *
1040 REM *                                                                   *
1050 REM *     J.A.D.BALFOUR                                                 *
1060 REM *                                                                   *
1070 REM *********************************************************************
1080 REM
1090 REM   ---  NOTE THAT VARIABLES ARE DEFINED IN APPENDIX A
1100 REM
1110 OPTION BASE 1
1120 PRINT : PRINT
1130 PRINT "-----------------------------------------------------------"
1140 PRINT
1150 PRINT "PROGRAM    M F O R C E . P F    TO CALCULATE THE MEMBER"
1160 PRINT "END FORCES FROM END DISPLACEMENTS FOR A PLANE FRAME MEMBER"
1170 PRINT
1180 PRINT "-----------------------------------------------------------"
1190 PRINT : PRINT
1200 INPUT "NUMBER OF MEMBERS      = ", NMEMB
1210 PRINT
1220 PRINT "ELEMENT      LOWER      HIGHER"
1230 PRINT "  TYPE       NODE        NODE"
1240 PRINT
1250 PRINT "    1        FIXED      FIXED"
1260 PRINT "    2        FIXED      PINNED"
1270 PRINT "    3        PINNED     FIXED"
1280 PRINT "    4        PINNED     PINNED"
1290 REM
1300 REM   ---  LOOP FOR ALL MEMBERS
1310 REM
1320 FOR K = 1 TO NMEMB
1330 PRINT : PRINT
1340 PRINT "ELEMENT "; K : PRINT
1350 INPUT "LOWER  NODE NUMBER     = ", IN
1360 INPUT "HIGHER NODE NUMBER     = ", JN
1370 PRINT
1380 PRINT          "+ + + + + + + + + + +"
1390 PRINT USING "+    ELEMENT ## ##   +"; IN, JN
1400 PRINT          "+ + + + + + + + + + +"
1410 PRINT
1420 INPUT "AREA                          = ", A
1430 INPUT "SECOND MOMENT OF AREA         = ", A2
1440 INPUT "LENGTH                        = ", L
1450 INPUT "ELASTIC CONSTANT              = ", E
1460 INPUT "ANGLE WITH THE GLOBAL X-AXIS  = ", ALPHA
1470 INPUT "MEMBER TYPE                   = ", MTYPE
1480 PRINT
1490 PRINT USING "X-DISPLACEMENT AT NODE ##   = "; IN;
1500 INPUT "", XI
1510 PRINT USING "Y-DISPLACEMENT AT NODE ##   = "; IN;
1520 INPUT "", YI
1530 PRINT USING "ROTATION          AT NODE ##   = "; IN;
1540 INPUT "", RI
1550 PRINT USING "X-DISPLACEMENT AT NODE ##   = "; JN;
```

```
1560 INPUT "", XJ
1570 PRINT USING "Y-DISPLACEMENT AT NODE ##    = "; JN;
1580 INPUT "", YJ
1590 PRINT USING "ROTATION        AT NODE ##    = "; JN;
1600 INPUT "", RJ
1610 REM
1620 REM  --- CONVERT THE ANGLE TO RADIANS
1630 REM
1640 AR = 4 * ATN(1) * ALPHA / 180
1650 C  = COS(AR)
1660 S  = SIN(AR)
1670 REM
1680 REM  --- CALCULATE AND OUTPUT THE MEMBER END FORCES
1690 REM
1700 GOSUB 1910
1710 FXI =      A*E* ( (XI-XJ)*C + (YI-YJ)*S )/L
1720 FYI =   K1*E*A2*( (XJ-XI)*S + (YI-YJ)*C )/L^3 + E*A2*(K2*RI + K3*RJ)/L^2
1730 FRI =           (-K2*XI - K5*XJ)*S + (K2*YI + K5*YJ)*C
1740 FRI =      E*A2*(FRI/L^2 + (K4*RI + K6*RJ)/L)
1750 FXJ = - FXI
1760 FYJ = - FYI
1770 FRJ = - FRI + L*FYI
1780 PRINT : PRINT
1790 PRINT "* * * * * * * * * * * *"
1800 PRINT "*   MEMBER END FORCES   *"
1810 PRINT "* * * * * * * * * * * *"
1820 PRINT
1830 PRINT "                 AXIAL        SHEAR        BENDING"
1840 PRINT "                 FORCE        FORCE        MOMENT"
1850 PRINT USING "NODE ##    "; IN;
1860 PRINT USING "#.####^^^^    #.####^^^^    #.####^^^^"; FXI, FYI, FRI
1870 PRINT USING "NODE ##    "; JN;
1880 PRINT USING "#.####^^^^    #.####^^^^    #.####^^^^"; FXJ, FYJ, FRJ
1890 NEXT K
1900 END
1910 REM * * * * * * * * * * * * * * * * * * * * * * * * * * * * * * * *
1920 REM *                                                              *
1930 REM *    SUBROUTINE TO SET UP ELEMENT STIFFNESS FACTORS FOR        *
1940 REM *    DIFFERENT END CONDITIONS                                  *
1950 REM *                                                              *
1960 REM * * * * * * * * * * * * * * * * * * * * * * * * * * * * * * * *
1970 REM
1980 REM
1990 REM --- ZERO STIFFNESS FACTORS (I.E. PINNED/PINNED MEMBER)
2000 REM
2010 K1 =  0 : K2 =  0 : K3 =  0 : K4 =  0
2020 K5 =  0 : K6 =  0 : K7 =  0 : K8 =  0
2030 ON MTYPE GOTO 2070, 2130, 2180, 2190
2040 REM
2050 REM --- FIXED/FIXED MEMBER
2060 REM
2070 K1 = 12 : K2 =  6 : K3 =  6 : K4 =  4
2080 K5 = -6 : K6 =  2 : K7 = -6 : K8 =  4
2090 GOTO 2190
2100 REM
2110 REM --- FIXED/PINNED MEMBER
2120 REM
2130 K1 =  3 : K2 =  3 : K4 =  3 : K5 = -3
2140 GOTO 2190
2150 REM
2160 REM --- PINNED/FIXED MEMBER
2170 REM
2180 K1 =  3 : K3 =  3 : K7 = -3 : K8 =  3
2190 RETURN
```

Sample Run In the following sample run the program MFORCE.PF is used to evaluate the member end forces for the structure analysed in example 7.5. Input from the keyboard is shown in bold typeface.

PLANE FRAMES

PROGRAM   M F O R C E . P F   TO CALCULATE THE MEMBER
END FORCES FROM END DISPLACEMENTS FOR A PLANE FRAME MEMBER

NUMBER OF MEMBERS   = 2

ELEMENT TYPE	LOWER NODE	HIGHER NODE
1	FIXED	FIXED
2	FIXED	PINNED
3	PINNED	FIXED
4	PINNED	PINNED

ELEMENT 1

LOWER NODE NUMBER = 1
HIGHER NODE NUMBER = 2

```
+ + + + + + + + + +
+   ELEMENT  1  2   +
+ + + + + + + + + +
```

AREA   = 9000
SECOND MOMENT OF AREA   = .25E9
LENGTH   = 3000
ELASTIC CONSTANT   = 200
ANGLE WITH THE GLOBAL X-AXIS   = 90
MEMBER TYPE   = 1

X-DISPLACEMENT AT NODE 1   = 0
Y-DISPLACEMENT AT NODE 1   = 0
ROTATION   AT NODE 1   = -2.16E-3
X-DISPLACEMENT AT NODE 2   = .792
Y-DISPLACEMENT AT NODE 2   = -.297
ROTATION   AT NODE 2   = .532E-3

```
* * * * * * * * * * * *
*   MEMBER END FORCES   *
* * * * * * * * * * * *
```

	AXIAL FORCE	SHEAR FORCE	BENDING MOMENT
NODE 1	0.1782E+03	-.3667E+02	-.9987E+05
NODE 2	-.1782E+03	0.3667E+02	-.1013E+05

ELEMENT 2

LOWER NODE NUMBER = 2
HIGHER NODE NUMBER = 3

```
+ + + + + + + + + +
+   ELEMENT  2  3   +
+ + + + + + + + + +
```

AREA   = 9000
SECOND MOMENT OF AREA   = .25E9
LENGTH   = 7071
ELASTIC CONSTANT   = 200
ANGLE WITH THE GLOBAL X-AXIS   = 8.13
MEMBER TYPE   = 1

```
X-DISPLACEMENT  AT  NODE   2    =   .792
Y-DISPLACEMENT  AT  NODE   2    =  -.297
ROTATION        AT  NODE   2    =   .532E-3
X-DISPLACEMENT  AT  NODE   3    =  0
Y-DISPLACEMENT  AT  NODE   3    =  0
ROTATION        AT  NODE   3    =  -.179E-3


* * * * * * * * * * * *
*   MEMBER END FORCES   *
* * * * * * * * * * * *

                AXIAL         SHEAR       BENDING
                FORCE         FORCE       MOMENT
NODE   2    0.1889E+03    0.1429E+01    0.1008E+05
NODE   3   -.1889E+03    -.1429E+01    0.2459E+02
```

8 Plane Frames - Further Topics

8.1 INTRODUCTION

The previous chapter dealt with the analysis of plane frames having rigidly jointed prismatic members, simple boundary conditions, and loads applied only at the nodes. In practice, however, plane frames are often subjected to more complex loading regimens involving member loads, temperature effects, settlement, and lack of fit. The analysis is likely to be futher complicated by inclined supports, elastic foundations, members with pins, and non-prismatic members.

This chapter shows how these effects can easily be incorporated into the stiffness method. The last section of this chapter presents a program for the automatic analysis of plane frames. This program allows loads to be applied to the members and will deal with members containing pins; thus demonstrating the practical implementation of some of the theory presented in this chapter.

8.2 MEMBER LOADS, TEMPERATURE EFFECTS, SETTLEMENT, AND LACK OF FIT - Program FIX.PF

So far, loads have been applied only at nodes. In practice, however, framed structures often carry loads along the length of their members. Such forces can be dealt with using the concepts of fixed end force, equivalent joint force, and superposition. The theory developed for member loads is easily extended to allow temperature effects, settlement, and lack of fit to be included in the analysis.

To illustrate, consider the structure shown in fig 8.1 which carries a uniformly distributed load along the length of member 2,3. Although a specific problem is considered here, the theory is presented in general terms to allow the same strategy to be adopted for member loads, temperature effects, settlement, and lack of fit.

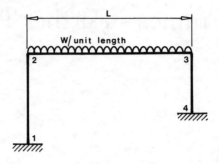

FIGURE 8.1   A PORTAL FRAME CARRYING MEMBER LOADS

The procedure is as follows.

1. Lock all joints against displacement (rotation and translation).

2. Apply member loads, temperature change, settlement, and force fit ill-fitting members.

3. Calculate the forces developed at the locked ends of the members (the *fixed end forces*).

4. Transform the fixed end forces into the nodal axes system.

5. Add the negative of the transformed fixed end forces (the *equivalent joint forces*) to the nodal loads and then analyse the structure.

6. The final member forces and displacements are found by superimposing the fixed end force system on the results of the analysis described in 5.

In the case of the portal shown in fig 8.1 the fixed end forces are as shown in fig 8.2.

It is convenient, in the first instance, to express the fixed end forces for each member in terms of the member axes system. For a member "i,j" affected by member loads, temperature change, settlement, or lack of fit, the vectors of fixed end force $\mathbf{f}^f_{ji}$ and $\mathbf{f}^f_{ij}$ at ends "i" and "j" of the member will take the form

PLANE FRAMES - FURTHER TOPICS

$$\mathbf{f}^f_{ij} = \begin{bmatrix} f^f_{ijx} \\ f^f_{ijy} \\ m^f_{ij} \end{bmatrix} \qquad \mathbf{f}^f_{ji} = \begin{bmatrix} f^f_{ijx} \\ f^f_{ijy} \\ m^f_{ji} \end{bmatrix}$$

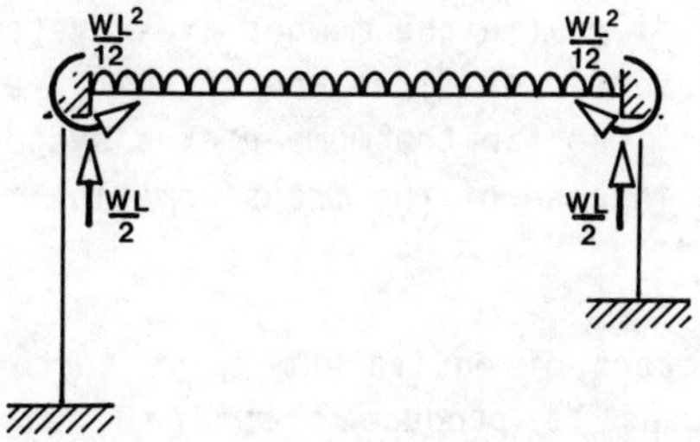

FIGURE 8.2   FIXED END FORCES

The fixed end forces are an artificial system of applied loads that serve to hold the nodal displacements at zero. To return to the true structural behaviour it is necessary to superimpose the negative of these forces (i.e. the equivalent joint forces), as illustrated in fig 8.3.

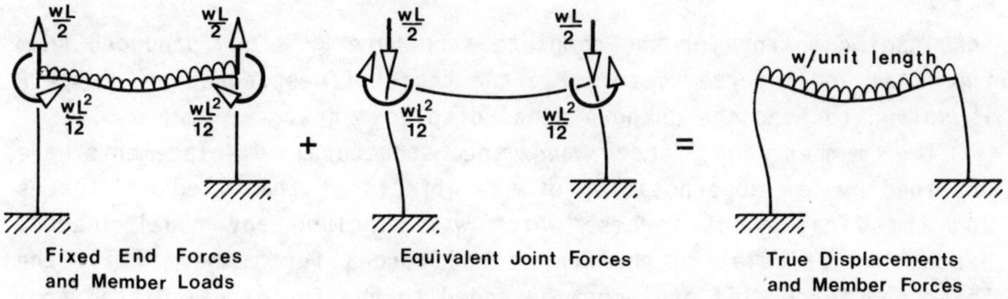

Fixed End Forces
and Member Loads

Equivalent Joint Forces

True Displacements
and Member Forces

FIGURE 8.3   SUPERPOSITION OF THE FIXED END FORCES AND THE EQUIVALENT JOINT FORCES.

Before the equivalent joint forces can be included in the equations of nodal equilibrium they must be expressed in the global axes system. Hence the vectors of equivalent joint force are obtained by transforming the negative of the fixed end force vectors into the global axes system.

i.e. $\quad \mathbf{P}^e_{ij} = -\mathbf{T}^{-1}_{ij} \mathbf{f}^f_{ij} \qquad$ and $\qquad \mathbf{P}^e_{ji} = -\mathbf{T}^{-1}_{ij} \mathbf{f}^f_{ji}$

where
- P^e_{ij} is the vector of equivalent joint forces at end "i" of member "i,j" (in the global axes system).
- P^e_{ji} is the vector of equivalent joint forces at end "j" of member "i,j" (in the global axes system).
- f^f_{ij} is the vector of fixed end forces at end "i" of member "i,j" (in the member axes system).
- f^f_{ji} is the vector of fixed end forces at end "j" of member "i,j" (in the member axes system).
- T^{-1}_{ij} is the inverse of the transformation matrix for member "i,j".

At each node the vectors of equivalent joint force are added to the vectors of nodal loads to produce the final nodal force vectors. Hence the final nodal force vector at node "i" takes the form

$$P_i = P^n_i + \sum P^e_{ij}$$

where
- P_i is the final force vector at node "i".
- P^n_i is the vector of nodal loads at node "i".
- $\sum P^e_{ij}$ is the sum of the vectors of equivalent joint force at end "i" of all of the members framing into joint "i".

The loading vector for the complete structure P is now produced from the final nodal force vectors P_i. and the stiffness equation $K \Delta = P$ is solved to find the unknown nodal displacements.

The member end forces and the structural displacements are obtained by the superposition of the effects of the fixed end forces and the final nodal forces (which will include any nodal loads). Hence when calculating the member end forces for member "i,j", the fixed end forces (if any) must be added to the forces calculated from the nodal displacements.

i.e. $f_{ij} = f^n_{ij} + f^f_{ij}$

$$f_{ij} = k^j_{ii} T_{ij} \Delta_i + k_{ij} T_{ij} \Delta_j + f^f_{ij} \qquad (8.1)$$

The solution of the structure stiffness equation using the final nodal loads yields the true nodal displacements. Away from the nodes the displacements caused by the fixed end force system (see fig 8.3) are not necessarily zero. and must be included if displacements are to be calculated away from the nodes. If, for example, the vertical displacement is required half way along the horizontal member of the portal shown in fig 8.3, then $wL^4/384EI$ (the displacement due to the fixed end force system) must be added to the displacement due to the final force system.

Member Loads

In practice the members of framed structures are often loaded along their length. If the member loads consist of point loads only, then by locating a node under each load the problem can be solved using the theory developed in Chapter 7. However, the introduction of additional nodes increases the problem size and hence the cost of the analysis. Furthermore, distributed loads cannot be satisfactorily catered for by the introduction of additional nodes, and the use of equivalent joint forces should be used whenever member loads are present. Additional nodes should only be introduced if displacement between joints is of interest.

Lack of Fit

Lack of fit is a problem associated with steel structures and usually arises from tolerance on the member sizes or inaccuracy in the position of column footings. During construction the structure is forced together causing a small change in the design geometry and an unintentional stressing of the members. Usually the lack of fit is small; the force fit being achieved during the tightening of bolted connections, and the resulting change in geometry is usually neglected for design purposes.

Temperature Effects

Whereas lack of fit normally affects only steel structures, temperature changes can affect all types of structure. In general temperature changes can cause extremely complex stress patterns. This chapter only considers the simple case where one or more of the members of a structure changes temperature uniformly relative to the rest of the structure. In such cases the temperature effects can be considered as a special case of lack of fit. When the temperature of a member changes, the member tries to change its length. If this change of length is resisted by the other members of the structure (or the restraints) then stress will be developed. To the structural analyst the problem is identical to that presented by lack of fit, only the "lack of fit" is due to temperature change, rather than structural tolerance. The "lack of fit" caused by temperature change is calculated using the coefficient of thermal expansion "c" which is, in effect, the strain induced per unit change in temperature. If an unrestrained member undergoes a change in temperature of Δt, then its length will change according to the following relationship:

$$\delta L = c L \Delta t$$

If, however, the ends of the member were fully restrained during a Δt change in temperature, then the axial force induced in the member would be

$$F = A \sigma$$
$$= A \varepsilon E = A \delta l E / L = A c \Delta t E \quad (8.2)$$

Equation (8.2) is, of course, the fixed end force for a change of temperature Δt in the member.

Finally it should be noted that statically determinate structures are not stressed by lack of fit or temperature changes of this type. For example, if member 2,4 of the structure shown in fig 8.4 is short (or cools relative to the other members) then the geometry of the structure changes, but there is no associated change of stress (if small displacement theory is assumed to be valid).

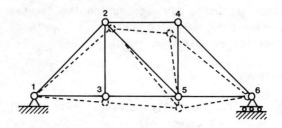

FIGURE 8.4   LACK OF FIT IN A STATICALLY DETERMINATE STRUCTURE

If the structure is made statically indeterminate by the introduction of a new member as shown in fig 8.5 then any lack of fit (or temperature change) of member 2,4 will stress the structure.

Note that the force in some of the members of the structure shown in fig 8.5 is statically determinate. Consequently lack of fit of these members (e.g. member 1,2) will not stress the structure. Conversely, any lack of fit of the members that are statically indeterminate (e.g. member 2,4) will cause stressing, but only in the statically indeterminate members of the structure.

PLANE FRAMES - FURTHER TOPICS

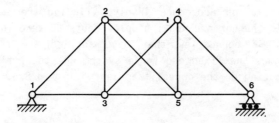

FIGURE 8.5   LACK OF FIT IN A STATICALLY INDETERMINATE STRUCTURE

Settlement

Before a civil engineering structure is designed a site investigation is usually conducted to determine the nature of the ground upon which it will be built. Often the site investigation will reveal soil conditions that indicate that the structure will continue to settle long after the construction is finished. If all of the foundations settle by the same amount then the settlement amounts to rigid body motion of the structure, and the designer has little need for concern as no stress is induced. However, because soil conditions often vary over the plan area of a structure and the footings usually carry different loadings, differential settlement frequently occurs. Calculation of settlements is a complex and inexact branch of civil engineering that is not within the scope of this book, but once the amount of settlement has been estimated the problem can, like temperature change, be regarded as a special case of lack of fit. Here the lack of fit applies not to the members, but to the structural boundaries.

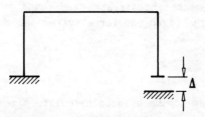

FIGURE 8.6   SETTLEMENT TO A FOOTING OF A PORTAL

Figure 8.6 shows a portal suffering from vertical settlement of the right-hand footing. The right-hand column has been severed to unstress the structure. In common with the other topics in this section the solution procedure involves finding the set of forces that will hold the unknown nodal displacements at zero (note that the displacements at boundaries that have settled are not unknown). Figure 8.7 shows the fixed end force system due to settlement of the right-hand footing of the portal frame shown in fig 8.6.

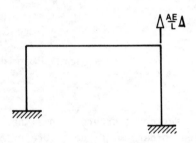

FIGURE 8.7   FIXED END FORCES

The structure is then analysed for the negative of the fixed end forces (the equivalent joint forces) as shown in fig 8.8.

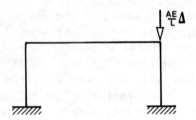

FIGURE 8.8   EQUIVALENT JOINT FORCES

The final displacements and stresses are found by the superposition of the effects of the fixed end force system and the equivalent joint force system.

Fixed End Forces

A variety of methods can be used to calculate fixed end forces, and a selection of the more popular methods is presented in most structural analysis textbooks. For convenience fixed end moments for some common loading cases are given in fig 8.9.

PLANE FRAMES - FURTHER TOPICS

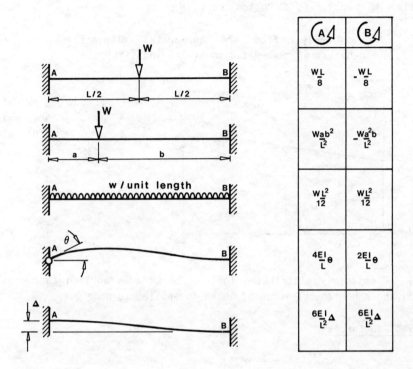

FIGURE 8.9   FIXED END MOMENTS FOR SOME COMMON LOADING CASES

Example 8.1

Find the member end forces for member 1,2 of the structure shown below if

(i) A uniformly distributed load of 10 kN/m is applied to member 2,3, and a horizontal force of 50 kN (to the right) is applied to node 2.

(ii) Member 1,2 undergoes a 20 degree centigrade rise in temperature. The structure is constructed from a material having a coefficient of thermal expansion of 1.1×10^{-5} per degree centigrade.

(iii) The left-hand support settles by 15 mm vertically downwards and 10 mm horizontally (to the right).

(iv) Member 2,3 is 20 mm too short.

(v) And finally, find the horizontal displacement at mid-height of member 1,2 under loading (iii).

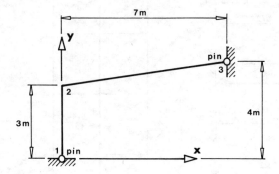

$E = 200 \text{ kN/mm}^2$
$A = 9000 \text{ mm}^2$
$I = 0.25 \times 10^9 \text{ mm}^4$

(i) A uniformly distributed load of 10 kN/m is applied to member 2,3, and a horizontal force of 50 kN is applied to node 2.

Fixed end forces

$wL/2 = (10 \times \sqrt{50})/2 = 35.4$ kN
$wL^2/12 = (10 \times 50)/12 = 41.7$ kN m

$$f_{23}^f = \begin{bmatrix} 0.0 \text{ kN} \\ 35.4 \text{ kN} \\ 41.7 \text{ kN m} \end{bmatrix} \qquad f_{32}^f = \begin{bmatrix} 0.0 \text{ kN} \\ 35.4 \text{ kN} \\ -41.7 \text{ kN m} \end{bmatrix}$$

Transform the fixed end force vectors into equivalent joint force vectors ($\alpha_{23} = 8.13°$).

$$P_{23}^e = - T_{23}^{-1} f_{23}^f$$

$$= - \begin{bmatrix} 0.990 & -0.141 & 0 \\ 0.141 & 0.990 & 0 \\ 0 & 0 & 1 \end{bmatrix} \begin{bmatrix} 0 \\ 35.4 \\ 41.7 \end{bmatrix} = \begin{bmatrix} 4.99 \\ -35.0 \\ -41.7 \end{bmatrix}$$

PLANE FRAMES - FURTHER TOPICS

Similarly

$$P^e_{32} = -T^{-1}_{32} f^f_{32} = \begin{bmatrix} 4.99 \\ -35.0 \\ 41.7 \end{bmatrix}$$

Calculate the vectors of final nodal force.

$$P_1 = P^n_1 + P^e_{12} = \begin{bmatrix} 0 \\ 0 \\ 0 \end{bmatrix} + \begin{bmatrix} 0 \\ 0 \\ 0 \end{bmatrix} = \begin{bmatrix} 0 \\ 0 \\ 0 \end{bmatrix}$$

$$P_2 = P^n_2 + P^e_{21} + P^e_{23} = \begin{bmatrix} 50 \\ 0 \\ 0 \end{bmatrix} + \begin{bmatrix} 0 \\ 0 \\ 0 \end{bmatrix} + \begin{bmatrix} 4.99 \\ -35.0 \\ -41.7 \end{bmatrix} = \begin{bmatrix} 55.0 \\ -35.0 \\ -41.7 \end{bmatrix}$$

$$P_3 = P^n_3 + P^e_{32} = \begin{bmatrix} 0 \\ 0 \\ 0 \end{bmatrix} + \begin{bmatrix} 4.99 \\ -35.0 \\ 41.7 \end{bmatrix} = \begin{bmatrix} 5.0 \\ -35.0 \\ 41.7 \end{bmatrix}$$

Discarding the forces associated with the boundary conditions (i.e. P_{1x}, P_{1y}, P_{3x}, and P_{3y}) and employing the structure stiffness found previously (this structure is identical to the one analysed in example 7.3) produces the final stiffness equation. Note that the units are kN and mm.

$$\begin{bmatrix} 66.7E6 & 33333 & 0 & 33.3E6 & 0 \\ 33333 & 271.7 & 35.40 & 32484 & -848.5 \\ 0 & 35.40 & 606.8 & 5940 & 5940 \\ 33.3E6 & 32484 & 5940 & 95.0E6 & 14.1E6 \\ 0 & -848.5 & 5940 & 14.1E6 & 28.3E6 \end{bmatrix} \begin{bmatrix} \theta_1 \\ \Delta_{2x} \\ \Delta_{2y} \\ \theta_2 \\ \theta_3 \end{bmatrix} = \begin{bmatrix} 0.0 \\ 55.0 \\ -35.1 \\ -41700 \\ 41700 \end{bmatrix}$$

which yields

$$\begin{bmatrix} \theta_1 \\ \Delta_{2x} \\ \Delta_{2y} \\ \theta_2 \\ \theta_3 \end{bmatrix} = \begin{bmatrix} 3.22 \times 10^{-4} \text{ rads} \\ 0.292 \text{ mm} \\ -0.085 \text{ mm} \\ -9.39 \times 10^{-4} \text{ rads} \\ 1.97 \times 10^{-3} \text{ rads} \end{bmatrix}$$

COMPUTER ANALYSIS OF STRUCTURAL FRAMEWORKS

The member end forces for element 1,2 are found by adding the forces associated with the above displacements to the fixed end forces:

$$f_{12} = f^n_{12} + f^f_{12}$$

$$= k^2_{11} T_{12} \Delta_1 + k_{12} T_{12} \Delta_2 + f^f_{12}$$

$$\begin{bmatrix} f_{12x} \\ f_{12y} \\ m_{12} \end{bmatrix} = \begin{bmatrix} EA/L & 0 & 0 \\ 0 & 12EI/L^3 & 6EI/L^2 \\ 0 & 6EI/L^2 & 4EI/L \end{bmatrix} \begin{bmatrix} \cos\alpha & \sin\alpha & 0 \\ -\sin\alpha & \cos\alpha & 0 \\ 0 & 0 & 1 \end{bmatrix} \begin{bmatrix} \Delta_{1x} \\ \Delta_{1y} \\ \Theta_1 \end{bmatrix} +$$

$$\begin{bmatrix} -EA/L & 0 & 0 \\ 0 & -12EI/L^3 & 6EI/L^2 \\ 0 & -6EI/L^2 & 2EI/L \end{bmatrix} \begin{bmatrix} \cos\alpha & \sin\alpha & 0 \\ -\sin\alpha & \cos\alpha & 0 \\ 0 & 0 & 1 \end{bmatrix} \begin{bmatrix} \Delta_{2x} \\ \Delta_{2y} \\ \Theta_2 \end{bmatrix} +$$

$$\begin{bmatrix} f^f_{12x} \\ f^f_{12y} \\ m^f_{12} \end{bmatrix}$$

$$= \begin{bmatrix} 600 & 0 & 0 \\ 0 & 22.2 & 33333 \\ 0 & 33333 & 66.7E6 \end{bmatrix} \begin{bmatrix} 0 & 1 & 0 \\ -1 & 0 & 0 \\ 0 & 0 & 1 \end{bmatrix} \begin{bmatrix} 0 \\ 0 \\ 3.22E-4 \end{bmatrix} +$$

$$\begin{bmatrix} -600 & 0 & 0 \\ 0 & -22.2 & 33333 \\ 0 & -33333 & 33.3E6 \end{bmatrix} \begin{bmatrix} 0 & 1 & 0 \\ -1 & 0 & 0 \\ 0 & 0 & 1 \end{bmatrix} \begin{bmatrix} 0.292 \\ -0.085 \\ -9.39E-4 \end{bmatrix} +$$

$$\begin{bmatrix} 0 \\ 0 \\ 0 \end{bmatrix}$$

$$= \begin{bmatrix} 0.0 \\ 10.7 \\ 21480 \end{bmatrix} + \begin{bmatrix} 51.0 \\ -24.8 \\ -21540 \end{bmatrix} + \begin{bmatrix} 0 \\ 0 \\ 0 \end{bmatrix}$$

$$= \begin{bmatrix} 51.0 \text{ kN} \\ -14.1 \text{ kN} \\ -60.0 \text{ kN mm} \end{bmatrix}$$

Similarly

$$f_{21} = f^n_{21} + f^f_{21}$$

PLANE FRAMES - FURTHER TOPICS

$$= k_{21} T_{12} \Delta_1 + k_{22}^1 T_{12} \Delta_2 + f_{21}^f$$

$$= \begin{bmatrix} 0.0 \\ -10.7 \\ 10720 \end{bmatrix} + \begin{bmatrix} -51.0 \\ 24.8 \\ -52900 \end{bmatrix} + \begin{bmatrix} 0 \\ 0 \\ 0 \end{bmatrix}$$

$$= \begin{bmatrix} -51.0 \text{ kN} \\ 14.1 \text{ kN} \\ -42180 \text{ kN mm} \end{bmatrix}$$

(ii) Member 1,2 undergoes a 20°C rise in temperature.

<u>Fixed end forces</u>

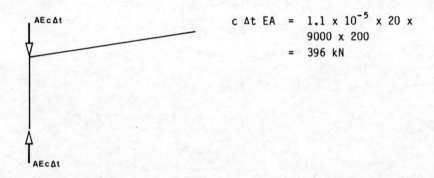

$$c \Delta t \, EA = 1.1 \times 10^{-5} \times 20 \times 9000 \times 200$$
$$= 396 \text{ kN}$$

Hence

$$f_{12}^f = \begin{bmatrix} 396 \\ 0 \\ 0 \end{bmatrix} \quad \text{and} \quad f_{21}^f = \begin{bmatrix} -396 \\ 0 \\ 0 \end{bmatrix}$$

Transform the fixed force vectors into equivalent joint force.

$$P_{12}^e = -T_{12}^{-1} f_{12}^f$$

$$= \begin{bmatrix} -0 & -1 & 0 \\ 1 & 0 & 0 \\ 0 & 0 & 1 \end{bmatrix} \begin{bmatrix} 396 \\ 0 \\ 0 \end{bmatrix} = \begin{bmatrix} 0 \\ -396 \\ 0 \end{bmatrix}$$

Similarly

$$P_{21}^e = \begin{bmatrix} 0 \\ 396 \\ 0 \end{bmatrix}$$

After calculating the nodal force vectors and applying the boundary conditions, the final stiffness equation is

$$\begin{bmatrix} 66.7E6 & 33333 & 0 & 33.3E6 & 0 \\ 33333 & 271.7 & 35.40 & 32484 & -848.5 \\ 0 & 35.40 & 606.8 & 5940 & 5940 \\ 33.3E6 & 32484 & 5940 & 95.0E6 & 14.1E6 \\ 0 & -848.5 & 5940 & 14.1E6 & 28.3E6 \end{bmatrix} \begin{bmatrix} \theta_1 \\ \Delta_{2x} \\ \Delta_{2y} \\ \theta_2 \\ \theta_3 \end{bmatrix} = \begin{bmatrix} 0 \\ 0 \\ 396 \\ 0 \\ 0 \end{bmatrix}$$

which yields

$$\begin{bmatrix} \theta_1 \\ \Delta_{2x} \\ \Delta_{2y} \\ \theta_2 \\ \theta_3 \end{bmatrix} = \begin{bmatrix} 0.490 \times 10^{-4} \text{ rads} \\ -0.0916 \text{ mm} \\ 0.659 \text{ mm} \\ -6.67 \times 10^{-6} \text{ rads} \\ -1.38 \times 10^{-4} \text{ rads} \end{bmatrix}$$

and the member end forces for element 1,2 are given by

$$f_{12} = f_{12}^n + f_{12}^f$$

$$= k_{11}^2 T_{12} \Delta_1 + k_{12} T_{12} \Delta_2 + f_{12}^f$$

$$= \begin{bmatrix} 0.0 \\ 1.633 \\ 3268 \end{bmatrix} + \begin{bmatrix} -395.4 \\ -2.256 \\ -3275 \end{bmatrix} + \begin{bmatrix} 396 \\ 0 \\ 0 \end{bmatrix}$$

$$= \begin{bmatrix} 0.6 \text{ kN} \\ -0.623 \text{ kN} \\ -7 \text{ kN mm} \end{bmatrix}$$

Similarly

$$f_{21} = f_{21}^n + f_{21}^f$$

$$= k_{21} T_{12} \Delta_1 + k_{22}^1 T_{12} \Delta_2 + f_{21}^f$$

$$= \begin{bmatrix} 0.0 \\ -1.633 \\ 1632 \end{bmatrix} + \begin{bmatrix} 395.4 \\ 2.256 \\ -3498 \end{bmatrix} + \begin{bmatrix} -396 \\ 0 \\ 0 \end{bmatrix}$$

$$= \begin{bmatrix} -0.6 \text{ kN} \\ 0.623 \text{ kN} \\ -1866 \text{ kN mm} \end{bmatrix}$$

PLANE FRAMES - FURTHER TOPICS

(iii) Support 1 settles 15 mm vertically (downwards) and 10 mm horizontally (to the right).

Following exactly the same procedures as were used in parts (i) and (ii) the fixed end force vectors can be shown to be

$$\mathbf{f}^f_{12} = \begin{bmatrix} -9000 \text{ kN} \\ -222 \text{ kN} \\ -333000 \text{ kN mm} \end{bmatrix} \qquad \mathbf{f}^f_{21} = \begin{bmatrix} 9000 \text{ kN} \\ 222 \text{ kN} \\ -333000 \text{ kN mm} \end{bmatrix}$$

Fixed end forces

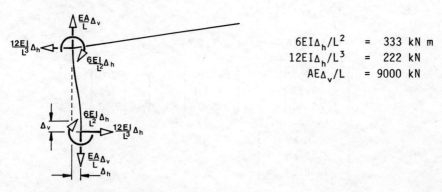

$6EI\Delta_h/L^2 = 333$ kN m
$12EI\Delta_h/L^3 = 222$ kN
$AE\Delta_v/L = 9000$ kN

After calculating the final force vectors and applying the boundary conditions the final stiffness equation is

$$\begin{bmatrix} 66.7E6 & 33333 & 0 & 33.3E6 & 0 \\ 33333 & 271.7 & 35.40 & 32484 & -848.5 \\ 0 & 35.40 & 606.8 & 5940 & 5940 \\ 33.3E6 & 32484 & 5940 & 95.0E6 & 14.1E6 \\ 0 & -848.5 & 5940 & 14.1E6 & 28.3E6 \end{bmatrix} \begin{bmatrix} \theta_1 \\ \Delta_{2x} \\ \Delta_{2y} \\ \theta_2 \\ \theta_3 \end{bmatrix} = \begin{bmatrix} 333000 \\ 222 \\ -9000 \\ 333000 \\ 0 \end{bmatrix}$$

which yields

$$\begin{bmatrix} \theta_1 \\ \Delta_{2x} \\ \Delta_{2y} \\ \theta_2 \\ \theta_3 \end{bmatrix} = \begin{bmatrix} 2.68 \times 10^{-3} \text{ rads} \\ 2.15 \text{ mm} \\ -15.0 \text{ mm} \\ 2.47 \times 10^{-3} \text{ rads} \\ 1.98 \times 10^{-3} \text{ rads} \end{bmatrix}$$

and the member forces for element 1,2 are given by

$$\mathbf{f}_{12} = \mathbf{f}^n_{12} + \mathbf{f}^f_{12}$$

$$= k_{11}^2 T_{12} \Delta_1 + k_{12} T_{12} \Delta_2 + f_{12}^f$$

$$= \begin{bmatrix} 0.0 \\ 89.33 \\ 178800 \end{bmatrix} + \begin{bmatrix} 9000 \\ 130.1 \\ 153900 \end{bmatrix} + \begin{bmatrix} -9000 \\ -222 \\ -333000 \end{bmatrix}$$

$$= \begin{bmatrix} 0 \text{ kN} \\ -2.57 \text{ kN} \\ -300 \text{ kN mm} \end{bmatrix}$$

Similarly

$$f_{21} = f_{21}^n + f_{21}^f$$

$$= k_{21} T_{12} \Delta_1 + k_{22}^1 T_{12} \Delta_2 + f_{21}^f$$

$$= \begin{bmatrix} 0.0 \\ -89.33 \\ 89240 \end{bmatrix} + \begin{bmatrix} -9000 \\ -130.1 \\ 236400 \end{bmatrix} + \begin{bmatrix} 9000 \\ 222 \\ -333000 \end{bmatrix}$$

$$= \begin{bmatrix} 0 \text{ kN} \\ 2.57 \text{ kN} \\ -7360 \text{ kN mm} \end{bmatrix}$$

Note that the member end forces due to nodal displacements and the fixed end forces both contain large numbers of approximately equal magnitude. In these conditions more significant figures could be usefully employed in the calculations.

(iv) Member 2,3 is 20 mm too short.

Fixed end forces

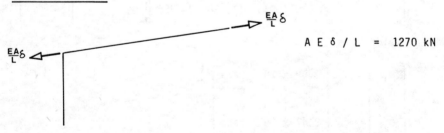

$A E \delta / L = 1270$ kN

From here the solution follows exactly the same procedures as were used in (i), (ii) and (iii), yielding the following equivalent joint force and solution vectors:

PLANE FRAMES - FURTHER TOPICS

$$P^e_{23} = \begin{bmatrix} 1260 \text{ kN} \\ 180 \text{ kN} \\ 0 \text{ kN mm} \end{bmatrix} \quad P^e_{32} = \begin{bmatrix} -1260 \text{ kN} \\ -180 \text{ kN} \\ 0 \text{ kN mm} \end{bmatrix}$$

$$\begin{bmatrix} \theta_1 \\ \Delta_{2x} \\ \Delta_{2y} \\ \theta_2 \\ \theta_3 \end{bmatrix} = \begin{bmatrix} -0.193 \times 10^{-2} \text{ rads} \\ 5.013 \text{ mm} \\ 0.837 \times 10^{-2} \text{ mm} \\ -0.114 \times 10^{-2} \text{ rads} \\ 0.718 \times 10^{-3} \text{ rads} \end{bmatrix}$$

Finding the member end forces as before produces

$$f_{12} = f^n_{12} + f^f_{12}$$

$$= k^2_{11} T_{12} \Delta_1 + k_{12} T_{12} \Delta_2 + f^f_{12}$$

$$= \begin{bmatrix} 0.0 \\ -64.33 \\ -128700 \end{bmatrix} + \begin{bmatrix} -5.022 \\ 73.29 \\ 129100 \end{bmatrix} + \begin{bmatrix} 0 \\ 0 \\ 0 \end{bmatrix}$$

$$= \begin{bmatrix} -5.02 \text{ kN} \\ 8.96 \text{ kN} \\ 400 \text{ kN mm} \end{bmatrix}$$

Similarly

$$f_{21} = f^n_{21} + f^f_{21}$$

$$= k_{21} T_{12} \Delta_1 + k^1_{22} T_{12} \Delta_2 + f^f_{21}$$

$$= \begin{bmatrix} 0.0 \\ 64.33 \\ -64270 \end{bmatrix} + \begin{bmatrix} 5.022 \\ -73.29 \\ 91060 \end{bmatrix} + \begin{bmatrix} 0 \\ 0 \\ 0 \end{bmatrix}$$

$$= \begin{bmatrix} 5.02 \text{ kN} \\ -8.96 \text{ kN} \\ 26790 \text{ kN mm} \end{bmatrix}$$

(v) The displacement at mid-height of member 1,2 under loading (iii) can be found as follows.

The deformed shape of a prismatic beam loaded only at its ends is a cubic polynomial (see section 2.11):

$$y = ax^3 + bx^2 + cx + d$$

COMPUTER ANALYSIS OF STRUCTURAL FRAMEWORKS

The boundary conditions are (from the solution vector found in (iii)):

at   x = 0   y   =   0
                          dy/dx = 0.00268
at   x = 3000   y   =   -2.15
                          dy/dx = 0.00247

Substitution into the expressions for y and dy/dx yields

        a = 0.729E-9   b = -3.32E-6
        c = 2.680E-3   d = 0.0

Hence
    $y = 0.729E{-9}\ x^3 - 3.32E{-6}\ x^2 + 2.680E{-3}\ x$

consequently, y = -0.99 mm when x = 1500 mm.

Note that this is the displacement at mid-height of element 1,2 in terms of the shape function axes under the action of the equivalent joint forces. As the shape function "y" axis lies at 180 degrees to the global "x" axis, the above displacement represents a horizontal displacement of 0.99 mm to the right. The final displacement is found by adding the displacement from the fixed end forces (i.e. half of the horizontal displacement at node 1) to the above displacement.

Final displacement   =   0.99 + 10.0/2 = 5.99 mm

Program FIX.PF

This program calculates the fixed end forces and the equivalent joint forces for a number of common loading conditions on plane frame members with a variety of end connections.

Input The number of loading conditions for which fixed end forces and equivalent joint forces are to be found is input to allow the program to loop for all load cases. Then the loading condition plus details of the element are input, thus allowing the fixed end forces and equivalent joint forces to be calculated.

Comments on the Algorithm The fixed end forces are first evaluated (in the member axes system). The equivalent joint forces are then found by transforming the fixed end forces into the global axes system. Note that members with pins are dealt with in sections 8.5 and 8.7.

PLANE FRAMES - FURTHER TOPICS

Output The fixed end forces (in the member axes system) and the equivalent joint forces (in the global axes system) are output.

Listing

```
1000 REM  ***********************************************************************
1010 REM  *                                                                     *
1020 REM  *     PROGRAM     F I X . P F     TO CALCULATE THE FIXED END          *
1030 REM  *     AND THE EQUIVALENT JOINT FORCES FOR A PLANE FRAME MEMBER        *
1040 REM  *                                                                     *
1050 REM  *            J.A.D.BALFOUR                                            *
1060 REM  *                                                                     *
1070 REM  ***********************************************************************
1080 REM
1090 REM  --- NOTE THAT VARIABLES ARE DEFINED IN APPENDIX A
1100 REM
1110 OPTION BASE 1
1120 PRINT : PRINT
1130 PRINT "----------------------------------------------------"
1140 PRINT
1150 PRINT "PROGRAM      F I X . P F     TO CALCULATE THE FIXED END"
1160 PRINT "AND EQUIVALENT JOINT FORCES FOR A PLANE FRAME MEMBER"
1170 PRINT
1180 PRINT "----------------------------------------------------"
1190 PRINT : PRINT
1200 INPUT "NUMBER OF LOADINGS TO BE CONSIDERED  =  ", NLOAD
1210 PRINT
1220 PRINT "ELEMENT        LOWER           HIGHER"
1230 PRINT " TYPE          NODE            NODE"
1240 PRINT
1250 PRINT "   1          FIXED           FIXED"
1260 PRINT "   2          FIXED           PINNED"
1270 PRINT "   3          PINNED          FIXED"
1280 PRINT "   4          PINNED          PINNED"
1290 FOR K = 1 TO NLOAD
1300 REM
1310 REM  --- INPUT THE ELEMENT DATA
1320 REM
1330 GOSUB 1440
1340 REM
1350 REM  --- CALCULATE THE FIXED END FORCES AND EQUIVALENT JOINT FORCES
1360 REM
1370 GOSUB 1940
1380 REM
1390 REM  --- OUTPUT RESULTS
1400 REM
1410 GOSUB 3070
1420 NEXT K
1430 END
1440 REM  * * * * * * * * * * * * * * * * * * * * * * * * * * * * * * * *
1450 REM  *                                                             *
1460 REM  *      SUBROUTINE TO INPUT THE ELEMENT DATA                   *
1470 REM  *                                                             *
1480 REM  * * * * * * * * * * * * * * * * * * * * * * * * * * * * * * * *
1490 REM
1500 PRINT : PRINT
1510 PRINT "LOAD CASE"; K : PRINT
1520 INPUT "LOWER  NODE NUMBER OF ELEMENT  =  ", IN
1530 INPUT "HIGHER NODE NUMBER OF ELEMENT  =  ", JN
1540 PRINT
1550 PRINT           "+ + + + + + + + + + +"
1560 PRINT USING "+    ELEMENT ## ##   +"; IN, JN
1570 PRINT           "+ + + + + + + + + + +"
1580 PRINT
1590 PRINT "SELECT LOADING TYPE"
1600 PRINT "(1) UNIFORMLY DISTRIBUTED"
1610 PRINT "(2) POINT LOAD"
1620 PRINT "(3) LACK OF FIT"
1630 PRINT "(4) THERMAL EXPANSION"
1640 PRINT "(5) SETTLEMENT"
```

```
1650 INPUT LTYPE
1660 REM
1670 REM  ---  INPUT THE NECESSARY MEMBER DATA
1680 REM
1690 IF LTYPE > 2 THEN GOTO 1730
1700 PRINT
1710 PRINT "NOTE THAT A POSITIVE MEMBER LOAD ACTS"
1720 PRINT "IN THE DIRECTION OF THE MEMBER Y-AXIS"
1730 PRINT
1740 IF LTYPE = 1 THEN GOTO 1810
1750 IF LTYPE = 2 THEN GOTO 1810
1760 INPUT "AREA                              = ", A
1770 IF LTYPE = 3 THEN GOTO 1810
1780 IF LTYPE = 4 THEN L = 1
1790 IF LTYPE = 4 THEN GOTO 1840
1800 INPUT "SECOND MOMENT OF AREA             = ", A2
1810 INPUT "LENGTH                            = ", L
1820 IF LTYPE = 1 THEN GOTO 1850
1830 IF LTYPE = 2 THEN GOTO 1850
1840 INPUT "ELASTIC CONSTANT                  = ", E
1850 INPUT "ANGLE WITH THE GLOBAL X-AXIS      = ", ALPHA
1860 INPUT "ELEMENT TYPE                      = ", MTYPE
1870 REM
1880 REM  ---  CONVERT THE ANGLE TO RADIANS
1890 REM
1900 AR = 4 * ATN(1) * ALPHA / 180
1910 C  = COS(AR)
1920 S  = SIN(AR)
1930 RETURN
1940 REM * * * * * * * * * * * * * * * * * * * * * * * * * * * * * * * *
1950 REM *                                                              *
1960 REM *    SUBROUTINE TO CALCULATE THE FIXED END FORCES AND THE      *
1970 REM *    EQUIVALENT JOINT FORCES                                   *
1980 REM *                                                              *
1990 REM * * * * * * * * * * * * * * * * * * * * * * * * * * * * * * * *
2000 REM
2010 ON LTYPE GOTO 2050, 2160, 2300, 2410, 2530
2020 REM
2030 REM  ---  UNIFORMLY DISTRIBUTED LOAD
2040 REM
2050 INPUT "LOAD INTENSITY                    = ", W
2060 FEF(1) =    0
2070 FEF(2) = - W * L / 2
2080 FEF(3) = - W * L^2 / 12
2090 FEF(4) =    0
2100 FEF(5) =    FEF(2)
2110 FEF(6) = - FEF(3)
2120 GOTO 2830
2130 REM
2140 REM  ---  POINT LOAD
2150 REM
2160 PRINT "DISTANCE - LOAD TO NODE"; IN; "   = ";
2170 INPUT "", X
2180 INPUT "MAGNITUDE OF THE LOAD             = ", P
2190 Y      =   L - X
2200 FEF(1) =   0
2210 FEF(2) = - P * (Y/L)^2 * (1 + 2*X/L)
2220 FEF(3) = - P * X * Y^2 / L^2
2230 FEF(4) =   0
2240 FEF(5) = - P * (X/L)^2 * (1 + 2*Y/L)
2250 FEF(6) =   P * Y * X^2 / L^2
2260 GOTO 2830
2270 REM
2280 REM  ---  LACK OF FIT
2290 REM
2300 INPUT "LACK OF FIT (SHORT IS + VE)       = ", X
2310 FEF(1) = - X * A * E / L
2320 FEF(2) =   0
2330 FEF(3) =   0
2340 FEF(4) = - FEF(1)
2350 FEF(5) =   0
2360 FEF(6) =   0
2370 GOTO 2830
```

```
2380 REM
2390 REM    ---  THERMAL EXPANSION
2400 REM
2410 INPUT "COEFF. OF THERMAL EXPANSION    = ", CO
2420 INPUT "TEMPERATURE RISE               = ", X
2430 FEF(1) =    A * CO * X * E
2440 FEF(2) =    0
2450 FEF(3) =    0
2460 FEF(4) = -  FEF(1)
2470 FEF(5) =    0
2480 FEF(6) =    0
2490 GOTO 2830
2500 REM
2510 REM    ---  SETTLEMENT
2520 REM
2530 PRINT "DOES SETTLEMENT TAKE PLACE AT"
2540 PRINT "    (1) NODE", IN
2550 PRINT "    (2) NODE", JN
2560 INPUT NS
2570 PRINT
2580 INPUT "SETTLEMENT GLOBAL X-DIRECTION  = ", X
2590 INPUT "SETTLEMENT GLOBAL Y-DIRECTION  = ", Y
2600 INPUT "ROTATION (DEGREES)             = ", R
2610 REM
2620 REM    ---  CONVERT SETTLEMENTS TO LOCAL AXES THEN CALC F.E.F.S
2630 REM
2640 R       =    4 * ATN(1) * R / 180
2650 CC      =    COS(AR)
2660 SS      =    SIN(AR)
2670 IF NS   = 1 THEN GOTO 2700
2680 CC      =    COS(AR + 4*ATN(1))
2690 SS      =    SIN(AR + 4*ATN(1))
2700 FEF(1) =         E * A  * ( X*CC + Y*SS) / L
2710 FEF(2) =    12 * E * A2 * (-X*SS + Y*CC) / L^3  +  6 * E * A2 * R / L^2
2720 FEF(3) =     6 * E * A2 * (-X*SS + Y*CC) / L^2  +  4 * E * A2 * R / L
2730 FEF(4) = -  FEF(1)
2740 FEF(5) = -  FEF(2)
2750 FEF(6) = -  FEF(3) + L*FEF(2)
2760 IF NS   = 1 THEN GOTO 2830
2770 X       =    FEF(3)
2780 FEF(3) = FEF(6)
2790 FEF(6) = X
2800 REM
2810 REM    ---  TAKE ACCOUNT OF ANY PINS
2820 REM
2830 ON MTYPE GOTO 3000, 2890, 2840, 2840
2840 FEF(2) = FEF(2) - 1.5*FEF(3)/L
2850 FEF(5) = FEF(5) + 1.5*FEF(3)/L
2860 FEF(6) = FEF(6) - FEF(3)/2
2870 FEF(3) = 0
2880 IF MTYPE = 3 THEN GOTO 3000 ELSE GOTO 2940
2890 FEF(2) = FEF(2) - 1.5*FEF(6)/L
2900 FEF(3) = FEF(3) - FEF(6)/2
2910 FEF(5) = FEF(5) + 1.5*FEF(6)/L
2920 FEF(6) = 0
2930 GOTO 3000
2940 FEF(2) = FEF(2) - FEF(6)/L
2950 FEF(5) = FEF(5) + FEF(6)/L
2960 FEF(6) = 0
2970 REM
2980 REM    ---  CONVERT FIXED END FORCES TO EQUIVALENT JOINT FORCES
2990 REM
3000 EJF(1) = -  (FEF(1)*C - FEF(2)*S)
3010 EJF(2) = -  (FEF(1)*S + FEF(2)*C)
3020 EJF(3) = -   FEF(3)
3030 EJF(4) = -  (FEF(4)*C - FEF(5)*S)
3040 EJF(5) = -  (FEF(4)*S + FEF(5)*C)
3050 EJF(6) = -   FEF(6)
3060 RETURN
```

```
3070 REM ****************************************
3080 REM *                                      *
3090 REM *   SUBROUTINE TO OUTPUT THE FIXED END FORCES AND THE  *
3100 REM *       EQUIVALENT JOINT FORCES                        *
3110 REM *                                      *
3120 REM ****************************************
3130 REM
3140 PRINT : PRINT
3150 PRINT "* * * * * * * * * * * * * *"
3160 PRINT "*     FIXED END FORCES       *"
3170 PRINT "*    (MEMBER AXES SYSTEM)    *"
3180 PRINT "* * * * * * * * * * * * * *"
3190 PRINT
3200 PRINT USING "              NODE ##         NODE ##"; IN, JN
3210 IF MTYPE = 1 THEN PRINT "        (FIXED)          (FIXED)"
3220 IF MTYPE = 2 THEN PRINT "        (FIXED)          (PINNED)"
3230 IF MTYPE = 3 THEN PRINT "        (PINNED)         (FIXED)"
3240 IF MTYPE = 4 THEN PRINT "        (PINNED)         (PINNED)"
3250 PRINT USING "X-FORCE    #.####^^^^    #.####^^^^"; FEF(1), FEF(4)
3260 PRINT USING "Y-FORCE    #.####^^^^    #.####^^^^"; FEF(2), FEF(5)
3270 PRINT USING "MOMENT     #.####^^^^    #.####^^^^"; FEF(3), FEF(6)
3280 PRINT
3290 PRINT "* * * * * * * * * * * * * *"
3300 PRINT "*    EQUIVALENT JOINT FORCES  *"
3310 PRINT "*     (GLOBAL AXES SYSTEM)    *"
3320 PRINT "* * * * * * * * * * * * * *"
3330 PRINT
3340 PRINT USING "              NODE ##         NODE ##"; IN, JN
3350 IF MTYPE = 1 THEN PRINT "        (FIXED)          (FIXED)"
3360 IF MTYPE = 2 THEN PRINT "        (FIXED)          (PINNED)"
3370 IF MTYPE = 3 THEN PRINT "        (PINNED)         (FIXED)"
3380 IF MTYPE = 4 THEN PRINT "        (PINNED)         (PINNED)"
3390 PRINT USING "X-FORCE    #.####^^^^    #.####^^^^"; EJF(1), EJF(4)
3400 PRINT USING "Y-FORCE    #.####^^^^    #.####^^^^"; EJF(2), EJF(5)
3410 PRINT USING "MOMENT     #.####^^^^    #.####^^^^"; EJF(3), EJF(6)
3420 RETURN
```

Sample Run The following sample shows program FIX.PF being used to evaluate the fixed end forces and the equivalent joint forces for example 8.1. Input from the keyboard is shown in bold typeface.

```
----------------------------------------------------
PROGRAM    F I X . P F    TO CALCULATE THE FIXED END
AND EQUIVALENT JOINT FORCES FOR A PLANE FRAME MEMBER
----------------------------------------------------

NUMBER OF LOADINGS TO BE CONSIDERED  =  4

ELEMENT     LOWER       HIGHER
 TYPE       NODE        NODE

   1        FIXED       FIXED
   2        FIXED       PINNED
   3        PINNED      FIXED
   4        PINNED      PINNED

LOAD CASE 1

LOWER  NODE NUMBER OF ELEMENT  =  2
HIGHER NODE NUMBER OF ELEMENT  =  3

+ + + + + + + + + + +
+    ELEMENT  2 3    +
+ + + + + + + + + + +

SELECT LOADING TYPE
```

PLANE FRAMES - FURTHER TOPICS

```
(1) UNIFORMLY DISTRIBUTED
(2) POINT LOAD
(3) LACK OF FIT
(4) THERMAL EXPANSION
(5) SETTLEMENT
? 1

NOTE THAT A POSITIVE MEMBER LOAD ACTS
IN THE DIRECTION OF THE MEMBER Y-AXIS

LENGTH                        =  7071
ANGLE WITH THE GLOBAL X-AXIS  =  8.13
ELEMENT TYPE                  =  1
LOAD INTENSITY                = -0.01


* * * * * * * * * * * * * * *
*     FIXED END FORCES      *
*    (MEMBER AXES SYSTEM)   *
* * * * * * * * * * * * * * *

                NODE  2         NODE  3
                (FIXED)         (FIXED)
X-FORCE       0.0000E+00      0.0000E+00
Y-FORCE       0.3536E+02      0.3536E+02
MOMENT        0.4167E+05     -.4167E+05


* * * * * * * * * * * * * * *
*   EQUIVALENT JOINT FORCES *
*    (GLOBAL AXES SYSTEM)   *
* * * * * * * * * * * * * * *

                NODE  2         NODE  3
                (FIXED)         (FIXED)
X-FORCE       0.5000E+01      0.5000E+01
Y-FORCE      -.3500E+02      -.3500E+02
MOMENT       -.4167E+05      0.4167E+05


LOAD CASE 2

LOWER  NODE NUMBER OF ELEMENT =  1
HIGHER NODE NUMBER OF ELEMENT =  2

+ + + + + + + + + + +
+    ELEMENT  1  2  +
+ + + + + + + + + + +

SELECT LOADING TYPE
(1) UNIFORMLY DISTRIBUTED
(2) POINT LOAD
(3) LACK OF FIT
(4) THERMAL EXPANSION
(5) SETTLEMENT
? 4

AREA                          =  9000
ELASTIC CONSTANT              =  200
ANGLE WITH THE GLOBAL X-AXIS  =  90
ELEMENT TYPE                  =  1
COEFF. OF THERMAL EXPANSION   =  1.1E-5
TEMPERATURE RISE              =  20

* * * * * * * * * * * * * * *
*     FIXED END FORCES      *
*    (MEMBER AXES SYSTEM)   *
* * * * * * * * * * * * * * *

                NODE  1         NODE  2
                (FIXED)         (FIXED)
X-FORCE       0.3960E+03     -.3960E+03
Y-FORCE       0.0000E+00      0.0000E+00
MOMENT        0.0000E+00      0.0000E+00
```

****
*   EQUIVALENT JOINT FORCES   *
*   (GLOBAL AXES SYSTEM)   *
****

```
                 NODE 1          NODE 2
                (FIXED)         (FIXED)
X-FORCE         0.1483E-03     -.1483E-03
Y-FORCE        -.3960E+03      0.3960E+03
MOMENT          0.0000E+00     0.0000E+00
```

LOAD CASE 3

LOWER NODE NUMBER OF ELEMENT = 1
HIGHER NODE NUMBER OF ELEMENT = 2

+ + + + + + + + + + +
+   ELEMENT 1 2   +
+ + + + + + + + + + +

SELECT LOADING TYPE
(1) UNIFORMLY DISTRIBUTED
(2) POINT LOAD
(3) LACK OF FIT
(4) THERMAL EXPANSION
(5) SETTLEMENT
? 5

```
AREA                          = 9000
SECOND MOMENT OF AREA         = .25E9
LENGTH                        = 3000
ELASTIC CONSTANT              = 200
ANGLE WITH THE GLOBAL X-AXIS  = 90
ELEMENT TYPE                  = 1
```
DOES SETTLEMENT TAKE PLACE AT
    (1) NODE   1
    (2) NODE   2
? 1

```
SETTLEMENT GLOBAL X-DIRECTION = 10
SETTLEMENT GLOBAL Y-DIRECTION = -15
ROTATION (DEGREES)            = 0
```

****
*   FIXED END FORCES   *
*   (MEMBER AXES SYSTEM)   *
****

```
                 NODE 1          NODE 2
                (FIXED)         (FIXED)
X-FORCE        -.9000E+04      0.9000E+04
Y-FORCE        -.2222E+03      0.2222E+03
MOMENT         -.3333E+06     -.3333E+06
```

****
*   EQUIVALENT JOINT FORCES   *
*   (GLOBAL AXES SYSTEM)   *
****

```
                 NODE 1          NODE 2
                (FIXED)         (FIXED)
X-FORCE        -.2222E+03      0.2222E+03
Y-FORCE         0.9000E+04    -.9000E+04
MOMENT          0.3333E+06     0.3333E+06
```

```
LOAD CASE 4

LOWER  NODE NUMBER OF ELEMENT   =  2
HIGHER NODE NUMBER OF ELEMENT   =  3

+ + + + + + + + + + +
+    ELEMENT  2  3   +
+ + + + + + + + + + +

SELECT LOADING TYPE
(1) UNIFORMLY DISTRIBUTED
(2) POINT LOAD
(3) LACK OF FIT
(4) THERMAL EXPANSION
(5) SETTLEMENT
? 3

AREA                            =  9000
LENGTH                          =  7071
ELASTIC CONSTANT                =  200
ANGLE WITH THE GLOBAL X-AXIS    =  8.13
ELEMENT TYPE                    =  1
LACK OF FIT (SHORT IS + VE)     =  5


* * * * * * * * * * * * *
*   FIXED END FORCES    *
*   (MEMBER AXES SYSTEM) *
* * * * * * * * * * * * *

                 NODE  2        NODE  3
                (FIXED)         (FIXED)
X-FORCE        -.1273E+04     0.1273E+04
Y-FORCE        0.0000E+00     0.0000E+00
MOMENT         0.0000E+00     0.0000E+00


* * * * * * * * * * * * *
*  EQUIVALENT JOINT FORCES *
*  (GLOBAL AXES SYSTEM)    *
* * * * * * * * * * * * *

                 NODE  2        NODE  3
                (FIXED)         (FIXED)
X-FORCE        0.1260E+04    -.1260E+04
Y-FORCE        0.1800E+03    -.1800E+03
MOMENT         0.0000E+00     0.0000E+00
```

8.3 INCLINED SUPPORTS

Chapter 3 (section 3.6) outlined how local axes are used to deal with inclined supports. If at any node the nodal axes (i.e. the axes used to describe nodal force and displacement) point in the directions of the coordinate axes then that nodal axes system is referred to as a global axes system. The solution procedure is greatly simplified by the use of global axes at all nodes. There are occasions, however, when it is either inconvenient or impossible to describe the boundary conditions in terms of the global axes, and local axes must be introduced. Figure 8.10(a) shows a structure where the engineer might prefer to introduce local axes at node 1 rather than use coordinate axes orientated as shown in fig 8.10(b).

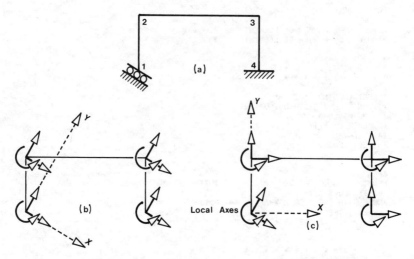

FIGURE 8.10   TWO POSSIBLE CHOICES OF COORDINATE AXES

The advantage of the coordinate axes system shown in fig 8.10(b) is that it allows the analysis to proceed without the complication of local axes. Its disadvantages are that the calculation of nodal coordinates would be more difficult and error-prone and, if the nodal loads are horizontal and vertical, then they would have to be transformed into the inclined global axes systems prior to analysis. In general, if the available computer program allows the use of local axes then the choice of the coordinate axes system should be governed by the ease of data preparation.

Some structures, such as that shown in fig 8.11, can only be analysed through the use of local axes as no single axes can be used to describe the boundary conditions.

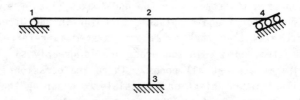

FIGURE 8.11   EXAMPLE OF A STRUCTURE THAT DEMANDS THE USE OF LOCAL AXES

Figure 8.12 shows nodal axes systems that could be used in the analysis of the structure shown in fig 8.11.

PLANE FRAMES - FURTHER TOPICS

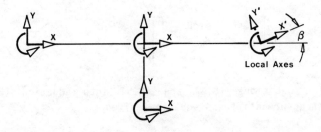

FIGURE 8.12   AN EXAMPLE OF THE USE OF LOCAL AXES

Prior to the application of the boundary conditions the stiffness relationship can be expressed solely in terms of the global axes. The initial stiffness relationship (in submatrices) for the structure shown in fig 8.12 is

$$\begin{bmatrix} K_{11} & K_{12} & 0 & 0 \\ K_{21} & K_{22} & K_{23} & K_{24} \\ 0 & K_{32} & K_{33} & 0 \\ 0 & K_{42} & 0 & K_{44} \end{bmatrix} \begin{bmatrix} \Delta_1 \\ \Delta_2 \\ \Delta_3 \\ \Delta_4 \end{bmatrix} = \begin{bmatrix} P_1 \\ P_2 \\ P_3 \\ P_4 \end{bmatrix} \qquad (8.3)$$

The boundary conditions at nodes 1 and 3 can be easily expressed in terms of the global axes.

$$\Delta_{1y} = \Delta_{3x} = \Delta_{3y} = \theta_3 = 0$$

In terms of the global axes the boundary conditions at node 4 are as follows:

$$\Delta_{4y} / \Delta_{4x} = \tan\beta$$

Obviously the first boundary condition cannot be applied directly, and in order to obtain a solution the equation of equilibrium for node 4 must be expressed in the local (primed) axes system.

The problem is to rewrite the initial structure stiffness equation with the vectors of nodal force and displacement at node 4 expressed in terms of the local axes system. If the local axes system has been rotated by β anticlockwise from the global axes system as shown in fig 8.12, then the global vectors of force and displacement can be expressed in the local axes system as follows:

$$\Delta_4 = T_4^{-1} \Delta_4' \qquad (8.4)$$

$$P_4 = T_4^{-1} P_4' \qquad (8.5)$$

where

$$T_4^{-1} = \begin{bmatrix} \cos\beta & -\sin\beta & 0 \\ \sin\beta & \cos\beta & 0 \\ 0 & 0 & 1 \end{bmatrix}$$

Introducing equations (8.4) and (8.5) into equation (8.3) (the initial structure stiffness equation) gives

$$\begin{bmatrix} K_{11} & K_{12} & 0 & 0 \\ K_{21} & K_{22} & K_{23} & K_{24} \\ 0 & K_{32} & K_{33} & 0 \\ 0 & K_{42} & 0 & K_{44} \end{bmatrix} \begin{bmatrix} \Delta_1 \\ \Delta_2 \\ \Delta_3 \\ T_4^{-1}\Delta_4' \end{bmatrix} = \begin{bmatrix} P_1 \\ P_2 \\ P_3 \\ T_4^{-1}P_4' \end{bmatrix}$$

Now premultiplication of Δ_4' by T_4^{-1} is the same as postmultiplication of the fourth column of the initial structure stiffness matrix by T_4^{-1}.

$$\begin{bmatrix} K_{11} & K_{12} & 0 & 0 \\ K_{21} & K_{22} & K_{23} & K_{24}T_4^{-1} \\ 0 & K_{32} & K_{33} & 0 \\ 0 & K_{42} & 0 & K_{44}T_4^{-1} \end{bmatrix} \begin{bmatrix} \Delta_1 \\ \Delta_2 \\ \Delta_3 \\ \Delta_4' \end{bmatrix} = \begin{bmatrix} P_1 \\ P_2 \\ P_3 \\ T_4^{-1}P_4' \end{bmatrix}$$

If row 4 is premultiplied by the inverse of T_4^{-1} (i.e. T_4) then the preceding equation becomes

$$\begin{bmatrix} K_{11} & K_{12} & 0 & 0 \\ K_{21} & K_{22} & K_{23} & K_{24}T_4^{-1} \\ 0 & K_{32} & K_{33} & 0 \\ 0 & T_4 K_{42} & 0 & T_4 K_{44} T_4^{-1} \end{bmatrix} \begin{bmatrix} \Delta_1 \\ \Delta_2 \\ \Delta_3 \\ \Delta_4' \end{bmatrix} = \begin{bmatrix} P_1 \\ P_2 \\ P_3 \\ P_4' \end{bmatrix} \quad (8.6)$$

as

$$T_4 T_4^{-1} P_4' = P_4'$$

Note also that, as the structure stiffness matrix is symmetrical

$$[K_{24}T_4^{-1}]^T = T_4 K_{42}$$

Hence, by applying additional transformations to the stiffness submatrices associated with node 4, the nodal force and displacement vectors have been expressed in terms of the local axes, thus allowing the boundary conditions to be applied directly.

If element "i,j" has a local axes system at end "j" then the element stiffness matrix that is added to the structure stiffness matrix is one that expresses the element stiffness in terms of the global axes at end "i" and the local axes at end "j" as follows:

$$\begin{bmatrix} F_{ij} \\ \hline F'_{ji} \end{bmatrix} = \begin{bmatrix} K_{ii}^j & | & K_{ij}T_j^{-1} \\ \hline T_j K_{ji} & | & T_j K_{jj}^i T_j^{-1} \end{bmatrix} \begin{bmatrix} \Delta_{ij} \\ \hline \Delta'_{ji} \end{bmatrix}$$

where
$$K_{ii}^j = T_{ij}^{-1} k_{ii}^j T_{ij} \quad \text{etc.}$$

Had the local axes system been at "i" then the corresponding equation would be

$$\begin{bmatrix} F'_{ij} \\ \hline F_{ji} \end{bmatrix} = \begin{bmatrix} T_i K_{ii}^j T_i^{-1} & | & T_i K_{ij} \\ \hline K_{ji} T_i^{-1} & | & K_{jj}^i \end{bmatrix} \begin{bmatrix} \Delta'_{ij} \\ \hline \Delta_{ji} \end{bmatrix}$$

Once the boundary conditions have been applied then the solution of the remaining equations yields the unknown displacements in terms of the nodal axes systems. For example, equation (8.6) yields the displacements at nodes 1 and 2 in the global axes system, and the displacements at node 4 in the local axes system. When using local axes the fact that some displacements are found in terms of the local axes system means that the equation used to find member end forces requires modification. The member end forces for member "i,j" have been previously found using the following equations:

$$f_{ij} = k_{ii}^j T_{ij} \Delta_i + k_{ij} T_{ij} \Delta_j + f_{ij}^f$$

$$f_{ji} = k_{ji} T_{ij} \Delta_i + k_{jj}^i T_{ij} \Delta_j + f_{ji}^f$$

However, if end "i" is connected to an inclined support then the displacements at "i" will have been found in a local axes system and will have to be transformed into the global axes system using equation (8.4).

$$f_{ij} = k_{ii}^j T_{ij} T_i^{-1} \Delta'_i + k_{ij} T_{ij} \Delta_j + f_{ij}^f \qquad (8.7)$$

$$f_{ji} = k_{ji} T_{ij} T_i^{-1} \Delta'_i + k_{jj}^i T_{ij} \Delta_j + f_{ji}^f \qquad (8.8)$$

where T_i^{-1} is the matrix which transforms vectors of force and and displacement at node "i" from the local to the global axes system.

If the local axes are at end "j" then the member end forces are found using the following equations:

$$f_{ij} = k_{ii}^j T_{ij} \Delta_i + k_{ij} T_{ij} T_j^{-1} \Delta'_j + f_{ij}^f \qquad (8.9)$$

and
$$f_{ji} = k_{ji} T_{ij} \Delta_i + k_{jj}^i T_{ij} T_j^{-1} \Delta_j^i + f_{ji}^f \qquad (8.10)$$

Example 8.2

Find the displacements for the structure shown in the following diagram. Find also the member end forces for member 2,4.

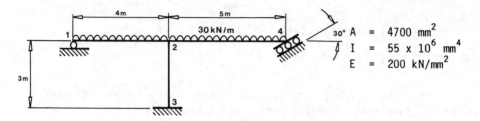

The following diagram shows the unrestrained nodal displacements and the freedom numbers.

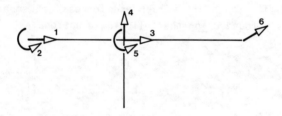

Generate the freedom vector. Note that force and displacement at node 4 will be expressed in terms of the local axes system shown above.

Zero the freedom vector $\begin{bmatrix} 0 \\ 0 \\ 0 \\ 0 \\ 0 \\ 0 \\ 0 \\ 0 \\ 0 \\ 0 \\ 0 \\ 0 \end{bmatrix}$ Apply the boundary conditions $\begin{bmatrix} 0 \\ 1 \\ 0 \\ 0 \\ 0 \\ 0 \\ 1 \\ 1 \\ 1 \\ 0 \\ 1 \\ 1 \end{bmatrix}$ Number the freedoms $\begin{bmatrix} 1 \\ 0 \\ 2 \\ 3 \\ 4 \\ 5 \\ 0 \\ 0 \\ 0 \\ 6 \\ 0 \\ 0 \end{bmatrix}$

PLANE FRAMES - FURTHER TOPICS

Fixed end forces (in the member axes systems)

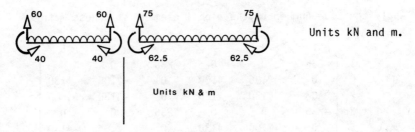

Units kN and m.

Units kN & m

Equivalent joint forces (in the global axes systems).

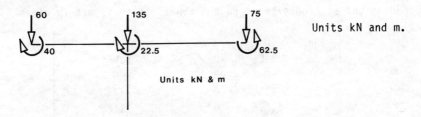

Units kN and m.

Units kN & m

Transform the equivalent joint forces at node 4 into the local axes system (note that the units are kN and mm).

$$P^{e'}_{42} = T_4 \, P^{e}_{42}$$

$$\begin{bmatrix} P^{e'}_{4x} \\ P^{e'}_{4y} \\ M^{e'}_{4} \end{bmatrix} = \begin{bmatrix} 0.866 & 0.5 & 0 \\ -0.5 & 0.866 & 0 \\ 0 & 0 & 1 \end{bmatrix} \begin{bmatrix} 0 \\ -75 \\ 62500 \end{bmatrix} = \begin{bmatrix} -37.5 \\ -65 \\ 62500 \end{bmatrix}$$

Final loadings (in the nodal axes systems).

Units kN & m

Now generate the final structure stiffness matrix.

Member 1,2

$A = 4700 \text{ mm}^2$, $I = 55 \times 10^6 \text{ mm}^4$, $L = 4000 \text{ mm}$, $E = 200 \text{ kN/mm}^2$,

COMPUTER ANALYSIS OF STRUCTURAL FRAMEWORKS

$\alpha = 90°$,   code number = [1 0 2 3 4 5].

Apply the code number to the global element stiffness matrix.

$$\begin{array}{c} \\ 1 \\ 0 \\ 2 \\ 3 \\ 4 \\ 5 \end{array} \begin{array}{c} 102345 \\ \left[\begin{array}{cccccc} 235 & 0 & 0 & -235 & 0 & 0 \\ 0 & 2.06 & 4130 & 0 & -2.06 & 4130 \\ 0 & 4130 & 11E6 & 0 & -4130 & 5.5E6 \\ -235 & 0 & 0 & 235 & 0 & 0 \\ 0 & -2.06 & -4130 & 0 & 2.06 & -4130 \\ 0 & 4130 & 5.5E6 & 0 & -4130 & 11E6 \end{array}\right] \end{array}$$

Adding the stiffness terms to the final stiffness matrix gives

$$\left[\begin{array}{cccccc} 235 & 0 & -235 & 0 & 0 & 0 \\ 0 & 11E6 & 0 & -4130 & 5.5E6 & 0 \\ -235 & 0 & 235 & 0 & 0 & 0 \\ 0 & -4130 & 0 & 2.06 & -4130 & 0 \\ 0 & 5.5E6 & 0 & -4130 & 11E6 & 0 \\ 0 & 0 & 0 & 0 & 0 & 0 \end{array}\right]$$

Member 2,3

$A = 4700$ mm^2, $I = 55 \times 10^6$ mm^4, $L = 3000$ mm, $E = 200$ kN/mm^2, $\alpha = 270°$,   code number = [3 4 5 0 0 0].

Apply the code number to the global element stiffness matrix.

$$\begin{array}{c} \\ 3 \\ 4 \\ 5 \\ 0 \\ 0 \\ 0 \end{array} \begin{array}{c} 345000 \\ \left[\begin{array}{cccccc} 4.89 & 0 & 7330 & -4.89 & 0 & 7330 \\ 0 & 313 & 0 & 0 & -313 & 0 \\ 7330 & 0 & 14.7E6 & -7330 & 0 & 7.4E6 \\ -4.89 & 0 & -7330 & 4.89 & 0 & -7330 \\ 0 & -313 & 0 & 0 & 313 & 0 \\ 7330 & 0 & 7.4E6 & -7330 & 0 & 14.7E6 \end{array}\right] \end{array}$$

Adding the stiffness terms to the final stiffness matrix gives

$$\left[\begin{array}{cccccc} 235 & 0 & -235 & 0 & 0 & 0 \\ 0 & 11E6 & 0 & -4130 & 5.5E6 & 0 \\ -235 & 0 & 240 & 0 & 7330 & 0 \\ 0 & -4130 & 0 & 315 & -4130 & 0 \\ 0 & 5.5E6 & 7330 & -4130 & 25.7E6 & 0 \\ 0 & 0 & 0 & 0 & 0 & 0 \end{array}\right]$$

PLANE FRAMES - FURTHER TOPICS

Member 2,4

$A = 4700$ mm^2, $\quad I = 55 \times 10^6$ mm^4, $\quad L = 3000$ mm, $\quad E = 200$ kN/mm^2,
$\alpha = 0°$, $\quad$ code number = [3 4 5 6 0 0].

$$K_{22}^4 = \begin{bmatrix} 188 & 0 & 0 \\ 0 & 1.06 & 2640 \\ 0 & 2640 & 8.8E6 \end{bmatrix}$$

The matrix to transform force and displacement vectors from the global axes to the local axes at node 4 is ($\beta = 30°$).

$$T_4 = \begin{bmatrix} 0.866 & 0.5 & 0 \\ -0.5 & 0.866 & 0 \\ 0 & 0 & 1 \end{bmatrix}$$

and

$$K_{24}\, T_4^{-1} = \begin{bmatrix} -163 & 94.0 & 0.0 \\ -0.53 & -0.92 & 2640 \\ -1320 & -2286 & 4.4E6 \end{bmatrix}$$

$$T_4\, K_{42} = \begin{bmatrix} -163 & -0.53 & -1320 \\ 94 & -0.92 & -2286 \\ 0.0 & 2640 & 4.4E6 \end{bmatrix}$$

$$T_4\, K_{44}^2\, T_4^{-1} = \begin{bmatrix} 141 & -80.9 & -1320 \\ -80.9 & 47.8 & -2286 \\ -1320 & -2286 & 8.8E6 \end{bmatrix}$$

Apply the code number to the element stiffness matrix.

	3	4	5	6	0	0
3	188	0	0	-163	94.0	0
4	0	1.06	2640	-0.53	-0.92	2640
5	0	2640	8.8E6	-1320	-2286	4.4E6
6	-163	-0.53	-1320	141	-80.9	-1320
0	94	-0.92	-2286	-80.9	47.8	-2286
0	0	2640	4.4E6	-1320	-2286	8.8E6

Note that the preceding element stiffness matrix relates forces and displacements at the ends of member 2,4 where the forces and displacements at end 2 are expressed in the global axes system and at end 4 in the local axes system. Adding the stiffness of member 2,4 to the structure stiffness matrix allows the complete structure stiffness equation to be written:

COMPUTER ANALYSIS OF STRUCTURAL FRAMEWORKS

$$\begin{bmatrix} 235 & 0 & -235 & 0 & 0 & 0 \\ 0 & 11E6 & 0 & -4130 & 5.5E6 & 0 \\ -235 & 0 & 428 & 0 & 7330 & -163 \\ 0 & -4130 & 0 & 316 & -1490 & -0.53 \\ 0 & 5.5E6 & 7330 & -1490 & 34.5E6 & -1320 \\ 0 & 0 & -163 & -0.53 & -1320 & 141 \end{bmatrix} \begin{bmatrix} \Delta_{1x} \\ \Theta_1 \\ \Delta_{2x} \\ \Delta_{2y} \\ \Theta_2 \\ \Delta'_{4x} \end{bmatrix} = \begin{bmatrix} 0 \\ -40000 \\ 0 \\ -135 \\ -22500 \\ -37.5 \end{bmatrix}$$

Solution of these equations yields the displacements

$$\begin{bmatrix} \Delta_{1x} \\ \Theta_1 \\ \Delta_{2x} \\ \Delta_{2y} \\ \Theta_2 \\ \Delta'_{4x} \end{bmatrix} = \begin{bmatrix} -12.32 \text{ mm} \\ -0.00491 \text{ rads} \\ -12.32 \text{ mm} \\ -0.505 \text{ mm} \\ 0.00217 \text{ rads} \\ -14.49 \text{ mm} \end{bmatrix}$$

The member end forces for member 2,4 are found using the equations (8.9) and (8.10).

End 2

$$f_{24} = k_{22}^4 T_{24} \Delta_2 + k_{24} T_{24} T_4^{-1} \Delta'_4 + f_{24}^f$$

$$= \begin{bmatrix} 188 & 0 & 0 \\ 0 & 1.06 & 2640 \\ 0 & 2640 & 8.8E6 \end{bmatrix} \begin{bmatrix} 1 & 0 & 0 \\ 0 & 1 & 0 \\ 0 & 0 & 1 \end{bmatrix} \begin{bmatrix} -12.32 \\ -0.505 \\ 0.00217 \end{bmatrix} +$$

$$\begin{bmatrix} -188 & 0 & 0 \\ 0 & -1.06 & 2640 \\ 0 & -2640 & 4.4E6 \end{bmatrix} \begin{bmatrix} 1 & 0 & 0 \\ 0 & 1 & 0 \\ 0 & 0 & 1 \end{bmatrix} \begin{bmatrix} 0.866 & -0.500 & 0 \\ 0.500 & 0.866 & 0 \\ 0 & 0 & 1 \end{bmatrix} \begin{bmatrix} -14.49 \\ 0 \\ 0 \end{bmatrix}$$

$$+ \begin{bmatrix} 0 \\ 75 \\ 62500 \end{bmatrix}$$

$$= \begin{bmatrix} -2322 \\ 5.19 \\ 17600 \end{bmatrix} + \begin{bmatrix} 2358 \\ 7.68 \\ 19140 \end{bmatrix} + \begin{bmatrix} 0 \\ 75 \\ 62500 \end{bmatrix} = \begin{bmatrix} 36 \\ 87.9 \\ 99240 \end{bmatrix}$$

End 4

$$f_{42} = k_{42} T_{24} \Delta_2 + k_{44}^2 T_{24} T_4^{-1} \Delta'_4 + f_{42}^f$$

$$= \begin{bmatrix} 2322 \\ -5.19 \\ 8215 \end{bmatrix} + \begin{bmatrix} -2358 \\ -7.68 \\ 19140 \end{bmatrix} + \begin{bmatrix} 0 \\ 75 \\ -62500 \end{bmatrix} = \begin{bmatrix} -36 \\ 62.1 \\ -35140 \end{bmatrix}$$

8.4 ELASTIC SUPPORTS

Elastic supports can be used to model "imperfect" boundary conditions (e.g. when the foundations are expected to undergo significant movement due to the elastic deformation of the subsoil). In some structures elastic supports are deliberately introduced in order to reduce peak reactions and stresses.

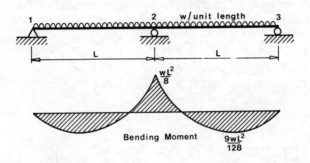

FIGURE 8.13   AN EXAMPLE OF RIGID SUPPORTS

For example, if the supports to the continuous beam shown in fig 8.13 are rigid then the hogging moment over the central support will exceed the sagging moment in the span by 78%. Also the magnitude of the central reaction will be more than three times that of the end reactions.

If the rigid central support is replaced by an elastic support of suitable stiffness as shown in fig 8.14, then the hogging and sagging moments can be equalised, and some of the central reaction will be distributed to the end supports.

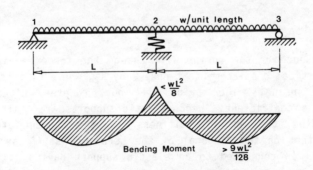

FIGURE 8.14   ELASTIC CENTRAL SUPPORT

COMPUTER ANALYSIS OF STRUCTURAL FRAMEWORKS

Of course if the spring is too weak then the sagging moment will exceed the hogging moment. It should also be noted that although the ground is usually regarded as rigid for the purposes of structural analysis it will, in fact, always deform under the loads from the foundations. Occasionally the engineer may have to regard the ground as an elastic support in order to represent satisfactorily the true structural behaviour. The structure shown in fig 8.14 will be used to illustrate how elastic supports can be incorporated into the stiffness method. The initial stiffness equation for the structure shown in fig 8.14, in terms of submatrices, is

$$\begin{bmatrix} K_{11} & K_{12} & 0 \\ K_{21} & K_{22} & K_{23} \\ 0 & K_{32} & K_{33} \end{bmatrix} \begin{bmatrix} \Delta_1 \\ \Delta_2 \\ \Delta_3 \end{bmatrix} = \begin{bmatrix} P_1 \\ P_2 \\ P_3 \end{bmatrix}$$

Expanding and applying the boundary conditions

$$\begin{bmatrix} \cancel{K_{11}} & \cancel{K_{12}} & \cancel{K_{13}} & \cancel{K_{14}} & \cancel{K_{15}} & \cancel{K_{16}} & 0 & \cancel{0} & 0 \\ \cancel{K_{21}} & \cancel{K_{22}} & \cancel{K_{23}} & \cancel{K_{24}} & \cancel{K_{25}} & \cancel{K_{26}} & 0 & \cancel{0} & 0 \\ K_{31} & K_{32} & K_{33} & K_{34} & K_{35} & K_{36} & 0 & 0 & 0 \\ K_{41} & K_{42} & K_{43} & K_{44} & K_{45} & K_{46} & K_{47} & K_{48} & K_{49} \\ K_{51} & K_{52} & K_{53} & K_{54} & K_{55} & K_{56} & K_{57} & K_{58} & K_{59} \\ K_{61} & K_{62} & K_{63} & K_{64} & K_{65} & K_{66} & K_{67} & K_{68} & K_{69} \\ 0 & 0 & 0 & K_{74} & K_{75} & K_{76} & K_{77} & K_{78} & K_{79} \\ \cancel{0} & \cancel{0} & \cancel{0} & \cancel{K_{84}} & \cancel{K_{85}} & \cancel{K_{86}} & \cancel{K_{87}} & \cancel{K_{88}} & \cancel{K_{89}} \\ 0 & 0 & 0 & K_{94} & K_{95} & K_{96} & K_{97} & K_{98} & K_{99} \end{bmatrix} \begin{bmatrix} \cancel{\Delta_{1x}} \\ \cancel{\Delta_{1y}} \\ \theta_1 \\ \Delta_{2x} \\ \Delta_{2y} \\ \theta_2 \\ \Delta_{3x} \\ \cancel{\Delta_{3y}} \\ \theta_3 \end{bmatrix} = \begin{bmatrix} \cancel{P_{1x}} \\ \cancel{P_{1y}} \\ M_1 \\ P_{2x} \\ P_{2y} \\ M_2 \\ P_{3x} \\ \cancel{P_{3y}} \\ M_3 \end{bmatrix}$$

Note that had the central support been rigid then Δ_{2y} would have been another zero displacement.

If the spring at node 2 did not exist then the "vertical" stiffness of node 2 would be

$$K_{55} = 12EI/L_{12}^3 + 12EI/L_{23}^3$$

The presence of the spring will change the force required to cause unit vertical displacement at node 2 by K_2, where K_2 is the stiffness of the spring. The spring will not, however, alter any other stiffness coefficient. Hence elastic supports are dealt with simply by adding the spring stiffness to the appropriate stiffness coefficient on the diagonal of the structure stiffness matrix. If node "i" is connected to an elastic support having stiffnesses K_{ix} and K_{iy} against translation and stiffness K_{ir} against rotation, then the elastic support can be regarded as a "special" element whose only contribution to the structure stiffness matrix is the submatrix K_{ii}^s

PLANE FRAMES - FURTHER TOPICS

where

$$K^s_{ii} = \begin{bmatrix} K_{ix} & 0 & 0 \\ 0 & K_{iy} & 0 \\ 0 & 0 & K_{ir} \end{bmatrix}$$

These stiffnesses are, of course, expressed in terms of the nodal axes system at node "i", and the above submatrix is added to the structure stiffness matrix in exactly the same way as the element stiffness submatrices K^j_{ii} and K^I_{jj}.

Example 8.3

Determine the bending moment diagram for the structure shown below if the spring has a stiffness of 40 kN/mm. Demonstrate the equilibrium of node 2.

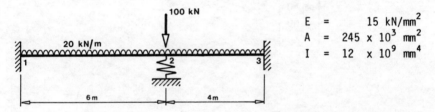

The initial structure stiffness equation, in submatrices, is as follows:

$$\begin{bmatrix} K_{11} & K_{12} & 0 \\ K_{21} & K_{22} & K_{23} \\ 0 & K_{32} & K_{33} \end{bmatrix} \begin{bmatrix} \Delta_1 \\ \Delta_2 \\ \Delta_3 \end{bmatrix} = \begin{bmatrix} P_1 \\ P_2 \\ P_3 \end{bmatrix}$$

Applying the boundary conditions $\Delta_1 = \Delta_3 = 0$ gives the final stiffness equation

$$K_{22} \Delta_2 = P_2$$

where

$$K_{22} = K^1_{22} + K^3_{22} + K^s_{22}$$

$$= \begin{bmatrix} 613 & 0 & 0 \\ 0 & 10 & -30000 \\ 0 & -30000 & 120E6 \end{bmatrix} + \begin{bmatrix} 919 & 0 & 0 \\ 0 & 33.8 & 67500 \\ 0 & 67500 & 180E6 \end{bmatrix} + \begin{bmatrix} 0 & 0 & 0 \\ 0 & 40 & 0 \\ 0 & 0 & 0 \end{bmatrix}$$

Loading

$$P_{21}^e = -T_{12}\, f_{21}^f$$

$$= -\begin{bmatrix} 1 & 0 & 0 \\ 0 & 1 & 0 \\ 0 & 0 & 1 \end{bmatrix} \begin{bmatrix} 0 \\ 60 \\ -60000 \end{bmatrix} = \begin{bmatrix} 0 \\ -60 \\ 60000 \end{bmatrix}$$

similarly

$$P_{23}^e = \begin{bmatrix} 0 \\ -40 \\ -26667 \end{bmatrix}$$

and the vector of nodal loads at node 2 is

$$P_2^n = \begin{bmatrix} 0 \\ -100 \\ 0 \end{bmatrix}$$

hence

$$P_2 = P_{21}^e + P_{23}^e + P_2^n$$

$$= \begin{bmatrix} 0 \\ -60 \\ 60000 \end{bmatrix} + \begin{bmatrix} 0 \\ -40 \\ -26667 \end{bmatrix} + \begin{bmatrix} 0 \\ -100 \\ 0 \end{bmatrix}$$

$$= \begin{bmatrix} 0 \\ -200 \\ 33333 \end{bmatrix}$$

The final stiffness equation is

$$\begin{bmatrix} 1532 & 0 & 0 \\ 0 & 83.8 & 37500 \\ 0 & 37500 & 300E6 \end{bmatrix} \begin{bmatrix} \Delta_{2x} \\ \Delta_{2y} \\ \theta_2 \end{bmatrix} = \begin{bmatrix} 0 \\ -200 \\ 33333 \end{bmatrix}$$

the solution to which is

$$\begin{bmatrix} \Delta_{2x} \\ \Delta_{2y} \\ \theta_2 \end{bmatrix} = \begin{bmatrix} 0 \text{ mm} \\ -2.58 \text{ mm} \\ 0.434E\text{-}3 \text{ rads} \end{bmatrix}$$

PLANE FRAMES - FURTHER TOPICS

The force at end 1 of member 1,2 is found using the now familiar equation

$$f_{12} = k_{11}^2 T_{12} \Delta_1 + k_{12} T_{12} \Delta_2 + f_{12}^f$$

$$\begin{bmatrix} f_{12x} \\ f_{12y} \\ m_{12} \end{bmatrix} = \begin{bmatrix} 613 & 0 & 0 \\ 0 & 10 & 30000 \\ 0 & 30000 & 120E6 \end{bmatrix} \begin{bmatrix} 1 & 0 & 0 \\ 0 & 1 & 0 \\ 0 & 0 & 1 \end{bmatrix} \begin{bmatrix} 0 \\ 0 \\ 0 \end{bmatrix} +$$

$$\begin{bmatrix} -613 & 0 & 0 \\ 0 & -10 & 30000 \\ 0 & -30000 & 60E6 \end{bmatrix} \begin{bmatrix} 1 & 0 & 0 \\ 0 & 1 & 0 \\ 0 & 0 & 1 \end{bmatrix} \begin{bmatrix} 0 \\ -2.58 \\ 0.434E-3 \end{bmatrix} + \begin{bmatrix} 0 \\ 60 \\ 60000 \end{bmatrix}$$

$$= \begin{bmatrix} 0 \\ 0 \\ 0 \end{bmatrix} + \begin{bmatrix} 0 \\ 38.8 \\ 103400 \end{bmatrix} + \begin{bmatrix} 0 \\ 60 \\ 60000 \end{bmatrix}$$

$$= \begin{bmatrix} 0 \\ 98.8 \\ 163400 \end{bmatrix}$$

Similar calculation shows the other member end forces to be

$$\begin{bmatrix} f_{21x} \\ f_{21y} \\ m_{21} \end{bmatrix} = \begin{bmatrix} 0 \\ 21.2 \\ 69500 \end{bmatrix} \quad \begin{bmatrix} f_{23x} \\ f_{23y} \\ m_{23} \end{bmatrix} = \begin{bmatrix} 0 \\ -17.9 \\ -69500 \end{bmatrix} \quad \begin{bmatrix} f_{32x} \\ f_{32y} \\ m_{32} \end{bmatrix} = \begin{bmatrix} 0 \\ 97.9 \\ -161700 \end{bmatrix}$$

The freebody diagram for node 2 is shown in the following diagram

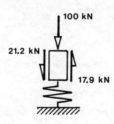

For vertical equilibrium of the element, the force in the spring must be 103.3 kN, and the force in the spring calculated from the vertical deflection of node 2 is 40 x 2.58 = 103.2 kN.

8.5 ELEMENTS WITH PINS

In Chapters 5 and 6 structures consisting exclusively of pin ended members were considered, and so far this chapter has considered structures consisting entirely of rigidly connected members. Although most civil engineering structures can be regarded as falling into one of these categories, there are times when the engineer has to deal with hybrid structures containing both rigidly jointed and pinned connections. Figure 8.15 gives three examples of such structures.

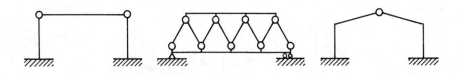

FIGURE 8.15   STRUCTURES CONTAINING BOTH RIGIDLY JOINTED AND PIN ENDED MEMBERS

Consider the left-hand structure shown in fig 8.15. Obviously the structure consists of three members. The problem facing the structural analyst is where to locate the nodes.

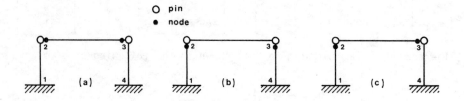

FIGURE 8.16   STRUCTURAL IDEALISATIONS

Figure 8.16 shows three possible idealisations. Where the nodes are located affects the character of the members as shown in the following table.

Idealisation	Member 1,2	Member 2,3	Member 3,4
(a)	fixed/pinned	fixed/fixed	pinned/fixed
(b)	fixed/fixed	pinned/pinned	fixed/fixed
(c)	fixed/fixed	pinned/fixed	pinned/fixed

PLANE FRAMES - FURTHER TOPICS

Note that although the way in which the structure is idealised affects the character of the individual elements, the character of the structure as a whole is unaffected (i.e. the deformed shape of the structure under a given system of loads will be the same for all three idealisations). The nodal rotations found during the analysis would, however, be different because of the discontinuity of rotation at the pins.

If the connection between a joint and a member is assumed pinned, then the end of the member connected to that joint can rotate relative to the joint. It is important that the structure is idealised such that the behaviour of the actual, "as built" structure is realistically represented. Consider a joint "i" which has three members framing into it. Figure 8.17 shows ways in which the joint might be idealised.

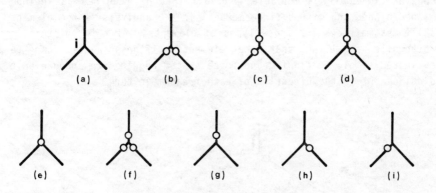

FIGURE 8.17   JOINT IDEALISATIONS

(a) shows the idealisation that should be used if all three members are rigidly connected.

(b), (c), and (d) show idealisations that could be used to represent a pinned connection. All three connections are structurally equivalent (i.e. all three members can rotate relative to each other). Note that the use of this idealisation throughout the structure results in a pin jointed frame.

(e) and (f) show idealisations that result in the joint having no rotational stiffness (i.e. the joint becomes a mechanism). If the analysis is concerned with joint rotations, e.g. plane frame analysis, then such idealisations lead to a breakdown in the solution

271

procedure as the joint rotations become indeterminate.

Idealisation (e) is frequently used to represent the joints of a pin jointed frame. Pin jointed frame analysis is unconcerned with joint rotations; hence the fact that the joints have no rotational stiffness is unimportant.

(g), (h), and (i) show idealisations where one member framing into the joint has a pinned connection, and the other two are rigidly connected. Note that these idealisations are *not* structurally equivalent.

The theory already developed has been based upon members that are rigidly connected to the joints at their ends. That theory is, however, generally applicable, and all that is required to include pinned/pinned and pinned/fixed members in the analysis is the element stiffness matrices for these types of elements. At this point it is worthwhile recalling that the element stiffness matrix can be evaluated simply by finding the force system required to produce unit displacement in the direction of each freedom in turn.

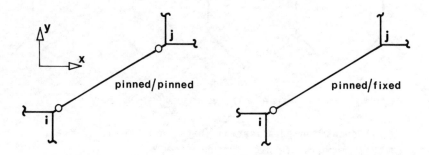

FIGURE 8.18   PINNED/PINNED AND PINNED/FIXED MEMBERS

Figure 8.18 shows a pinned/pinned and a pinned/fixed member running between nodes "i" and "j". It should be obvious that both members will affect the translational stiffness of joints "i" and "j". Further it should be noted that the pinned/pinned member adds no rotational stiffness to either joint, and that the pinned/fixed member adds rotational stiffness only to joint "j". Figure 8.19 shows unit displacement in the "x" direction and illustrates how moment will be developed at joint "j" in the pinned/fixed case, but not in the pinned/pinned case.

PLANE FRAMES - FURTHER TOPICS

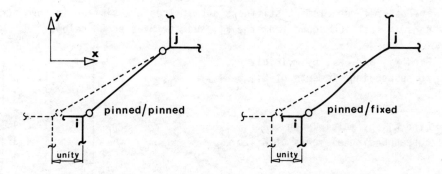

FIGURE 8.19   UNIT DISPLACEMENT IN THE "X" DIRECTION

Element Pinned at Both Ends

The element stiffness matrix in the member axes system for a pinned/pinned member can be written down by inspection and partitioned in the normal way to produce submatrices:

$$\begin{bmatrix} f_{ijx} \\ f_{ijy} \\ m_{ij} \\ \hline f_{jix} \\ f_{jiy} \\ m_{ji} \end{bmatrix} = \begin{bmatrix} EA/L & 0 & 0 & -EA/L & 0 & 0 \\ 0 & 0 & 0 & 0 & 0 & 0 \\ 0 & 0 & 0 & 0 & 0 & 0 \\ \hline -EA/L & 0 & 0 & EA/L & 0 & 0 \\ 0 & 0 & 0 & 0 & 0 & 0 \\ 0 & 0 & 0 & 0 & 0 & 0 \end{bmatrix} \begin{bmatrix} \delta_{ijx} \\ \delta_{ijy} \\ \theta_{ij} \\ \hline \delta_{jix} \\ \delta_{jiy} \\ \theta_{ji} \end{bmatrix}$$

i.e.
$$f_{ij} = k_{ii}^j \delta_{ij} + k_{ij} \delta_{ji}$$

and
$$f_{ji} = k_{ji} \delta_{ij} + k_{jj}^i \delta_{ji}$$

The element stiffness submatrices in the global axes system are found using the familiar triple multiplication

where
$$K_{ii}^j = T_{ij}^{-1} k_{ii}^j T_{ij}$$

$$T_{ij} = \begin{bmatrix} \cos\alpha & \sin\alpha & 0 \\ -\sin\alpha & \cos\alpha & 0 \\ 0 & 0 & 1 \end{bmatrix}$$

hence
$$K_{ii}^j = \begin{bmatrix} EA \cos^2\alpha / L & EA \cos\alpha \sin\alpha / L & 0 \\ EA \cos\alpha \sin\alpha / L & EA \cos^2\alpha / L & 0 \\ 0 & 0 & 0 \end{bmatrix} \quad (8.11)$$

COMPUTER ANALYSIS OF STRUCTURAL FRAMEWORKS

The other three element stiffness submatrices are similar in form and can, in fact, be found from the K_{ii}^j submatrix as shown below.

Find K_{ij} and K_{ji} by multiplying corresponding elements of K_{ii}^j by
$$\begin{bmatrix} -1 & -1 & 1 \\ -1 & -1 & 1 \\ 1 & 1 & 1 \end{bmatrix}$$

Find K_{jj}^i by multiplying corresponding elements of K_{ii}^j by
$$\begin{bmatrix} 1 & 1 & 1 \\ 1 & 1 & 1 \\ 1 & 1 & 1 \end{bmatrix}$$

As might be expected, these element stiffness submatrices are identical to the submatrices for a standard plane frame element having no flexural stiffness. By using the above submatrices members with pins at both ends can be included in the analysis using exactly the same procedures as were used for rigidly connected members.

Element Pinned at End "i"

The procedure to deal with elements pinned at only one end is identical to that used for elements pinned at both ends. The first stage is to determine the element stiffness matrix. Here the element will be assumed to be pinned at node "i" and rigidly connected at node "j", as shown in fig 8.20.

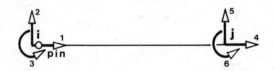

FIGURE 8.20   ELEMENT PINNED AT END "i"

The axial stiffness of the member is unaffected by the pin, hence

$$K_{11} = EA/L \qquad K_{41} = -EA/L \qquad K_{21} = K_{31} = K_{51} = K_{61} = 0$$

There are many ways in which the remaining stiffness terms can be found. Here use will again be made of the shape function. To illustrate, consider the stiffness terms associated with freedom 2 (see fig 8.21).

274

PLANE FRAMES - FURTHER TOPICS

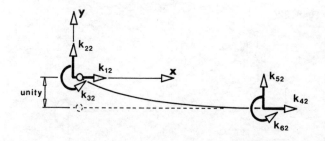

FIGURE 8.21   UNIT DISPLACEMENT AT FREEDOM 2

Section 2.11 showed that the deformed shape of a prismatic line element loaded only at its ends is described exactly by a cubic polynomial.

$$y = ax^3 + bx^2 + cx + d$$

$$dy/dx = 3ax^2 + 2bx + c$$

$$d^2y/dx^2 = 6ax + 2b$$

The boundary conditions associated with unit displacement in the direction of freedom 2 are

at $x = 0$, $y = 0$ therefore $0 = 0 + 0 + 0 + d$
             $d^2y/dx^2 = 0$ therefore $0 = 0 + 2b$
at $x = L$, $y = -1$ therefore $-1 = aL^3 + bL^2 + cL + d$
             $dy/dx = 0$ therefore $0 = 3aL^2 + 2bL + c$

hence
    $a = 1/2L^3$, $b = 0$, $c = -3/2L$, $d = 0$
giving
    $y = x^3/2L^3 - 3x/2L$
and
    $M = EI\, d^2y/dx^2$ (i.e. bending moment = EI x curvature)

       $= 3EIx/L^3$
therefore

    $M = 0$ when $x = 0$
    $M = 3EI/L^2$ when $x = L$

The sign means that the applied bending moment causes positive curvature at the right-hand end of the beam. Hence the force system associated with unit displacement in the direction of freedom 2 is as

275

shown in fig 8.22.

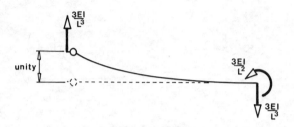

FIGURE 8.22   FORCE SYSTEM ASSOCIATED WITH UNIT DISPLACEMENT AT FREEDOM 2

Note that the forces $3EI/L^3$ are necessary to maintain the equilibrium of the element, and the stiffness coefficients associated with freedom 2 are

$$K_{12} = K_{32} = K_{42} = 0 \quad K_{22} = 3EI/L^3 \quad K_{52} = -3EI/L^3 \quad K_{62} = 3EI/L^2$$

Treating the other freedoms similarly the element stiffness relationship can be shown to be

$$\begin{bmatrix} f_{ijx} \\ f_{ijy} \\ m_{ij} \\ f_{jix} \\ f_{jiy} \\ m_{ji} \end{bmatrix} = \begin{bmatrix} EA/L & 0 & 0 & -EA/L & 0 & 0 \\ 0 & 3EI/L^3 & 0 & 0 & -3EI/L^3 & 3EI/L^2 \\ 0 & 0 & 0 & 0 & 0 & 0 \\ -EA/L & 0 & 0 & EA/L & 0 & 0 \\ 0 & -3EI/L^3 & 0 & 0 & 3EI/L^3 & -3EI/L^2 \\ 0 & 3EI/L^2 & 0 & 0 & -3EI/L^2 & 3EI/L \end{bmatrix} \begin{bmatrix} \delta_{ijx} \\ \delta_{ijy} \\ \theta_{ij} \\ \delta_{jix} \\ \delta_{jiy} \\ \theta_{ji} \end{bmatrix}$$

(8.12)

The above equation can be written in terms of the submatrices indicated by the broken lines as follows:

$$\begin{bmatrix} f_{ij} \\ f_{ji} \end{bmatrix} = \begin{bmatrix} k_{ii}^j & k_{ij} \\ k_{ji} & k_{jj}^i \end{bmatrix} \begin{bmatrix} \delta_{ij} \\ \delta_{ji} \end{bmatrix}$$

The global element stiffness submatrices, calculated using the triple matrix multiplication $T^{-1} k T$, are

$$K_{ii}^j = T_{ij}^{-1} k_{ii}^j T_{ij}$$

$$= \begin{bmatrix} (EA\cos^2\alpha/L + 3EI\sin^2\alpha/L^3) & (EA/L - 3EI/L^3)\sin\alpha\cos\alpha & 0 \\ (EA/L - 3EI/L^3)\sin\alpha\cos\alpha & (EA\sin^2\alpha/L + 3EI\cos^2\alpha/L^3) & 0 \\ 0 & 0 & 0 \end{bmatrix}$$

PLANE FRAMES - FURTHER TOPICS

$$\mathbf{K}_{ij} = \mathbf{T}_{ij}^{-1} \mathbf{k}_{ij} \mathbf{T}_{ij}$$

$$= \begin{bmatrix} -(EA\cos^2\alpha/L + 3EI\sin^2\alpha/L^3) & -(EA/L - 3EI/L^3)\sin\alpha\cos\alpha & -3EI\sin\alpha/L^2 \\ -(EA/L - 3EI/L^3)\sin\alpha\cos\alpha & -(EA\sin^2\alpha/L + 3EI\cos^2\alpha/L^3) & 3EI\cos\alpha/L^2 \\ 0 & 0 & 0 \end{bmatrix}$$

$$\mathbf{K}_{ji} = \mathbf{K}_{ij}^T$$

$$\mathbf{K}_{jj}^i = \mathbf{T}_{ij}^{-1} \mathbf{k}_{jj}^i \mathbf{T}_{ij}$$

$$= \begin{bmatrix} (EA\cos^2\alpha/L + 3EI\sin^2\alpha/L^3) & (EA/L - 3EI/L^3)\sin\alpha\cos\alpha & 3EI\sin\alpha/L^2 \\ (EA/L - 3EI/L^3)\sin\alpha\cos\alpha & (EA\sin^2\alpha/L + 3EI\cos^2\alpha/L^3) & -3EI\cos\alpha/L^2 \\ 3EI\sin\alpha/L^2 & -3EI\cos\alpha/L^2 & 3EI/L \end{bmatrix}$$

(8.13)

For hand calculation it is best to first determine the $\mathbf{K}_{jj}^i$ submatrix and then obtain the other submatrices as follows:

Find $\mathbf{K}_{ii}^j$ by multiplying corresponding elements of $\mathbf{K}_{jj}^i$ by
$$\begin{bmatrix} 1 & 1 & 0 \\ 1 & 1 & 0 \\ 0 & 0 & 0 \end{bmatrix}$$

Find $\mathbf{K}_{ij}$ by multiplying corresponding elements of $\mathbf{K}_{jj}^i$ by
$$\begin{bmatrix} -1 & -1 & -1 \\ -1 & -1 & -1 \\ 0 & 0 & 0 \end{bmatrix}$$

Find $\mathbf{K}_{ji}$ by multiplying corresponding elements of $\mathbf{K}_{jj}^i$ by
$$\begin{bmatrix} -1 & -1 & 0 \\ -1 & -1 & 0 \\ -1 & -1 & 0 \end{bmatrix}$$

(8.14)

Note that this is not matrix multiplication but simply multiplication of corresponding elements.

Member Pinned at End "j"

$$\begin{bmatrix} f_{ijx} \\ f_{ijy} \\ m_{ij} \\ f_{jix} \\ f_{jiy} \\ m_{ji} \end{bmatrix} = \begin{bmatrix} EA/L & 0 & 0 & -EA/L & 0 & 0 \\ 0 & 3EI/L^3 & 3EI/L^2 & 0 & -3EI/L^3 & 0 \\ 0 & 3EI/L^2 & 3EI/L & 0 & -3EI/L^2 & 0 \\ -EA/L & 0 & 0 & EA/L & 0 & 0 \\ 0 & -3EI/L^3 & -3EI/L^2 & 0 & 3EI/L^3 & 0 \\ 0 & 0 & 0 & 0 & 0 & 0 \end{bmatrix} \begin{bmatrix} \delta_{ijx} \\ \delta_{ijy} \\ \Theta_{ij} \\ \delta_{jix} \\ \delta_{jiy} \\ \Theta_{ji} \end{bmatrix}$$

COMPUTER ANALYSIS OF STRUCTURAL FRAMEWORKS

The techniques used for a member with a pin at "i" can be used to show that the element stiffness equation (in the member axes system) for a member with a pin at "j" and rigidly jointed at "i" is as shown in the preceding equation. The global element stiffness submatrices are developed in the usual way as follows:

$$K_{ii}^j = T_{ij}^{-1} k_{ii}^j T_{ij}$$

$$= \begin{bmatrix} (EA\cos^2\alpha/L + 3EI\sin^2\alpha/L^3) & (EA/L - 3EI/L^3)\sin\alpha\cos\alpha & -3EI\sin\alpha/L^2 \\ (EA/L - 3EI/L^3)\sin\alpha\cos\alpha & (EA\sin^2\alpha/L + 3EI\cos^2\alpha/L^3) & 3EI\cos\alpha/L^2 \\ -3EI\sin\alpha/L^2 & 3EI\cos\alpha/L^2 & 3EI/L \end{bmatrix}$$

(8.15)

$$K_{ij} = T_{ij}^{-1} k_{ij} T_{ij}$$

$$= \begin{bmatrix} -(EA\cos^2\alpha/L + 3EI\sin^2\alpha/L^3) & -(EA/L - 3EI/L^3)\sin\alpha\cos\alpha & 0 \\ -(EA/L - 3EI/L^3)\sin\alpha\cos\alpha & -(EA\sin^2\alpha/L + 3EI\cos^2\alpha/L^3) & 0 \\ 3EI\sin\alpha/L^2 & -3EI\cos\alpha/L^2 & 0 \end{bmatrix}$$

$$K_{ji} = K_{ij}^T$$

$$K_{jj}^i = T_{ij}^{-1} k_{jj}^i T_{ij}$$

$$= \begin{bmatrix} (EA\cos^2\alpha/L + 3EI\sin^2\alpha/L^3) & (EA/L - 3EI/L^3)\sin\alpha\cos\alpha & 0 \\ (EA/L - 3EI/L^3)\sin\alpha\cos\alpha & (EA\sin^2\alpha/L + 3EI\cos^2\alpha/L^3) & 0 \\ 0 & 0 & 0 \end{bmatrix}$$

For hand calculation it is best to first determine the K_{ii}^j submatrix and then obtain the other submatrices as follows:

Find K_{jj}^i by multiplying corresponding elements of K_{ii}^j by
$$\begin{bmatrix} 1 & 1 & 0 \\ 1 & 1 & 0 \\ 0 & 0 & 0 \end{bmatrix}$$

Find K_{ij} by multiplying corresponding elements of K_{ii}^j by
$$\begin{bmatrix} -1 & -1 & 0 \\ -1 & -1 & 0 \\ -1 & -1 & 0 \end{bmatrix}$$

Find K_{ji} by multiplying corresponding elements of K_{ii}^j by
$$\begin{bmatrix} -1 & -1 & -1 \\ -1 & -1 & -1 \\ 0 & 0 & 0 \end{bmatrix}$$

(8.16)

Note again that this is not matrix multiplication but simply multiplication of corresponding elements.

Member Loads and Member End Forces

If a member is pinned at one or more ends then the pin (or pins) must be taken into account when determining the fixed end and the equivalent joint forces.

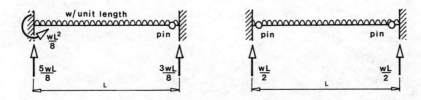

FIGURE 8.23   FIXED END FORCES FOR MEMBERS CONTAINING PINS

Figure 8.23 shows the fixed end forces due to a uniformly distributed load acting on members containing pins. The fixed end forces for the member pinned at both ends can be obtained by inspection. The member pinned at one end behaves, under "vertical" loads, as a propped cantilever, and fixed end forces for common loading cases are to be found in many textbooks. It is worthwhile noting that if the fixed end forces for a fixed/fixed member are known, then the fixed end forces for the fixed/pinned case can easily be found by superposition as illustrated in fig 8.24.

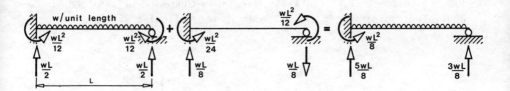

FIGURE 8.24   FIXED END FORCES FOR A FIXED/PINNED MEMBER

Once the nodal displacements have been found the now familiar matrix equations are used to find the member end forces.

$$f_{ij} = k_{ii}^{j} T_{ij} \Delta_i + k_{ij} T_{ij} \Delta_j$$

$$f_{ji} = k_{ji} T_{ij} \Delta_i + k_{jj}^{i} T_{ij} \Delta_j$$

COMPUTER ANALYSIS OF STRUCTURAL FRAMEWORKS

The stiffness submatrices used in the above equations must, of course, be consistent with the end conditions of the element (e.g. if the member has a pinned connection at end "i" and a rigid connection at end "j" the submatrices must be as shown in equation (8.12)).

Example 8.4

For the structure shown prove that the reactive forces at node 1 are independent of which member is assumed to contain the pin.

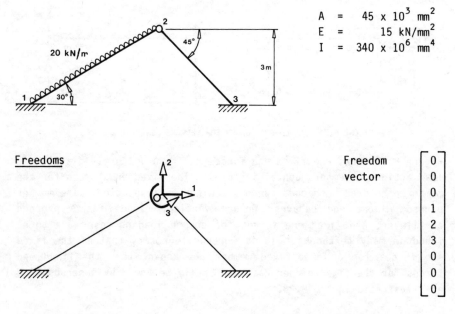

$A = 45 \times 10^3 \text{ mm}^2$
$E = 15 \text{ kN/mm}^2$
$I = 340 \times 10^6 \text{ mm}^4$

(i) If element 1,2 contains the pin, then the fixed end and equivalent joint forces are as shown in the following diagram.

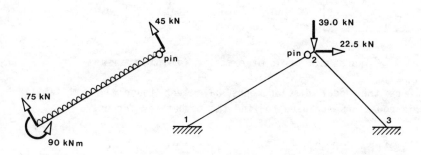

PLANE FRAMES - FURTHER TOPICS

Member 1,2

$A = 45000$ mm^2, $\quad I = 340 \times 10^6$ mm^4, $\quad L = 6000$ mm, $\quad E = 15$ kN/mm^2, $\alpha = 30°$, $\quad$ code number $= [0\ 0\ 0\ 1\ 2\ 3]$.

This is a fixed/pinned member, therefore equation (8.15) yields the K_{ii}^j global element stiffness submatrix, and the remaining submatrices are found using transformations (8.16). Apply the code number to the global element stiffness matrix.

$$\begin{array}{c} \\ 0 \\ 0 \\ 0 \\ 1 \\ 2 \\ 3 \end{array} \begin{array}{cccccc} 0 & 0 & 0 & 1 & 2 & 3 \\ \left[\begin{array}{cccccc} 84.4 & 48.7 & -213 & -84.4 & -48.7 & 0 \\ 48.7 & 28.2 & 368 & -48.7 & -28.2 & 0 \\ -213 & 368 & 2.55E6 & 213 & -368 & 0 \\ -84.4 & -48.7 & 213 & 84.4 & 48.7 & 0 \\ -48.7 & -28.2 & -368 & 48.7 & 28.2 & 0 \\ 0 & 0 & 0 & 0 & 0 & 0 \end{array}\right] \end{array}$$

Add the element stiffness to the structure stiffness matrix.

$$\begin{bmatrix} 84.4 & 48.7 & 0 \\ 48.7 & 28.2 & 0 \\ 0 & 0 & 0 \end{bmatrix}$$

Member 2,3

$A = 45000$ mm^2, $\quad I = 340 \times 10^6$ mm^4, $\quad L = 4243$ mm, $\quad E = 15$ kN/mm^2, $\alpha = 315°$, $\quad$ code number $= [1\ 2\ 3\ 0\ 0\ 0]$.

This is a fixed/fixed (or standard) element. Apply the code number to the global element stiffness matrix found using equation (7.11) and the transformations given by equations (7.12).

$$\begin{array}{c} \\ 1 \\ 2 \\ 3 \\ 0 \\ 0 \\ 0 \end{array} \begin{array}{cccccc} 1 & 2 & 3 & 0 & 0 & 0 \\ \left[\begin{array}{cccccc} 80.0 & -79.2 & 1200 & -80.0 & 79.2 & 1200 \\ -79.2 & 80.0 & 1200 & 79.2 & -80.0 & 1200 \\ 1200 & 1200 & 4.81E6 & -1200 & -1200 & 2.40E6 \\ -80.0 & 79.2 & -1200 & 80.0 & -79.2 & -1200 \\ 79.2 & -80.0 & -1200 & -79.2 & 80.0 & -1200 \\ 1200 & 1200 & 2.40E6 & -1200 & -1200 & 4.81E6 \end{array}\right] \end{array}$$

COMPUTER ANALYSIS OF STRUCTURAL FRAMEWORKS

Adding the element stiffness completes the generation of the final structure stiffness matrix, and the final structure stiffness equation is

$$\begin{bmatrix} 164.4 & -30.5 & 1200 \\ -30.5 & 108.2 & 1200 \\ 1200 & 1200 & 4.81E6 \end{bmatrix} \begin{bmatrix} \Delta_{2x} \\ \Delta_{2y} \\ \theta_2 \end{bmatrix} = \begin{bmatrix} 22.5 \\ -39.0 \\ 0 \end{bmatrix}$$

which yields

$$\begin{bmatrix} \Delta_{2x} \\ \Delta_{2y} \\ \theta_2 \end{bmatrix} = \begin{bmatrix} 0.0732 & mm \\ -0.341 & mm \\ 0.667 \times 10^{-4} & rads \end{bmatrix}$$

Calculate the reactive forces at node 1 from the member end forces.

Member 1,2 (fixed/pinned), end 1

$$f_{12} = k^2_{11} T_{12} \Delta_1 + k_{12} T_{12} \Delta_2 + f^f_{12}$$

$$= k_{12} T_{12} \Delta_2 + f^f_{12} \quad as \quad \Delta_1 = 0$$

hence

$$\begin{bmatrix} f_{12x} \\ f_{12y} \\ m_{12} \end{bmatrix} = \begin{bmatrix} EA/L & 0 & 0 \\ 0 & -3EI/L^3 & 0 \\ 0 & -3EI/L^2 & 0 \end{bmatrix} \begin{bmatrix} \cos\alpha & \sin\alpha & 0 \\ -\sin\alpha & \cos\alpha & 0 \\ 0 & 0 & 1 \end{bmatrix} \begin{bmatrix} \Delta_{2x} \\ \Delta_{2y} \\ \theta_2 \end{bmatrix} + \begin{bmatrix} f^f_{12x} \\ f^f_{12y} \\ m^f_{12} \end{bmatrix}$$

$$= \begin{bmatrix} -113 & 0 & 0 \\ 0 & -0.071 & 0 \\ 0 & -425 & 0 \end{bmatrix} \begin{bmatrix} 0.866 & 0.5 & 0 \\ -0.5 & 0.866 & 0 \\ 0 & 0 & 1 \end{bmatrix} \begin{bmatrix} 0.0732 \\ -0.341 \\ 667E-7 \end{bmatrix} + \begin{bmatrix} 0 \\ 75 \\ 90000 \end{bmatrix}$$

$$= \begin{bmatrix} 12.1 \\ 0.0236 \\ 141 \end{bmatrix} + \begin{bmatrix} 0 \\ 75 \\ 90000 \end{bmatrix} = \begin{bmatrix} 12.1 \\ 75 \\ 90140 \end{bmatrix}$$

$$R_1 = \sum T^{-1}_{1j} f_{1j} - P_1 = T^{-1}_{12} f_{12}$$

Therefore

$$\begin{bmatrix} R_{1x} \\ R_{1y} \\ R_{1m} \end{bmatrix} = - \begin{bmatrix} 0.866 & -0.5 & 0 \\ 0.5 & 0.866 & 0 \\ 0 & 0 & 1 \end{bmatrix} \begin{bmatrix} 12.1 \\ 75 \\ 90140 \end{bmatrix} = \begin{bmatrix} -27.0 \text{ kN} \\ 71.0 \text{ kN} \\ -90140 \text{ kN mm} \end{bmatrix}$$

PLANE FRAMES - FURTHER TOPICS

(ii) If element 2,3 contains the pin, then the fixed end and equivalent joint forces are

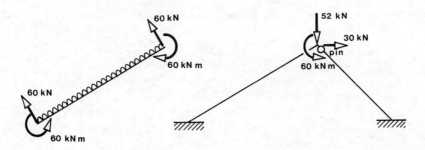

Member 1,2

$A = 45000$ mm^2, $\quad I = 340 \times 10^6$ mm^4, $\quad L = 6000$ mm, $\quad E = 15$ kN/mm^2,
$\alpha = 30°$, $\quad$ code number = [0 0 0 1 2 3].

This is a fixed/fixed member, therefore the global element stiffness matrix is found using equation (7.11) and the transformations given by equations (7.12).

Apply the code number to the global element stiffness matrix.

$$\begin{array}{c} \\ 0 \\ 0 \\ 0 \\ 1 \\ 2 \\ 3 \end{array} \begin{array}{c} 0 0 0 1 2 3 \\ \begin{bmatrix} 84.4 & 48.6 & -425 & -84.4 & -48.6 & -425 \\ 48.6 & 28.3 & 736 & -48.7 & -28.3 & 736 \\ -425 & 736 & 3.4E6 & 425 & -736 & 1.7E6 \\ -84.4 & -48.7 & 425 & 84.4 & 48.6 & 425 \\ -48.6 & -28.3 & -736 & 48.6 & 28.3 & -736 \\ -425 & 736 & 1.7E6 & 425 & -736 & 3.4E6 \end{bmatrix} \end{array}$$

Adding the stiffness terms to the final stiffness matrix gives

$$\begin{bmatrix} 84.4 & 48.6 & 425 \\ 48.6 & 28.3 & -736 \\ 425 & -736 & 3.4E6 \end{bmatrix}$$

Member 2,3

$A = 45000$ mm^2, $\quad I = 340 \times 10^6$ mm^4, $\quad L = 4243$ mm, $\quad E = 15$ kN/mm^2,
$\alpha = 315°$, $\quad$ code number = [1 2 3 0 0 0].

This ia a pinned/fixed member, therefore equation (8.13) yields the

COMPUTER ANALYSIS OF STRUCTURAL FRAMEWORKS

K_{jj}^i global element stiffness submatrix, and the remaining submatrices are found using transformations (8.14).

Apply the code number to the global element stiffness matrix.

$$\begin{array}{c} \\ 1 \\ 2 \\ 3 \\ 0 \\ 0 \\ 0 \end{array} \begin{array}{c} 123000 \\ \begin{bmatrix} 79.6 & -79.4 & 0 & -79.6 & 79.4 & 601 \\ -79.4 & 79.6 & 0 & 79.4 & -79.6 & 601 \\ 0 & 0 & 0 & 0 & 0 & 0 \\ -79.6 & 79.4 & 0 & 79.6 & -79.4 & -601 \\ 79.4 & -79.6 & 0 & -79.4 & 79.6 & -601 \\ 601 & 601 & 0 & -601 & -601 & 3.61E6 \end{bmatrix} \end{array}$$

Adding the element stiffness completes the generation of the final structure stiffness matrix, and the final structure stiffness equation is

$$\begin{bmatrix} 164.0 & -30.8 & 425 \\ -30.8 & 107.9 & -736 \\ 425 & -736 & 3.4E6 \end{bmatrix} \begin{bmatrix} \Delta_{2x} \\ \Delta_{2y} \\ \theta_2 \end{bmatrix} = \begin{bmatrix} 30.0 \\ -52.0 \\ 60000 \end{bmatrix}$$

which yields

$$\begin{bmatrix} \Delta_{2x} \\ \Delta_{2y} \\ \theta_2 \end{bmatrix} = \begin{bmatrix} 0.0733 \text{ mm} \\ -0.341 \text{ mm} \\ 0.0176 \text{ rads} \end{bmatrix}$$

Calculate the reactions at node 1 from the member end forces.

Member 1,2 (fixed/fixed)

$$f_{12} = k_{11}^2 T_{12} \Delta_1 + k_{12} T_{12} \Delta_2 + f_{12}^f$$

$$= k_{12} T_{12} \Delta_2 + f_{12}^f \quad \text{as} \quad \Delta_1 = 0$$

hence

$$\begin{bmatrix} f_{12x} \\ f_{12y} \\ m_{12} \end{bmatrix} = \begin{bmatrix} EA/L & 0 & 0 \\ 0 & -12EI/L^3 & 6EI/L^2 \\ 0 & -6EI/L^2 & 2EI/L \end{bmatrix} \begin{bmatrix} \cos\alpha & \sin\alpha & 0 \\ -\sin\alpha & \cos\alpha & 0 \\ 0 & 0 & 1 \end{bmatrix} \begin{bmatrix} \Delta_{2x} \\ \Delta_{2y} \\ \theta_2 \end{bmatrix} + \begin{bmatrix} f_{12x}^f \\ f_{12y}^f \\ m_{12}^f \end{bmatrix}$$

$$= \begin{bmatrix} -113 & 0 & 0 \\ 0 & -0.283 & 850 \\ 0 & -850 & 1.7E6 \end{bmatrix} \begin{bmatrix} 0.866 & 0.5 & 0 \\ -0.5 & 0.866 & 0 \\ 0 & 0 & 1 \end{bmatrix} \begin{bmatrix} 0.0733 \\ -0.341 \\ 0.0176 \end{bmatrix} + \begin{bmatrix} 0 \\ 60 \\ 60000 \end{bmatrix}$$

PLANE FRAMES - FURTHER TOPICS

$$= \begin{bmatrix} 12.1 \\ 15.1 \\ 30200 \end{bmatrix} + \begin{bmatrix} 0 \\ 60 \\ 60000 \end{bmatrix} = \begin{bmatrix} 12.1 \\ 75.1 \\ 90200 \end{bmatrix}$$

$$R_1 = \sum T_{1j}^{-1} f_{1j} - P_1 = T_{12}^{-1} f_{12}$$

$$\begin{bmatrix} R_{1x} \\ R_{1y} \\ R_{1m} \end{bmatrix} = \begin{bmatrix} 0.866 & -0.5 & 0 \\ 0.5 & 0.866 & 0 \\ 0 & 0 & 1 \end{bmatrix} \begin{bmatrix} 12.1 \\ 75.1 \\ 90200 \end{bmatrix} = \begin{bmatrix} -27.1 \text{ kN} \\ 71.1 \text{ kN} \\ 90200 \text{ kN mm} \end{bmatrix}$$

These figures are, given the limited number of significant figures employed, the same as when the pin was assumed to be in member 1,2.

8.6 NON-PRISMATIC MEMBERS

All of the members considered so far have been straight and of uniform cross-section, i.e. prismatic members. If, however, a member is curved or has section properties which vary along its length, then it is known as a non-prismatic member. Figure 8.25 shows four examples of non-prismatic members.

       (a)   is a stepped member.
       (b)   is a tapered member.
       (c)   is a curved member.
       (d)   is a curved and tapered member.

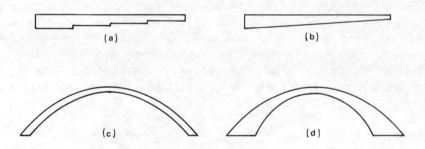

FIGURE 8.25   EXAMPLES OF NON-PRISMATIC MEMBERS

One way to deal with such members is to evaluate their stiffness either analytically or numerically and then treat them as standard members (as was done with members containing pins). A variety of techniques can be used to evaluate the stiffness of non-prismatic members. The majority of problems involving non-prismatic members are, however, most easily tackled by treating the non-prismatic members as a series of standard prismatic beam elements as illustrated in fig 8.26.

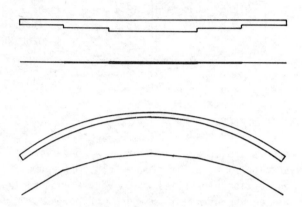

FIGURE 8.26   NON-PRISMATIC MEMBERS TREATED AS A SERIES OF PRISMATIC MEMBERS

Note that the disadvantage of this approach is that additional nodes are introduced. This increases the size of the problem and hence the cost of the analysis. In the case of the straight non-prismatic member illustrated in fig 8.26 an additional four nodes have been introduced which increases the degree of freedom by 12. Usually, however, the disadvantage of increased problem size is more than outweighed by the ease of analysis. Only when there are many non-prismatic members, or when the sub-division of non-prismatic members makes the problem too expensive or too large for the available computer should the engineer consider the alternative method of calculating the actual stiffness of the non-prismatic members.

Example 8.5

The program PFRAME.PF presented later in this chapter has been used to compare the end deflection of the tapered member shown below when it is idealised as 2, 4, and 8 prismatic members. The cross-section is rectangular with a breadth of 400 mm and $E = 200$ kN/mm^2.

PLANE FRAMES - FURTHER TOPICS

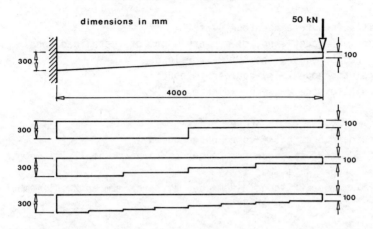

The results are as follows:

Number of elements	Downwards end deflection (mm)
2	25.4
4	13.5
8	12.5

Notice how the end deflection quickly approaches the true value of 12.8 mm as the number of elements increases.

8.7 PUTTING SOME OF IT TOGETHER - Program PFRAME.PF

This chapter and the previous chapter have now covered the procedures necessary for the analysis of plane frames using the stiffness method. In this section the short computer programs to generate the final stiffness matrix (FSTIFF.PF) and to calculate member forces (MFORCE.PF) are combined with the Gaussian elimination equation solver (GAUSS.EQ) to produce an automatic plane frame analysis program (PFRAME.PF). To illustrate how the theory covered in this chapter might be implemented in an automatic plane frame analysis program PFRAME.PF has been given the ability to deal with member loads and members containing pins.

Input The only piece of information the user supplies to the program from the keyboard is the name of the file containing the data. This data file can be constructed using either a text editor or the data preprocessor program PRE.MA/PRE.01/PRE.02 described in section 5.11. If using a text editor the data file should be constructed to the following format:

```
<job title (not to exceed one line)>
3,2,6,<no. of nodes>,<no. of members>,<no. of restrained nodes>,
<no. of nodal loads>, <no. of member loads>
<x-coordinate node 1>,<y-coordinate node 1>
<x-coordinate node 2>,<y-coordinate node 2>
          .              .              .
          .              .              .
<x-coordinate node n>,<y-coordinate node n>
<node "i" member 1>,<node "j" member 1>,<area member 1>,<I member 1>,
<member 1 type>,<E member 1>
<node "i" member 2>,<node "j" member 2>,<area member 2>,<I member 2>,
<member 2 type>,<E member 2>
          .              .              .
```

        note - "i" must be less then "j"

        member types are

TYPE	NODE I	NODE J
1	fixed	fixed
2	fixed	pinned
3	pinned	fixed
4	pinned	pinned

```
<node "i" member n>,<node "j" member n>,<area member n>,<I member n>,
<member n type>,<E member n>
<restrained node no. 1 node number>,<direction(s)>
<restrained node no. 2 node number>,<direction(s)>
<restrained node no. 3 node number>,<direction(s)>
```

      note - restraint directions are as follows
          restraint in the global x-direction = 1
          restraint in the global y-direction = 2
          rotational restraint   = 3

      Input restraints as a composite number (see section 7.8)

```
<restrained node no. n node number>,<direction(s)>
<nodal load no. 1  node number>,<direction>,<magnitude>
<nodal load no. 2  node number>,<direction>,<magnitude>
<nodal load no. 3  node number>,<direction>,<magnitude>
```

    note - load directions are as follows
        load in the global x-direction = 1
        load in the global y-direction = 2
        moment   = 3

PLANE FRAMES - FURTHER TOPICS

```
       .             .             .
       .             .             .
       .             .             .
<nodal load no. n  node number>,<direction>,<magnitude>
<member load 1 - node "i">,<member load 1 - node "j">,<load type>,
<load intensity>,<distance from node "i">
<member load 2 - node "i">,<member load 2 - node "j">,<load type>,
<load intensity>,<distance from node "i">
```
         .   .   .
  note - load types are
         1 uniformly distributed (intensity = load/unit length)
         2 point load   (intensity = load magnitude)

       - member loads are described in the member axes system,
         and a positive member load acts perpendicular to the
         member in the positive "y" direction.

       - distance from node "i" is immaterial for load type 1
         .   .   .
         .   .   .
```
<member load n - node "i">,<member load n - node "j">,<load type>,
<load intensity>,<distance from node "i">
```

Comments on the Algorithm   The length of each member and the cosine and sine of the angle it makes with the global "x" axis are calculated from the member node numbers and the nodal coordinates. Only the upper semi-band of the final stiffness matrix is generated. Members containing pins are dealt with by using the variables $K_1 - K_8$ in the element stiffness matrix as shown in the following equation.

$$\begin{bmatrix} f_{ijx} \\ f_{ijy} \\ m_{ij} \\ f_{jix} \\ f_{jiy} \\ m_{ji} \end{bmatrix} = \begin{bmatrix} EA/L & 0 & 0 & -EA/L & 0 & 0 \\ & K_1EI/L^3 & K_2EI/L^2 & 0 & -K_1EI/L^3 & K_3EI/L^2 \\ & & K_4EI/L & 0 & K_5EI/L^2 & K_6EI/L \\ & \text{symmetrical} & & EA/L & 0 & 0 \\ & & & & K_1EI/L^3 & K_7EI/L^2 \\ & & & & & K_8EI/L \end{bmatrix} \begin{bmatrix} \delta_{ijx} \\ \delta_{ijy} \\ \theta_{ij} \\ \delta_{jix} \\ \delta_{jiy} \\ \theta_{ji} \end{bmatrix}$$

MEMBER TYPE	NODE "i"	NODE "j"	K_1	K_2	K_3	K_4	K_5	K_6	K_7	K_8
1	Fixed	Fixed	12	6	6	4	-6	2	-6	4
2	Fixed	Pinned	3	3	0	3	-3	0	0	0
3	Pinned	Fixed	3	0	3	0	0	0	-3	3
4	Pinned	Pinned	0	0	0	0	0	0	0	0

COMPUTER ANALYSIS OF STRUCTURAL FRAMEWORKS

Uniformly distributed loads and point loads acting perpendicular to the members are catered for through the use of fixed end forces and equivalent joint forces. Initially the fixed end forces are calculated on the assumption that the member is rigidly connected to the nodes at its ends. The moments developed at pinned connections are then released to produce fixed end forces that are consistent with the end connections. Once the unknown nodal displacements have been found they are used to calculate the member end forces. Note that the stiffness factors K_1 - K_8 are used to facilitate the calculation of the member end forces. The reactions are evaluated by summing the contributions from the members connected to the restrained nodes.

Output The data from the data file is echoed and messages are output to indicate the beginning of each solution phase. After solution of the stiffness equation nodal displacements are output, the axial force, shear force, and bending moment at each end of each member are then output. Finally the components of reaction (in the global axes system) at each restrained node are output.

Listing

```
1000 REM  ***********************************************************************
1010 REM  *                                                                     *
1020 REM  *    PROGRAM   P F R A M E . P F   FOR THE AUTOMATIC ANALYSIS         *
1030 REM  *    OF PLANE FRAMES                                                  *
1040 REM  *                                                                     *
1050 REM  *    DATA FROM DATA FILE GENERATED BY A TEXT EDITOR OR BY THE         *
1060 REM  *    PREPROCESSOR PROGRAM P R E . M A / P R E . 0 1 / P R E . 0 2     *
1070 REM  *                                                                     *
1080 REM  *    J.A.D.BALFOUR                                                    *
1090 REM  *                                                                     *
1100 REM  ***********************************************************************
1110 REM
1120 REM  ---  NOTE THAT THE VARIABLES ARE DEFINED IN APPENDIX A
1130 REM
1140 OPTION BASE 1
1150 REM
1160 REM  ---  THE FOLLOWING DIMENSION STATEMENT SETS THE PROBLEM SIZE
1170 REM
1180 COMMON FILE$
1190 DIM NODE(20,2),    MEMB(40,9), REST(20,5), NLOD(40,3), MLOD(20,9)
1200 DIM FSTIFF(40,20), LOD(40),    FREE(60),   EFREE(6),   ESTIFF(6,6)
1210 DIM TEMP(12)
1220 PRINT : PRINT
1230 PRINT "----------------------------------------------"
1240 PRINT
1250 PRINT "PROGRAM   P F R A M E . P F   FOR THE AUTOMATIC"
1260 PRINT "ANALYSIS OF PLANE FRAMES"
1270 PRINT
1280 PRINT "----------------------------------------------"
1290 REM
1300 REM  ---  CHECK IF THIS PROGRAM HAS BEEN CHAINED FROM PRE.MA/PRE.01/PRE.02
1310 REM
1320 IF LEN(FILE$) <> 0 THEN GOTO 1350
1330 PRINT
```

PLANE FRAMES - FURTHER TOPICS

```
1340 INPUT          "NAME OF THE DATA FILE  =  ", FILE$
1350 OPEN "I", #1, FILE$
1360 PRINT
1370 PRINT          "READING DATA"
1380 GOSUB 1620
1390 REM
1400 REM            PRINT DATA
1410 GOSUB 2030
1420 PRINT
1430 PRINT          "GENERATING THE FREEDOM VECTOR"
1440 GOSUB 2720
1450 PRINT
1460 PRINT          "GENERATING THE STRUCTURE STIFFNESS MATRIX"
1470 GOSUB 3080
1480 PRINT
1490 PRINT          "GENERATING THE LOADING VECTOR"
1500 GOSUB 3970
1510 PRINT
1520 PRINT          "SOLVING EQUATIONS  -  NO. OF EQUATIONS  = "; NDOF
1530 PRINT          "SEMI-BANDWIDTH                          = "; BAND
1540 GOSUB 4790
1550 REM
1560 REM            OUTPUT NODAL DISPLACEMENTS
1570 GOSUB 5390
1580 REM
1590 REM            CALCULATE AND OUTPUT MEMBER FORCES AND REACTIONS
1600 GOSUB 5640
1610 END
1620 REM * * * * * * * * * * * * * * * * * * * * * * * * * * * * * * *
1630 REM *                                                            *
1640 REM *    SUBROUTINE TO READ DATA FROM THE INPUT FILE             *
1650 REM *                                                            *
1660 REM * * * * * * * * * * * * * * * * * * * * * * * * * * * * * * *
1670 REM
1680 INPUT #1, TITLE$
1690 INPUT #1, PTYPE, NCORD, NPROP, NNODE, NMEMB, NREST, NNLOD, NMLOD
1700 IF PTYPE = 3 THEN GOTO 1750
1710 PRINT : PRINT "NOT A PLANE FRAME DATA FILE" : END
1720 REM
1730 REM  ---  READ NODAL COORDINATES
1740 REM
1750 FOR I = 1 TO NNODE
1760 INPUT #1, NODE(I,1), NODE(I,2)
1770 NEXT I
1780 REM
1790 REM  ---  READ MEMBER DATA
1800 REM
1810 FOR I = 1 TO NMEMB
1820 INPUT #1, MEMB(I,1), MEMB(I,2), MEMB(I,3), MEMB(I,4), MEMB(I,5), MEMB(I,6)
1830 NEXT I
1840 REM
1850 REM  ---  READ RESTRAINT DATA
1860 REM
1870 FOR I = 1 TO NREST
1880 INPUT #1, REST(I,1), REST(I,2)
1890 NEXT I
1900 REM
1910 REM  ---  READ NODAL LOADS
1920 REM
1930 FOR I = 1 TO NNLOD
1940 INPUT #1, NLOD(I,1), NLOD(I,2), NLOD(I,3)
1950 NEXT I
1960 REM
1970 REM  ---  READ MEMBER LOADS
1980 REM
1990 FOR I = 1 TO NMLOD
2000 INPUT #1, MLOD(I,1), MLOD(I,2), MLOD(I,3), MLOD(I,4), MLOD(I,5)
2010 NEXT I
2020 RETURN
```

```
2030 REM ******************************************
2040 REM *                                          *
2050 REM *     SUBROUTINE TO PRINT DATA             *
2060 REM *                                          *
2070 REM ******************************************
2080 REM
2090 PRINT
2100 PRINT "+ + + + + + + +"
2110 PRINT "+    JOB TITLE   +    -   "; TITLE$
2120 PRINT "+ + + + + + + +"
2130 PRINT
2140 PRINT "+ + + + + + + + + + + +"
2150 PRINT "+    NODAL COORDINATES   +"
2160 PRINT "+ + + + + + + + + + + +"
2170 PRINT
2180 PRINT "NODE             X               Y"
2190 PRINT "NUMBER"
2200 FOR I = 1 TO NNODE
2210 PRINT " "; I, NODE(I,1), NODE(I,2)
2220 NEXT I
2230 PRINT
2240 PRINT "+ + + + + + + + + + + +"
2250 PRINT "+    MEMBER PROPERTIES   +"
2260 PRINT "+ + + + + + + + + + + +"
2270 PRINT
2280 PRINT "MEMBER          A           I          TYPE          E"
2290 FOR I = 1 TO NMEMB
2300 PRINT USING "## ##      "; MEMB(I,1), MEMB(I,2);
2310 FOR J = 3 TO NPROP
2320 PRINT USING "    #.###^^^^"; MEMB(I,J);
2330 NEXT J
2340 PRINT
2350 NEXT I
2360 PRINT
2370 PRINT "+ + + + + + + +"
2380 PRINT "+   RESTRAINTS  +"
2390 PRINT "+ + + + + + + +"
2400 PRINT
2410 PRINT "NODE     DIRECTION(S)"
2420 PRINT "NUMBER"
2430 FOR I = 1 TO NREST
2440 PRINT USING "  ##           ###"; REST(I,1), REST(I,2)
2450 NEXT I
2460 IF NNLOD = 0 THEN GOTO 2570 ELSE PRINT
2470 PRINT "+ + + + + + + + +"
2480 PRINT "+   NODAL LOADS   +"
2490 PRINT "+ + + + + + + + +"
2500 PRINT
2510 PRINT       "NODE     DIRECTION    VALUE"
2520 PRINT       "NUMBER"
2530 FOR I = 1 TO NNLOD
2540 PRINT USING " ##         #"; NLOD(I,1), NLOD(I,2);
2550 PRINT USING "    #.###^^^^"; NLOD(I,3)
2560 NEXT I
2570 IF NMLOD = 0 THEN GOTO 2710
2580 PRINT
2590 PRINT "+ + + + + + + + + +"
2600 PRINT "+   MEMBER LOADS   +    -    UDL = 1,    POINT LOAD = 2"
2610 PRINT "+ + + + + + + + + +"
2620 PRINT
2630 PRINT "MEMBER    LOAD    MAGNITUDE    DISTANCE FROM"
2640 PRINT "          TYPE                 LOWER NODE"
2650 FOR I = 1 TO NMLOD
2660 PRINT USING "## ##        #"; MLOD(I,1), MLOD(I,2), MLOD(I,3);
2670 PRINT USING "    #.###^^^^"; MLOD(I,4);
2680 IF MLOD(I,3) = 2 THEN PRINT USING "    #.###^^^^"; MLOD(I,5)
2690 IF MLOD(I,3) <> 2 THEN PRINT "         N/A"
2700 NEXT I
2710 RETURN
```

PLANE FRAMES - FURTHER TOPICS

```
2720 REM ******************************************
2730 REM *                                          *
2740 REM *    SUBROUTINE TO GENERATE THE FREEDOM VECTOR   *
2750 REM *                                          *
2760 REM ******************************************
2770 REM
2780 REM --- ZERO THE FREEDOM VECTOR
2790 REM
2800 FOR I    = 1 TO 3*NNODE
2810 FREE(I) = 0
2820 NEXT I
2830 REM
2840 REM --- LOOP FOR ALL RESTRAINED NODES
2850 REM
2860 FOR I    = 1 TO NREST
2870 K        = REST(I,2)
2880 REM
2890 REM --- EVALUATE THE FREEDOM TO BE RESTRAINED FROM RIGHT-MOST DIGIT OF K
2900 REM
2910 J        = 3*REST(I,1) - 3 + K MOD 10
2920 FREE(J)  = 1
2930 K        = INT(K/10)
2940 IF K     > 0 THEN GOTO 2910
2950 NEXT I
2960 REM
2970 REM --- NUMBER THE FREEDOMS (RESTRAINTS SET TO ZERO)
2980 REM
2990 NDOF     = 0
3000 FOR I    = 1 TO 3*NNODE
3010 IF FREE(I) = 1 THEN GOTO 3050
3020 NDOF     = NDOF + 1
3030 FREE(I)  = NDOF
3040 GOTO 3060
3050 FREE(I)  = 0
3060 NEXT I
3070 RETURN
3080 REM ******************************************
3090 REM *                                          *
3100 REM *    SUBROUTINE TO GENERATE THE STRUCTURE STIFFNESS MATRIX   *
3110 REM *                                          *
3120 REM ******************************************
3130 REM
3140 REM --- ZERO THE STIFFNESS MATRIX AND LOAD VECTOR
3150 REM
3160 FOR I    = 1 TO NDOF
3170 FOR J    = 1 TO 20
3180 FSTIFF(I,J) = 0
3190 NEXT J
3200 LOD(I)   = 0
3210 NEXT I
3220 REM
3230 REM --- LOOP FOR EACH ELEMENT
3240 REM
3250 BAND     = 0
3260 PRINT "ADDING ELEMENT :- ";
3270 FOR K = 1 TO NMEMB
3280 IN = MEMB(K,1)
3290 JN = MEMB(K,2)
3300 PRINT K;
3310 REM
3320 REM --- CALCULATE ELEMENT LENGTH (STORE IN MEMB(K,7))
3330 REM
3340 MEMB(K,7) = SQR((NODE(JN,1)-NODE(IN,1))^2 + (NODE(JN,2)-NODE(IN,2))^2)
3350 REM
3360 REM --- CALCULATE THE COSINE AND SINE OF THE ANGLE THE MEMBER MAKES
3370 REM --- WITH THE COORDINATE X-AXIS (STORE IN MEMB(K,8) AND MEMB(K,9))
3380 REM
3390 MEMB(K,8) = (NODE(JN,1) - NODE(IN,1)) / MEMB(K,7)
3400 MEMB(K,9) = (NODE(JN,2) - NODE(IN,2)) / MEMB(K,7)
3410 REM
```

```
3420 REM --- GENERATE THE UPPER TRIANGLE OF THE ELEMENT STIFFNESS MATRIX
3430 REM
3440 A   = MEMB(K,3)
3450 A2  = MEMB(K,4)
3460 E   = MEMB(K,6)
3470 L   = MEMB(K,7)
3480 C   = MEMB(K,8)
3490 S   = MEMB(K,9)
3500 GOSUB 6670
3510 ESTIFF(1,1) =   E * A * C * C / L + K1 * E * A2 * S * S / L^3
3520 ESTIFF(1,2) =   E * A * S * C / L - K1 * E * A2 * S * C / L^3
3530 ESTIFF(1,3) = - K2 * E * A2 * S / L^2
3540 ESTIFF(1,4) = - ESTIFF(1,1)
3550 ESTIFF(1,5) = - ESTIFF(1,2)
3560 ESTIFF(1,6) = - K3 * E * A2 * S / L^2
3570 ESTIFF(2,2) =   E * A * S * S / L + K1 * E * A2 * C * C / L^3
3580 ESTIFF(2,3) =   K2 * E * A2 * C / L^2
3590 ESTIFF(2,4) = - ESTIFF(1,2)
3600 ESTIFF(2,5) = - ESTIFF(2,2)
3610 ESTIFF(2,6) =   K3 * E * A2 * C / L^2
3620 ESTIFF(3,3) =   K4 * E * A2 / L
3630 ESTIFF(3,4) = - K5 * E * A2 * S / L^2
3640 ESTIFF(3,5) =   K5 * E * A2 * C / L^2
3650 ESTIFF(3,6) =   K6 * E * A2 / L
3660 ESTIFF(4,4) =   ESTIFF(1,1)
3670 ESTIFF(4,5) =   ESTIFF(1,2)
3680 ESTIFF(4,6) = - K7 * E * A2 * S / L^2
3690 ESTIFF(5,5) =   ESTIFF(2,2)
3700 ESTIFF(5,6) =   K7 * E * A2 * C / L^2
3710 ESTIFF(6,6) =   K8 * E * A2 / L
3720 REM
3730 REM --- SET UP THE ELEMENT CODE NUMBER
3740 REM
3750 EFREE(1) = FREE(3*IN-2)
3760 EFREE(2) = FREE(3*IN-1)
3770 EFREE(3) = FREE(3*IN)
3780 EFREE(4) = FREE(3*JN-2)
3790 EFREE(5) = FREE(3*JN-1)
3800 EFREE(6) = FREE(3*JN)
3810 REM
3820 REM --- ADD THE ELEMENT STIFFNESS TERMS
3830 REM
3840 FOR I     = 1 TO 6
3850 IF EFREE(I) = 0 THEN GOTO 3930
3860 FOR J     = 1 TO 6
3870 IF EFREE(J) = 0 THEN GOTO 3920
3880 L         = EFREE(I)
3890 M         = EFREE(J) - L + 1
3900 IF M      > BAND THEN BAND = M
3910 FSTIFF(L,M) = FSTIFF(L,M) + ESTIFF(I,J)
3920 NEXT J
3930 NEXT I
3940 NEXT K
3950 PRINT
3960 RETURN
3970 REM * * * * * * * * * * * * * * * * * * * * * * * * * * * * * * *
3980 REM *                                                             *
3990 REM *     SUBROUTINE TO SET UP THE LOADING VECTOR                 *
4000 REM *                                                             *
4010 REM * * * * * * * * * * * * * * * * * * * * * * * * * * * * * * *
4020 REM
4030 REM --- ADD NODAL LOADS TO THE LOADING VECTOR
4040 REM
4050 FOR I = 1 TO NNLOD
4060 J     = 3*NLOD(I,1) - 3 + NLOD(I,2)
4070 J     = FREE(J)
4080 LOD(J) = NLOD(I,3)
4090 NEXT I
4100 REM
4110 REM --- ADD MEMBER LOADS TO THE LOADING VECTOR
4120 REM
```

```
4130 FOR I = 1 TO NMLOD
4140 IN   = MLOD(I,1)
4150 JN   = MLOD(I,2)
4160 FOR K       = 1 TO NMEMB
4170 IF MEMB(K,1) <> IN THEN GOTO 4190
4180 IF MEMB(K,2) =  JN THEN J = K
4190 NEXT K
4200 W = MLOD(I,4)
4210 L = MEMB(J,7)
4220 X = MLOD(I,5)
4230 Y = L - X
4240 C = MEMB(J,8)
4250 S = MEMB(J,9)
4260 REM
4270 REM --- CALC. AND STORE F.E.F. (INITIALLY ASSUME FIXED/FIXED MEMBER)
4280 REM
4290 ON MLOD(I,3) GOTO 4330, 4410
4300 REM
4310 REM --- UNIFORMLY DISTRIBUTED LOAD
4320 REM
4330 MLOD(I,6) = - W * L / 2
4340 MLOD(I,7) = - W * L^2 / 12
4350 MLOD(I,8) =   MLOD(I,6)
4360 MLOD(I,9) = - MLOD(I,7)
4370 GOTO 4480
4380 REM
4390 REM --- POINT LOAD
4400 REM
4410 MLOD(I,6) = - W * (Y/L)^2 * (1 + 2*X/L)
4420 MLOD(I,7) = - W * X * Y^2 / L^2
4430 MLOD(I,8) = - W * (X/L)^2 / L^2
4440 MLOD(I,9) =   W * X^2 * Y / L^2
4450 REM
4460 REM --- TAKE ACCOUNT OF ANY PINS
4470 REM
4480 ON MEMB(J,5) GOTO 4650, 4540, 4490, 4490
4490 MLOD(I,6) = MLOD(I,6) - 1.5*MLOD(I,7)/L
4500 MLOD(I,8) = MLOD(I,8) + 1.5*MLOD(I,7)/L
4510 MLOD(I,9) = MLOD(I,9) -     MLOD(I,7)/2
4520 MLOD(I,7) = 0
4530 IF MEMB(J,5) = 3 THEN GOTO 4650 ELSE GOTO 4590
4540 MLOD(I,6) = MLOD(I,6) - 1.5*MLOD(I,9)/L
4550 MLOD(I,7) = MLOD(I,7) -     MLOD(I,9)/2
4560 MLOD(I,8) = MLOD(I,8) + 1.5*MLOD(I,9)/L
4570 MLOD(I,9) = 0
4580 GOTO 4650
4590 MLOD(I,6) = MLOD(I,6) - MLOD(I,9)/L
4600 MLOD(I,8) = MLOD(I,8) + MLOD(I,9)/L
4610 MLOD(I,9) = 0
4620 REM
4630 REM --- ADD EQUIVALENT JOINT FORCES TO THE LOADING VECTOR
4640 REM
4650 J = FREE(3*IN - 2)
4660 IF J <> 0 THEN LOD(J) = LOD(J) + MLOD(I,6)*S
4670 J = FREE(3*IN - 1)
4680 IF J <> 0 THEN LOD(J) = LOD(J) - MLOD(I,6)*C
4690 J = FREE(3*IN)
4700 IF J <> 0 THEN LOD(J) = LOD(J) - MLOD(I,7)
4710 J = FREE(3*JN - 2)
4720 IF J <> 0 THEN LOD(J) = LOD(J) + MLOD(I,8)*S
4730 J = FREE(3*JN - 1)
4740 IF J <> 0 THEN LOD(J) = LOD(J) - MLOD(I,8)*C
4750 J = FREE(3*JN)
4760 IF J <> 0 THEN LOD(J) = LOD(J) - MLOD(I,9)
4770 NEXT I
4780 RETURN
4790 REM * * * * * * * * * * * * * * * * * * * * * * * * * * * * * * * *
4800 REM *                                                              *
4810 REM *    SUBROUTINE TO SOLVE LINEAR SIMULTANEOUS EQUATIONS BY      *
4820 REM *    GAUSS ELIMINATION WITH NO ROW INTERCHANGE                 *
4830 REM *                                                              *
4840 REM * * * * * * * * * * * * * * * * * * * * * * * * * * * * * * * *
```

```
4850 REM
4860 REM    ---  LOOP FOR ALL PIVOTS
4870 REM
4880 BB       = BAND
4890 PRINT "USING PIVOT     :- ";
4900 FOR I  = 1 TO NDOF
4910 PRINT I;
4920 REM
4930 REM    ---  CHECK IF IN THE UNUSED TRIANGLE
4940 REM
4950 IF I  > NDOF-BAND+1 THEN BB = NDOF-I+1
4960 PIVOT  = FSTIFF(I,1)
4970 REM
4980 REM    ---  NORMALISE
4990 REM
5000 FOR J  = 1 TO BB
5010 FSTIFF(I,J) = FSTIFF(I,J) / PIVOT
5020 NEXT J
5030 LOD(I) = LOD(I) / PIVOT
5040 REM
5050 REM    ---  CHECK IF LAST ROW
5060 REM
5070 IF BB  = 1 THEN GOTO 5260
5080 REM
5090 REM    ---  ELIMINATE (WITHIN BAND) FOR ALL ROWS ABOVE PIVOT
5100 REM
5110 FOR K  = 2 TO BB
5120 REM
5130 REM    ---  CALCULATE ROW NUMBER THEN EVALUATE MULTIPLIER
5140 REM
5150 L      = I + K - 1
5160 MULT   = FSTIFF(I,K) * PIVOT
5170 REM
5180 REM    ---  LOOP FOR ELEMENTS IN THE ELIMINATION ROW
5190 REM
5200 FOR J  = K TO BB
5210 M      = J - K + 1
5220 FSTIFF(L,M) = FSTIFF(L,M) - MULT * FSTIFF(I,J)
5230 NEXT J
5240 LOD(L) = LOD(L) - MULT*LOD(I)
5250 NEXT K
5260 NEXT I
5270 PRINT
5280 REM
5290 REM    ---  BACK SUBSTITUTE
5300 REM
5310 FOR I  = 1 TO NDOF-1
5320 BB     = I
5330 IF I  > BAND-1 THEN BB = BAND-1
5340 FOR J  = 1 TO BB
5350 LOD(NDOF-I) = LOD(NDOF-I) - FSTIFF(NDOF-I,J+1)*LOD(NDOF-I+J)
5360 NEXT J
5370 NEXT I
5380 RETURN
5390 REM  * * * * * * * * * * * * * * * * * * * * * * * * * * * * *
5400 REM  *                                                       *
5410 REM  *    SUBROUTINE TO OUTPUT THE DISPLACEMENTS             *
5420 REM  *                                                       *
5430 REM  * * * * * * * * * * * * * * * * * * * * * * * * * * * * *
5440 REM
5450 PRINT
5460 PRINT "* * * * * * * * * *"
5470 PRINT "*   DISPLACEMENTS   *        * :-  RESTRAINT"
5480 PRINT "* * * * * * * * * *"
5490 PRINT
5500 PRINT "NODE     X-DISPLACEMENT Y-DISPLACEMENT      ROTATION"
5510 PRINT
5520 FOR I = 1 TO NNODE
5530 PRINT USING "###     "; I;
5540 FOR J = 1 TO 3
5550 K      = FREE(3*I+J-3)
```

PLANE FRAMES - FURTHER TOPICS

```
5560 IF K  = 0 THEN GOTO 5590
5570 PRINT USING "      #.####^^^^"; LOD(K);
5580 GOTO 5600
5590 PRINT            "          *   ";
5600 NEXT J
5610 PRINT
5620 NEXT I
5630 RETURN
5640 REM  * * * * * * * * * * * * * * * * * * * * * * * * * * * * * * * * *
5650 REM  *                                                                *
5660 REM  *    SUBROUTINE TO CALCULATE AND OUTPUT THE MEMBER FORCES AND    *
5670 REM  *    REACTIONS                                                   *
5680 REM  *                                                                *
5690 REM  * * * * * * * * * * * * * * * * * * * * * * * * * * * * * * * * *
5700 REM
5710 REM  ---  ZERO THE REACTION COMPONENTS
5720 REM
5730 FOR I     = 1 TO NREST
5740 REST(I,3) = 0
5750 REST(I,4) = 0
5760 REST(I,5) = 0
5770 NEXT I
5780 FOR K = 1 TO NMEMB
5790 A  = MEMB(K,3)
5800 A2 = MEMB(K,4)
5810 E  = MEMB(K,6)
5820 L  = MEMB(K,7)
5830 C  = MEMB(K,8)
5840 S  = MEMB(K,9)
5850 I  = 3*MEMB(K,1) - 3
5860 REM
5870 REM  ---  EXTRACT THE NODAL DISPLACEMENTS - STORE TEMP(1)-(6)
5880 REM
5890 FOR M = 1 TO 6
5900 J     = FREE(I+M)
5910 IF J  = 0 THEN TEMP(M) = 0 ELSE TEMP(M) = LOD(J)
5920 IF M  = 3 THEN I = 3*MEMB(K,2) - 6
5930 NEXT M
5940 REM
5950 REM  ---  CALC. MEMBER END FORCES - STORE TEMP(7)-(12)
5960 REM
5970 GOSUB 6670
5980 TEMP(7)  =   A*E* ((TEMP(1) - TEMP(4))*C + (TEMP(2) - TEMP(5))*S)/L
5990 TEMP(8)  = K1*E*A2*((TEMP(4) - TEMP(1))*S + (TEMP(2) - TEMP(5))*C)/L^3
6000 TEMP(8)  = TEMP(8) + E*A2*(K2*TEMP(3) + K3*TEMP(6))/L^2
6010 TEMP(9)  = (-K2*TEMP(1) - K5*TEMP(4))*S + (K2*TEMP(2) + K5*TEMP(5))*C
6020 TEMP(9)  = E*A2*(TEMP(9)/L^2 + (K4*TEMP(3) + K6*TEMP(6))/L)
6030 TEMP(10) = - TEMP(7)
6040 TEMP(11) = - TEMP(8)
6050 TEMP(12) = - TEMP(9) + L*TEMP(8)
6060 REM
6070 REM  ---  ADD IN FIXED END FORCES
6080 REM
6090 IF NMLOD = 0 THEN GOTO 6180
6100 FOR I     = 1 TO NMLOD
6110 IF MEMB(K,1) <> MLOD(I,1) THEN GOTO 6170
6120 IF MEMB(K,2) <> MLOD(I,2) THEN GOTO 6170
6130 TEMP(8)  = TEMP(8)  + MLOD(I,6)
6140 TEMP(9)  = TEMP(9)  + MLOD(I,7)
6150 TEMP(11) = TEMP(11) + MLOD(I,8)
6160 TEMP(12) = TEMP(12) + MLOD(I,9)
6170 NEXT I
6180 PRINT
6190 PRINT            "* * * * * * * * * *"
6200 PRINT USING "*     ELEMENT ## ##    *"; MEMB(K,1), MEMB(K,2)
6210 PRINT            "* * * * * * * * * *"
6220 PRINT
6230 PRINT "              AXIAL         SHEAR        BENDING"
6240 PRINT "              FORCE         FORCE        MOMENT"
6250 PRINT USING "NODE ##"; MEMB(K,1);
6260 PRINT USING "     #.####^^^^"; TEMP(7), TEMP(8), TEMP(9)
```

```
6270 PRINT USING "NODE ##"; MEMB(K,2);
6280 PRINT USING "     #.####^^^^"; TEMP(10), TEMP(11), TEMP(12)
6290 REM
6300 REM  ---  ADD ANY CONTRIBUTION FROM THE MEMBER FORCES TO THE REACTIONS
6310 REM
6320 FOR I = 1 TO NREST
6330 J      = 7
6340 IF MEMB(K,1) = REST(I,1) THEN GOTO 6370
6350 J      = 10
6360 IF MEMB(K,2) <> REST(I,1) THEN GOTO 6400
6370 REST(I,3) = REST(I,3) + C*TEMP(J) - S*TEMP(J+1)
6380 REST(I,4) = REST(I,4) + S*TEMP(J) + C*TEMP(J+1)
6390 REST(I,5) = REST(I,5) + TEMP(J+2)
6400 NEXT I
6410 NEXT K
6420 REM
6430 REM  ---  ADD ANY APPLIED LOADS
6440 REM
6450 FOR I = 1 TO NREST
6460 FOR J = 1 TO NNLOD
6470 IF REST(I,1) <> NLOD(J,1) THEN GOTO 6500
6480 K      = NLOD(J,2)
6490 REST(I,2+K) = REST(I,2+K) - NLOD(J,3)
6500 NEXT J
6510 NEXT I
6520 REM
6530 REM  ---  OUTPUT THE REACTIONS
6540 REM
6550 PRINT
6560 PRINT "* * * * * * * * *"
6570 PRINT "*    REACTIONS  *"
6580 PRINT "* * * * * * * * *"
6590 PRINT
6600 PRINT "           X-COMPONENT     Y-COMPONENT          MOMENT"
6610 PRINT
6620 FOR I   = 1 TO NREST
6630 PRINT USING "NODE ##"; REST(I,1);
6640 PRINT USING "     #.####^^^^"; REST(I,3), REST(I,4), REST(I,5)
6650 NEXT I
6660 RETURN
6670 REM * * * * * * * * * * * * * * * * * * * * * * * * * * * * * * *
6680 REM *                                                            *
6690 REM *    SUBROUTINE TO SET UP ELEMENT STIFFNESS FACTORS FOR      *
6700 REM *       DIFFERENT END CONNECTIONS                            *
6710 REM *                                                            *
6720 REM * * * * * * * * * * * * * * * * * * * * * * * * * * * * * * *
6730 REM
6740 REM  ---  ZERO STIFFNESS FACTORS (I.E. PINNED/PINNED MEMBER)
6750 REM
6760 K1 =  0 : K2 =  0 : K3 =  0 : K4 =  0
6770 K5 =  0 : K6 =  0 : K7 =  0 : K8 =  0
6780 ON MEMB(K,5) GOTO 6820, 6880, 6930, 6940
6790 REM
6800 REM  ---  FIXED/FIXED MEMBER
6810 REM
6820 K1 = 12 : K2 =  6 : K3 =  6 : K4 =  4
6830 K5 = -6 : K6 =  2 : K7 = -6 : K8 =  4
6840 GOTO 6940
6850 REM
6860 REM  ---  FIXED/PINNED MEMBER
6870 REM
6880 K1 =  3 : K2 =  3 : K4 =  3 : K5 = -3
6890 GOTO 6940
6900 REM
6910 REM  ---  PINNED/FIXED MEMBER
6920 REM
6930 K1 =  3 : K3 =  3 : K7 = -3 : K8 =  3
6940 RETURN
```

PLANE FRAMES - FURTHER TOPICS

Data Generation The following printout shows program PRE.MA/PRE.O1/PRE.O2 being used to generate the data for example 7.5. Note that when the data is complete PFRAME.PF could have been run directly from PRE.O2. In that case the data file would have been stored on disk and then PFRAME.PF chained with the name of the data file being passed via COMMON.

```
-----------------------------------------------------------
PROGRAM    P R E . M A    TO PREPROCESS DATA FOR PROGRAMS
PTRUSS.PT, STRUSS.PT, PFRAME.PF, GRID.GD, AND SFRAME.SF
-----------------------------------------------------------

DO YOU WISH TO
(1) CREATE A NEW DATA FILE
(2) MODIFY AN EXISTING DATA FILE
? 1

NAME OF THE DATA FILE  =  DATA1.PF

WAIT - CHAINING PRE.O1

SELECT PROBLEM TYPE
(1) PLANE TRUSS
(2) SPACE TRUSS
(3) PLANE FRAME
(4) GRILLAGE
(5) SPACE FRAME
? 3

JOB TITLE  =  EXAMPLE 7.5  -  FILE DATA1.PF

NUMBER OF NODES  =  3

NODE 1

X-COORDINATE    =  0
Y-COORDINATE    =  0

NODE 2

X-COORDINATE    =  0
Y-COORDINATE    =  3000

NODE 3

X-COORDINATE    =  7000
Y-COORDINATE    =  4000

NUMBER OF MEMBERS        =  2

ELEMENT      LOWER       HIGHER
 TYPE        NODE         NODE

   1         FIXED        FIXED
   2         FIXED        PINNED
   3         PINNED       FIXED
   4         PINNED       PINNED

HIT CARRIAGE RETURN IF THE VALUE IS
THE SAME AS THE PREVIOUS MEMBER
```

MEMBER 1

LOWER NODE NUMBER   = 1
HIGHER NODE NUMBER   = 2
CROSS-SECTIONAL AREA = 9000
SECOND MOMENT OF AREA = .25E9
ELEMENT TYPE   = 1
ELASTIC CONSTANT   = 200

MEMBER 2

LOWER NODE NUMBER   = 2
HIGHER NODE NUMBER   = 3
CROSS-SECTIONAL AREA = 9000
SECOND MOMENT OF AREA = 2.5E+08
ELEMENT TYPE   = 1
ELASTIC CONSTANT   = 200

NUMBER OF RESTRAINED NODES = 2

RESTRAINT DIRECTIONS ARE
X-TRANSLATION = 1
Y-TRANSLATION = 2
ROTATION   = 3

ENTER RESTRAINED DIRECTIONS AS A COMPOSITE NUMBER
E.G. IF NODE IS RESTRAINED IN DIRECTIONS 1 AND 2 ENTER 12

HIT CARRIAGE RETURN IF THE VALUE IS
THE SAME AS THE PREVIOUS RESTRAINT
RESTRAINED NODE NUMBER 1

NODE TO BE RESTRAINED = 1
DIRECTION(S)   = 12

RESTRAINED NODE NUMBER 2

NODE TO BE RESTRAINED = 3
DIRECTION(S)   = 12

NUMBER OF NODAL LOADS = 3

LOAD DIRECTIONS ARE
FORCE IN THE X-DIRECTION = 1
FORCE IN THE Y-DIRECTION = 2
MOMENT   = 3

HIT CARRIAGE RETURN IF THE VALUE IS
THE SAME AS THE PREVIOUS LOAD

NODAL LOAD NUMBER 1

NODE CARRYING LOAD = 2
DIRECTION   = 1
MAGNITUDE OF LOAD = 150

NODAL LOAD NUMBER 2

NODE CARRYING LOAD = 2
DIRECTION   = 2
MAGNITUDE OF LOAD = -150

NODAL LOAD NUMBER 3

NODE CARRYING LOAD = 1
DIRECTION   = 3
MAGNITUDE OF LOAD = 100000

NUMBER OF MEMBER LOADS = 0

```
WAIT - CHAINING PRE.02

+ + + + + + + + +
+   JOB TITLE   +   -   EXAMPLE 7.5 - FILE DATA1.PF
+ + + + + + + + +

+ + + + + + + + + + + +
+   NODAL COORDINATES  +
+ + + + + + + + + + + +

NODE            X               Y
NUMBER
  1             0               0
  2             0               3000
  3             7000            4000

+ + + + + + + + + + + +
+   MEMBER PROPERTIES  +
+ + + + + + + + + + + +

MEMBER          A           I          PTYPE       E
NUMBER
 1  2      0.900E+04  0.250E+09  0.100E+01  0.200E+03
 2  3      0.900E+04  0.250E+09  0.100E+01  0.200E+03

+ + + + + + + + + +
+    RESTRAINTS   +
+ + + + + + + + + +

NODE      DIRECTION(S)
NUMBER
  1         12
  3         12

+ + + + + + + + + +
+   NODAL LOADS   +
+ + + + + + + + + +

  1         3        0.100E+06
  2         1        0.150E+03
  2         2       -.150E+03

DO YOU WISH TO (CR = CARRIAGE RETURN)
(1)  MODIFY DATA
(2)  LIST ALL THE CURRENT DATA
(3)  RUN DATA FILE WITH APPROPRIATE ANALYSIS PROGRAM
(CR) STOP
?                                            Carriage return entered
```

Sample Run 1 The following sample run shows program PFRAME.PF being used to analyse the structure analysed in example 7.5.

```
------------------------------------------------
PROGRAM   P F R A M E . P F   FOR THE AUTOMATIC
ANALYSIS OF PLANE FRAMES
------------------------------------------------

NAME OF THE DATA FILE = DATA1.PF

READING DATA

+ + + + + + + + +
+   JOB TITLE   +   -   EXAMPLE 7.5 - FILE DATA1.PF
+ + + + + + + + +
```

COMPUTER ANALYSIS OF STRUCTURAL FRAMEWORKS

```
+ + + + + + + + + + + +
+   NODAL COORDINATES   +
+ + + + + + + + + + + +
```

NODE NUMBER	X	Y
1	0	0
2	0	3000
3	7000	4000

```
+ + + + + + + + + + + +
+   MEMBER PROPERTIES   +
+ + + + + + + + + + + +
```

MEMBER		A	I	TYPE	E
1	2	0.900E+04	0.250E+09	0.100E+01	0.200E+03
2	3	0.900E+04	0.250E+09	0.100E+01	0.200E+03

```
+ + + + + + + + +
+   RESTRAINTS   +
+ + + + + + + + +
```

NODE NUMBER	DIRECTION(S)
1	12
3	12

```
+ + + + + + + + + +
+   NODAL LOADS   +
+ + + + + + + + + +
```

NODE NUMBER	DIRECTION	VALUE
1	3	-.100E+06
2	1	0.150E+03
2	2	-.150E+03

GENERATING THE FREEDOM VECTOR

GENERATING THE STRUCTURE STIFFNESS MATRIX
ADDING ELEMENT :- 1 2

GENERATING THE LOADING VECTOR

SOLVING EQUATIONS - NO. OF EQUATIONS = 5
SEMI-BANDWIDTH   = 4
USING PIVOT   :- 1 2 3 4 5

```
* * * * * * * * * * *
*   DISPLACEMENTS   *         * :- RESTRAINT
* * * * * * * * * * *
```

NODE	X-DISPLACEMENT	Y-DISPLACEMENT	ROTATION
1	*	*	-.2163E-02
2	0.7916E+00	-.2969E+00	0.5339E-03
3	*	*	-.1809E-03

```
* * * * * * * * * * *
*   ELEMENT  1  2   *
* * * * * * * * * * *
```

	AXIAL FORCE	SHEAR FORCE	BENDING MOMENT
NODE 1	0.1781E+03	-.3670E+02	-.1000E+06
NODE 2	-.1781E+03	0.3670E+02	-.1011E+05

```
* * * * * * * * * * *
*   ELEMENT  2  3   *
* * * * * * * * * * *
```

PLANE FRAMES - FURTHER TOPICS

```
                   AXIAL          SHEAR         BENDING
                   FORCE          FORCE         MOMENT
         NODE 2    0.1888E+03     0.1430E+01    0.1011E+05
         NODE 3   -.1888E+03     -.1430E+01     0.1953E-02

         * * * * * * * *
         *  REACTIONS   *
         * * * * * * * *

                   X-COMPONENT    Y-COMPONENT   MOMENT

         NODE 1    0.3670E+02     0.1781E+03    0.0000E+00
         NODE 3   -.1867E+03     -.2812E+02     0.1953E-02
```

Sample Run 2 The following sample run shows the program PFRAME.PF being used to analyse the structure shown in example 8.4 when the pin is assumed to be in member 1,2. The data was generated using the program PRE.MA/PRE.01/PRE.02 and was stored on file DATA2.PF.

```
-------------------------------------------------
PROGRAM   P F R A M E . P F   FOR THE AUTOMATIC
ANALYSIS OF PLANE FRAMES
-------------------------------------------------

NAME OF THE DATA FILE  =  DATA2.PF

READING DATA

+ + + + + + + + +
+  JOB TITLE   +    -   EXAMPLE 8.4 (PIN IN MEMBER 2 3)  -  DATA2.PF
+ + + + + + + + +


+ + + + + + + + + + + +
+  NODAL COORDINATES  +
+ + + + + + + + + + + +

NODE           X              Y
NUMBER
  1            0              0
  2            5196           3000
  3            8196           0

+ + + + + + + + + + + +
+  MEMBER PROPERTIES  +
+ + + + + + + + + + + +

MEMBER        A            I           TYPE          E
 1  2      0.450E+05    0.340E+09    0.200E+01    0.150E+02
 2  3      0.450E+05    0.340E+09    0.100E+01    0.150E+02

+ + + + + + + + +
+  RESTRAINTS   +
+ + + + + + + + +

NODE      DIRECTION(S)
NUMBER
  1         123
  3         123

+ + + + + + + + + +
+  MEMBER LOADS   +    -   UDL = 1,    POINT LOAD = 2
+ + + + + + + + + +

MEMBER     LOAD       MAGNITUDE     DISTANCE FROM
           TYPE                     LOWER NODE
 1  2       1         -.200E-01        N/A
```

COMPUTER ANALYSIS OF STRUCTURAL FRAMEWORKS

```
GENERATING THE FREEDOM VECTOR

GENERATING THE STRUCTURE STIFFNESS MATRIX
ADDING ELEMENT :- 1 2

GENERATING THE LOADING VECTOR

SOLVING EQUATIONS  -  NO. OF EQUATIONS  =  3
SEMI-BANDWIDTH                          =  3
USING PIVOT      :-  1  2  3

* * * * * * * * * *
*   DISPLACEMENTS   *          * :- RESTRAINT
* * * * * * * * * *

NODE      X-DISPLACEMENT  Y-DISPLACEMENT     ROTATION

  1              *              *                *
  2          0.7330E-01     -.3405E+00       0.6680E-04
  3              *              *                *

* * * * * * * * * *
*   ELEMENT  1  2  *
* * * * * * * * * *

              AXIAL           SHEAR          BENDING
              FORCE           FORCE          MOMENT
NODE  1    0.1201E+02      0.7502E+02      0.9014E+05
NODE  2   -.1201E+02       0.4498E+02     -.2136E-03

* * * * * * * * * *
*   ELEMENT  2  3  *
* * * * * * * * * *

              AXIAL           SHEAR          BENDING
              FORCE           FORCE          MOMENT
NODE  2    0.4655E+02     -.3785E-01      -.7248E-04
NODE  3   -.4655E+02      0.3785E-01      -.1606E+03

* * * * * * * *
*   REACTIONS   *
* * * * * * * *

           X-COMPONENT     Y-COMPONENT       MOMENT

NODE  1   -.2711E+02       0.7098E+02      0.9014E+05
NODE  3   -.3289E+02       0.3294E+02     -.1606E+03
```

9 Grillages

9.1 INTRODUCTION

Grillages, like plane frames, are rigidly jointed planar structures. The two types of structure differ only in the way in which they are loaded; plane frames are loaded in the plane of the structure, whereas grillages (grids) are loaded only perpendicular to the plane of the structure, as illustrated in fig 9.1. Structures such as floor systems, roof systems, and bridge decks are frequently analysed as grillages.

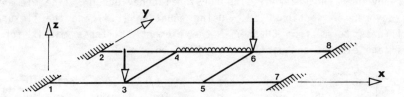

FIGURE 9.1   A SIMPLE GRILLAGE

Each unrestrained node of a grillage has three freedoms, and if the global axes are located as shown in fig 9.2, these freedoms are: translation in the "z" direction, rotation about the "x" axis, and rotation about the "y" axis. In other words grillage analysis is concerned with the out-of-plane displacements and rotations caused by out-of-plane loads applied to a plane structure. As rotation can occur out of the plane of the members, torsion (i.e. twisting along the length of the member) is possible. Take, for example, member 5,6 of the grillage shown in fig 9.1. Under the loading shown, differing degrees of rotation about the "y" axis at nodes 5 and 6 are likely to occur, resulting in torsion of the member.

305

9.2 THE ELEMENT STIFFNESS MATRIX IN THE MEMBER AXES SYSTEM

Figure 9.2 shows the member axes and freedoms adopted for grillage elements in this book. Further, the member axes will be orientated such that the member "z" axis always coincides with the global "z" axis, i.e. the load axis.

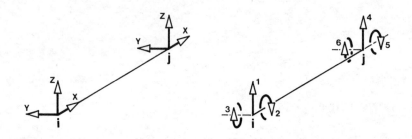

FIGURE 9.2   MEMBER AXES AND FREEDOMS FOR A GRILLAGE ELEMENT

The element stiffness coefficients are found, as before, by evaluation of the force systems required to produce unit displacement in the direction of each freedom in turn. With the exception of the torsional stiffness terms, all other stiffness coefficients have already been encountered. The torsional behaviour of line elements is covered in section 2.12. Using equation (2.4) and the flexural stiffness terms from Chapter 7 the element stiffness matrix for a grillage element can easily be shown to be.

$$\begin{bmatrix} f_{ijz} \\ m_{ijx} \\ m_{ijy} \\ f_{jiz} \\ m_{jix} \\ m_{jiy} \end{bmatrix} = \begin{bmatrix} 12EI/L^3 & 0 & -6EI/L^2 & -12EI/L^3 & 0 & -6EI/L^2 \\ 0 & GJ/L & 0 & 0 & -GJ/L & 0 \\ -6EI/L^2 & 0 & 4EI/L & 6EI/L^2 & 0 & 2EI/L \\ -12EI/L^3 & 0 & 6EI/L^2 & 12EI/L^3 & 0 & 6EI/L^2 \\ 0 & -GJ/L & 0 & 0 & GJ/L & 0 \\ -6EI/L^2 & 0 & 2EI/L & 6EI/L^2 & 0 & 4EI/L \end{bmatrix} \begin{bmatrix} \delta_{ijz} \\ \theta_{ijx} \\ \theta_{ijy} \\ \delta_{jiz} \\ \theta_{jix} \\ \theta_{jiy} \end{bmatrix}$$

(9.1)

When written in terms of the submatrices indicated by the broken lines, the above equation reduces to the well-known expressions

$$\begin{bmatrix} f_{ij} \\ f_{ji} \end{bmatrix} = \begin{bmatrix} k_{ii}^j & k_{ij} \\ k_{ji} & k_{jj}^i \end{bmatrix} \begin{bmatrix} \delta_{ij} \\ \delta_{ji} \end{bmatrix}$$

i.e

$$f_{ij} = k_{ii}^j \delta_{ij} + k_{ij} \delta_{ji} \tag{9.2}$$

and
$$f_{ji} = k_{ji}\delta_i + k^i_{jj}\delta_{ji} \tag{9.3}$$

9.3 TRANSFORMATION OF FORCE AND DISPLACEMENT

The member axes system adopted in the previous section demands that the member "z" axis coincides with the global "z" axis. Hence force and displacement in the z-direction are the same in both axes systems; rendering transformation unnecessary. However, transformation of rotation and moment between the member "x" and "y" axes and the global "x" and "y" axes is necessary. In general the member "x" axis will lie at an angle to the coordinate "x" axis. Let that angle be α, measured as shown in fig 9.3 (i.e. α is the clockwise rotation of member "i,j" about "i" that will make the member axes coincide with the global axes).

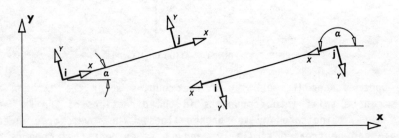

FIGURE 9.3 MEMBER ORIENTATION

Rotations and bending moments can be split into components in the same way as translations and forces. For instance, consider a bending moment about the global "x" axis as shown in fig 9.4(a).

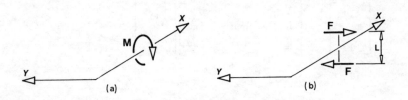

FIGURE 9.4 REPLACEMENT OF A MOMENT BY A COUPLE

The bending moment can be replaced by two equal and opposite forces equidistant from the "x" axis as shown in fig 9.4(b). Such a pair of forces is usually referred to as a *couple*, and M = F L (where "L" is the distance between forces).

Throughout this text moment and rotation about any axis are defined as positive if the sense is anticlockwise when observed looking down the axis towards the origin. Hence the direction of the upper force of positive couples about the "x" and "y" axes is as shown in fig 9.5.

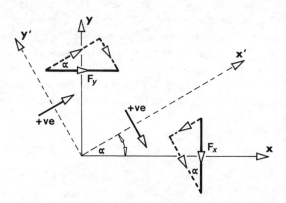

FIGURE 9.5 TRANSFORMATION OF A COUPLE

The upper forces (F_x and F_y) of the couples about the "x" and "y" axes can be split into components in the directions of the "x'" and "y'" axes as indicated by the broken lines. The lower force in each couple can be treated similarly, and it is evident that moments can be transformed in exactly the same way as forces. If the "x,y" axes system is regarded as the global axes system at node "i", and the "x',y'" axes system is regarded as the member axes system at end "i" of member "i,j", it is a simple matter to show that transformation of a nodal force vector from the global to the member axes system takes the following form:

$$\begin{bmatrix} f_{ijz} \\ m_{ijx} \\ m_{ijy} \end{bmatrix} = \begin{bmatrix} 1 & 0 & 0 \\ 0 & \cos\alpha & \sin\alpha \\ 0 & -\sin\alpha & \cos\alpha \end{bmatrix} \begin{bmatrix} F_{ijz} \\ M_{ijx} \\ M_{ijy} \end{bmatrix}$$

or $\quad f_{ij} = T_{ij} F_{ij}$

where f_{ij} is the member end force vector at end "i" of member "i,j" in the member axes system.

F_{ij} is the member end force vector at end "i" of member "i,j" in the global axes system.

T_{ij} is the transformation matrix for member "i,j".

GRILLAGES

Of course, the same transformation matrix will transform a nodal displacement vector from the global axes system to the member axes system, i.e.

$$\begin{bmatrix} \delta_{ijz} \\ \theta_{ijx} \\ \theta_{ijy} \end{bmatrix} = \begin{bmatrix} 1 & 0 & 0 \\ 0 & \cos\alpha & \sin\alpha \\ 0 & -\sin\alpha & \cos\alpha \end{bmatrix} \begin{bmatrix} \Delta_{ijz} \\ \Theta_{ijx} \\ \Theta_{ijy} \end{bmatrix}$$

The inverse of T_{ij} (shown below) effects the reverse transformation (i.e. from the member axes system to the global axes system):

$$T_{ij}^{-1} = \begin{bmatrix} 1 & 0 & 0 \\ 0 & \cos\alpha & -\sin\alpha \\ 0 & \sin\alpha & \cos\alpha \end{bmatrix}$$

9.4 THE SOLUTION ALGORITHM

By now it should be clear to the reader that one set of matrix equations governs the elastic behaviour of all types of skeletal structures. The relationship (in member axes) between member end forces and member end displacements is, in terms of submatrices

$$f_{ij} = k_{ii}^j \delta_{ij} + k_{ij} \delta_{ji}$$

and

$$f_{ji} = k_{ji} \delta_{ij} + k_{jj}^i \delta_{ji}$$

If the transformation matrix, T_{ij}, transforms vectors of nodal force and displacement at node "i" from the global axes system to the member axes system for member "i,j", then the equilibrium of node "i" is described by the following equation:

$$P_i = K_{ii}^a \Delta_i + K_{ia} \Delta_a + K_{ii}^b \Delta_i + K_{ib} \Delta_b +$$
$$K_{ii}^c \Delta_i + K_{ic} \Delta_c + \ldots\ldots + K_{ii}^n \Delta_i + K_{in} \Delta_n$$

$$= K_{ii} \Delta_i + K_{ia} \Delta_a + K_{ib} \Delta_b \ldots\ldots + K_{in} \Delta_n$$

where

$$K_{ii} = \sum K_{ii}^j$$

and

$$K_{ii}^j = T_{ij}^{-1} k_{ii}^j T_{ij}$$

$$= \begin{bmatrix} 1 & 0 & 0 \\ 0 & \cos\alpha & -\sin\alpha \\ 0 & \sin\alpha & \cos\alpha \end{bmatrix} \begin{bmatrix} 12EI/L^3 & 0 & -6EI/L^2 \\ 0 & GJ/L & 0 \\ -6EI/L^2 & 0 & 4EI/L \end{bmatrix} \begin{bmatrix} 1 & 0 & 0 \\ 0 & \cos\alpha & \sin\alpha \\ 0 & -\sin\alpha & \cos\alpha \end{bmatrix}$$

$$= \begin{bmatrix} 12EI/L^3 & 6EI\sin\alpha/L^2 & -6EI\cos\alpha/L^2 \\ 6EI\sin\alpha/L^2 & (GJ\cos^2\alpha/L + 4EI\sin^2\alpha/L) & (GJ/L - 4EI/L)\sin\alpha\cos\alpha \\ -6EI\cos\alpha/L^2 & (GJ/L - 4EI/L)\sin\alpha\cos\alpha & (GJ\sin^2\alpha/L + 4EI\cos^2\alpha/L) \end{bmatrix}$$

and

$$K_{ij} = T_{ij}^{-1} k_{ij} T_{ij}$$

$$= \begin{bmatrix} 1 & 0 & 0 \\ 0 & \cos\alpha & -\sin\alpha \\ 0 & \sin\alpha & \cos\alpha \end{bmatrix} \begin{bmatrix} -12EI/L^3 & 0 & -6EI/L^2 \\ 0 & -GJ/L & 0 \\ 6EI/L^2 & 0 & 2EI/L \end{bmatrix} \begin{bmatrix} 1 & 0 & 0 \\ 0 & \cos\alpha & \sin\alpha \\ 0 & -\sin\alpha & \cos\alpha \end{bmatrix}$$

$$= \begin{bmatrix} -12EI/L^3 & 6EI\sin\alpha/L^2 & -6EI\cos\alpha/L^2 \\ -6EI\sin\alpha/L^2 & -(GJ\cos^2\alpha/L - 2EI\sin^2\alpha/L) & -(GJ/L + 2EI/L)\sin\alpha\cos\alpha \\ 6EI\cos\alpha/L^2 & -(GJ/L + 2EI/L)\sin\alpha\cos\alpha & -(GJ\sin^2\alpha/L - 2EI\cos^2\alpha/L) \end{bmatrix}$$

The K_{ji} and K_{jj}^i submatrices can be found by a similar calculation. In practice only the K_{ii}^j and the K_{ij} submatrices need be evaluated for each element as the other submatrices can be found by multiplying the elements of the K_{ii}^j and K_{ij} submatrices as follows:

Find K_{jj}^i by multiplying corresponding elements of K_{ii}^j by $\begin{bmatrix} 1 & -1 & -1 \\ -1 & 1 & 1 \\ -1 & 1 & 1 \end{bmatrix}$

Find K_{ji} by multiplying corresponding elements of K_{ij} by $\begin{bmatrix} 1 & -1 & -1 \\ -1 & 1 & 1 \\ -1 & 1 & 1 \end{bmatrix}$

Note that this is not a matrix multiplication, but simply a multiplication of corresponding elements.

Writing the equation of nodal equilibrium for each node in turn produces simultaneous equations that are collectively known as the initial structure stiffness equation. Application of the boundary conditions eliminates a number of these equations and the solution of the reduced system of equations (known collectively as the final structure stiffness equation) yields the unknown nodal displacements. Once found, the nodal displacements are transformed into the member axes system and the member end forces are found using equations (9.2)

GRILLAGES

and (9.3).

Previous chapters have shown that the same matrix equations govern the behaviour of all types of structural frameworks. All that varies is the nature of the matrices involved. This will be demonstrated by using procedures developed in previous chapters to analyse the grillage shown in the following example.

Example 9.1

Find the nodal displacements and check the equilibrium of node 2 for the grillage shown below given that $I = 80 \times 10^9$ mm^4, $J = 40 \times 10^9$ mm^4, $G = 6$ kN/mm^2, and $E = 15$ kN/mm^2.

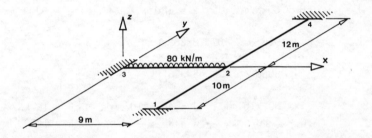

The initial stiffness equation, in terms of submatrices, is as follows:

$$\begin{bmatrix} K_{11} & K_{12} & 0 & 0 \\ K_{21} & K_{22} & K_{23} & K_{24} \\ 0 & K_{32} & K_{33} & 0 \\ 0 & K_{42} & 0 & K_{44} \end{bmatrix} \begin{bmatrix} \Delta_1 \\ \Delta_2 \\ \Delta_3 \\ \Delta_4 \end{bmatrix} = \begin{bmatrix} P_1 \\ P_2 \\ P_3 \\ P_4 \end{bmatrix}$$

Applying the boundary conditions $\Delta_1 = \Delta_3 = \Delta_4 = 0$, as shown, yields the final structure stiffness equation.

$$K_{22} \Delta_2 = P_2 \qquad \text{where} \qquad K_{22} = K_{22}^1 + K_{22}^3 + K_{22}^4$$

Loading

$$P_2 = P_2^n + P_{21}^e + P_{23}^e + P_{24}^e$$

$$= P_{23}^e = T_{23}^{-1} f_{23}^f$$

$$= \begin{bmatrix} 1 & 0 & 0 \\ 0 & -1 & 0 \\ 0 & 0 & -1 \end{bmatrix} \begin{bmatrix} 360 \\ 0 \\ -540000 \end{bmatrix} = \begin{bmatrix} -360 \\ 0 \\ -540000 \end{bmatrix}$$

311

Member 1,2 $I = 80 \times 10^9 \text{mm}^4$, $J = 40 \times 10^9 \text{mm}^4$, $E = 15 \text{ kN/mm}^2$,
              $G = 6 \text{ kN/mm}^2$, $\alpha = 90°$, $L = 10000 \text{ mm}$.

$$K_{22}^1 = \begin{bmatrix} 12EI/L^3 & -6EI\sin\alpha/L^2 & 6EI\cos\alpha/L^2 \\ -6EI\sin\alpha/L^2 & (GJ\cos^2\alpha/L + 4EI\sin^2\alpha/L) & (GJ/L - 4EI/L)\sin\alpha\cos\alpha \\ 6EI\cos\alpha/L^2 & (GJ/L - 4EI/L)\sin\alpha\cos\alpha & (GJ\sin^2\alpha/L + 4EI\cos^2\alpha/L) \end{bmatrix}$$

$$= \begin{bmatrix} 14.4 & -72000 & 0 \\ -72000 & 480E6 & 0 \\ 0 & 0 & 24E6 \end{bmatrix}$$

Member 2,3 $I = 80 \times 10^9 \text{mm}^4$, $J = 40 \times 10^9 \text{mm}^4$, $E = 15 \text{ kN/mm}^2$,
              $G = 6 \text{ kN/mm}^2$, $\alpha = 180°$, $L = 9000 \text{ mm}$.

$$K_{22}^3 = \begin{bmatrix} 19.8 & 0 & 88888 \\ 0 & 26.7E6 & 0 \\ 88888 & 0 & 533E6 \end{bmatrix}$$

Member 2,4 $I = 80 \times 10^9 \text{mm}^4$, $J = 40 \times 10^9 \text{mm}^4$, $E = 15 \text{ kN/mm}^2$,
              $G = 6 \text{ kN/mm}^2$, $\alpha = 90°$, $L = 12000 \text{ mm}$.

$$K_{22}^4 = \begin{bmatrix} 8.33 & 50000 & 0 \\ 50000 & 400E6 & 0 \\ 0 & 0 & 200E6 \end{bmatrix}$$

Substituting numerical values into the final stiffness equation gives

$$\begin{bmatrix} 42.5 & -22000 & 88888 \\ -22000 & 906.7E6 & 0 \\ 88888 & 0 & 577E6 \end{bmatrix} \begin{bmatrix} \Delta_{2z} \\ \theta_{2x} \\ \theta_{2y} \end{bmatrix} = \begin{bmatrix} -360 \\ 0 \\ -540000 \end{bmatrix}$$

the solution to which is

$$\begin{bmatrix} \Delta_{2z} \\ \theta_{2x} \\ \theta_{2y} \end{bmatrix} = \begin{bmatrix} -9.791 \text{ mm} \\ -2.376E-4 \text{ rads} \\ 5.724E-4 \text{ rads} \end{bmatrix}$$

Member end forces (at node 2)

$$f_{21} = k_{22}^1 \, T_{12} \, \Delta_2$$

GRILLAGES

$$= \begin{bmatrix} 14.4 & 0 & 72000 \\ 0 & 24E6 & 0 \\ 72000 & 0 & 480E6 \end{bmatrix} \begin{bmatrix} 1 & 0 & 0 \\ 0 & 0 & 1 \\ 0 & -1 & 0 \end{bmatrix} \begin{bmatrix} -9.791 \\ -2.376E-4 \\ 5.724E-4 \end{bmatrix}$$

$$= \begin{bmatrix} -124 \\ 13700 \\ -591000 \end{bmatrix}$$

$$f_{23} = k_{22}^3 \, T_{23} \, \Delta_2 + f_{23}^f$$

$$= \begin{bmatrix} 19.8 & 0 & -88888 \\ 0 & 26.7E6 & 0 \\ -88888 & 0 & 533E6 \end{bmatrix} \begin{bmatrix} 1 & 0 & 0 \\ 0 & -1 & 0 \\ 0 & 0 & -1 \end{bmatrix} \begin{bmatrix} -9.791 \\ -2.376E-4 \\ 5.724E-4 \end{bmatrix} +$$

$$\begin{bmatrix} 360 \\ 0 \\ -540000 \end{bmatrix}$$

$$= \begin{bmatrix} -142 \\ 6340 \\ 565000 \end{bmatrix} + \begin{bmatrix} 360 \\ 0 \\ -540000 \end{bmatrix} = \begin{bmatrix} 217 \\ 6340 \\ 25000 \end{bmatrix}$$

$$f_{24} = k_{22}^4 \, T_{24} \, \Delta_2$$

$$= \begin{bmatrix} 8.33 & 0 & -50000 \\ 0 & 20E6 & 0 \\ -50000 & 0 & 400E6 \end{bmatrix} \begin{bmatrix} 1 & 0 & 0 \\ 0 & 0 & 1 \\ 0 & -1 & 0 \end{bmatrix} \begin{bmatrix} -9.791 \\ -2.376E-4 \\ 5.724E-4 \end{bmatrix}$$

$$= \begin{bmatrix} -93.4 \\ -11400 \\ 585000 \end{bmatrix}$$

Noting that these forces are in the member axes systems, and that the forces experienced by the node are equal and opposite to the member end forces, allows the freebody diagram for node 2 to be drawn as shown (units are kN and m).

Σ Force in the global z-direction = 124 - 217 + 93.4 $\approx$ 0
Σ Moments about the global "x" axis = 6.3 - 591 + 585 $\approx$ 0
Σ Moments about the global "y" axis = 25.0 - 13.7 - 11.4 $\approx$ 0

9.5 GRILLAGE PROGRAM - GRID.GD

In this book different programs are presented for different types of structural frameworks. This has been done for the sake of clarity. The reader should already have noticed the striking similarity between the programs presented thus far. Indeed, the same procedures have been used in the programs for the analysis of plane trusses, space trusses, and plane frames. The program for the automatic analysis of grillages, GRID.GD, is a modification of the plane frame analysis program PFRAME.PF. The facility to deal with member loads (which are very common on grillages) has been maintained, but the provision for members containing pins has been dropped.

Input The only piece of information the user supplies to the program from the keyboard is the name of the data file to be read by the program. This data file can be constructed using either a text editor or the data preprocessor program PRE.MA/PRE.01/PRE.02 described in section 5.11. If using a text editor the data file should be constructed to the following format:

```
<job title (not to exceed one line)>
4,2,6,<no. of nodes>,<no. of members>,<no. of restrained nodes>,
<no. of nodal loads>, <no. of member loads>
<x-coordinate node 1>,<y-coordinate node 1>
<x-coordinate node 2>,<y-coordinate node 2>
<x-coordinate node 3>,<y-coordinate node 3>
        .              .              .
      .              .              .
<x-coordinate node n>,<y-coordinate node n>
<node "i" member 1>,<node "j" member 1>,<I member 1>,<J member 1>,
<G member 1>,<E member 1>
<node "i" member 2>,<node "j" member 2>,<I member 2>,<J member 2>,
<G member 2>,<E member 2>
          .              .              .
      note - "i" must be less then "j"     .
            .              .              .
            .              .              .
```

GRILLAGES

```
<node "i" member n>,<node "j" member n>,<I member n>,<J member n>,
<G member n>,<E member n>
<restrained node no. 1 node number>,<direction(s)>
<restrained node no. 2 node number>,<direction(s)>
<restrained node no. 3 node number>,<direction(s)>
     .                 .                  .
        note - restraint directions are as follows
               restraint in the global z-direction        = 1
               rotational restraint about the global x-axis = 2
               rotational restraint about the global y-axis = 3
     .                 .                  .
               Input restraints as a composite number
     .                 .                  .
<restrained node no. n node number>,<direction(s)>
<nodal load no. 1  node number>,<direction>,<magnitude>
<nodal load no. 2  node number>,<direction>,<magnitude>
<nodal load no. 3  node number>,<direction>,<magnitude>
     .                 .                  .
        note - load directions are as follows
               load in the global z-direction = 1
               moment about the global x-axis = 2
               moment about the global y-axis = 3
     .                 .                  .
<nodal load no. n node number>,<direction>,<magnitude>
<node "i" member load 1>,<node "j" member load 1>,<load type>,
<load intensity>,<distance from node "i">
<node "i" member load 2>,<node "j" member load 2>,<load type>,
<load intensity>,<distance from node "i">
     .                 .                  .
        note - load types are
               1  uniformly distributed (intensity = load/unit length)
               2  point load            (intensity = load magnitude)
             - member loads act perpendicular to the members in the
               positive z-direction        .
     .                 .                  .
<node "i" member load n>,<node "j" member load n>,<load type>,
<load intensity>,<distance from node "i">
```

Comments on the Algorithm The length of each member and the cosine and sine of the angle it makes with the global "x" axis are calculated from the member node numbers and the nodal coordinates. Only the upper semi-band of the final stiffness matrix is generated. Once the unknown nodal displacements have been found they are used to calculate the member end forces. The reactions are evaluated by summing the contributions from the members connected to the restrained nodes. If node "i" is restrained then the contribution, R_{ij}, from member "i,j" to the reactive forces (in the global system) at node "i" is as follows:

$$R_{ij} = T_{ij}^{-1} f_{ij}$$

$$\begin{bmatrix} R_{ijz} \\ RM_{ijx} \\ RM_{ijy} \end{bmatrix} = \begin{bmatrix} 1 & 0 & 0 \\ 0 & \cos\alpha & -\sin\alpha \\ 0 & \sin\alpha & \cos\alpha \end{bmatrix} \begin{bmatrix} f_{ijz} \\ m_{ijx} \\ m_{ijx} \end{bmatrix}$$

and the final reaction at node "i" is given by

$$R_i = \sum R_{ij} - P_1$$

where the summation is for all of the members attached to node "i".

Output The data from the data file is echoed and messages are output to indicate the beginning of each solution phase. After solution of the final structure stiffness equation nodal displacements are output. The shear force, bending moment, and torsional moment at each end of each member are then calculated and output. Finally the components of reaction (in the global axes system) at each restrained node are output.

Listing

```
1000 REM  ****************************************************************
1010 REM  *                                                                *
1020 REM  *    PROGRAM      G R I D . G D    FOR THE AUTOMATIC ANALYSIS OF *
1030 REM  *    GRILLAGES                                                   *
1040 REM  *                                                                *
1050 REM  *    DATA FROM DATA FILE GENERATED BY A TEXT EDITOR OR BY THE    *
1060 REM  *    PREPROCESSOR PROGRAM P R E . M A / P R E . 0 1 / P R E . 0 2 *
1070 REM  *                                                                *
1080 REM  *    J.A.D.BALFOUR                                               *
1090 REM  *                                                                *
1100 REM  ****************************************************************
1110 REM
1120 REM  --- NOTE THAT VARIABLES ARE DEFINED IN APPENDIX A
1130 REM
1140 OPTION BASE 1
1150 REM
1160 REM  --- THE FOLLOWING DIMENSION STATEMENT SETS THE PROBLEM SIZE
1170 REM
1180 COMMON FILE$
1190 DIM NODE(20,2),    MEMB(40,9), REST(20,5), NLOD(40,3), MLOD(20,9)
1200 DIM FSTIFF(40,20), LOD(40),    FREE(60),   EFREE(6),   ESTIFF(6,6)
1210 DIM TEMP(12)
1220 PRINT : PRINT
```

GRILLAGES

```
1230 PRINT "-----------------------------------------------------"
1240 PRINT
1250 PRINT "PROGRAM    G R I D . G D   FOR THE AUTOMATIC ANALYSIS"
1260 PRINT "OF GRILLAGES"
1270 PRINT
1280 PRINT "-----------------------------------------------------"
1290 REM
1300 REM  ---  CHECK IF THIS PROGRAM HAS BEEN CHAINED FROM PRE.MA/PRE.01/PRE.02
1310 REM
1320 IF LEN(FILE$) <> 0 THEN GOTO 1350
1330 PRINT
1340 INPUT           "NAME OF THE DATA FILE  =  ", FILE$
1350 OPEN "I", #1, FILE$
1360 PRINT
1370 PRINT           "READING DATA"
1380 GOSUB 1620
1390 REM
1400 REM             PRINT DATA
1410 GOSUB 2040
1420 PRINT
1430 PRINT           "GENERATING THE FREEDOM VECTOR"
1440 GOSUB 2720
1450 PRINT
1460 PRINT           "GENERATING THE STRUCTURE STIFFNESS MATRIX"
1470 GOSUB 3040
1480 PRINT
1490 PRINT           "GENERATING THE LOADING VECTOR"
1500 GOSUB 3930
1510 PRINT
1520 PRINT           "SOLVING EQUATIONS  -  NO. OF EQUATIONS  = "; NDOF
1530 PRINT           "SEMI-BANDWIDTH                          = "; BAND
1540 GOSUB 4600
1550 REM
1560 REM             OUTPUT NODAL DISPLACEMENTS
1570 GOSUB 5200
1580 REM
1590 REM             OUTPUT MEMBER FORCES AND REACTIONS
1600 GOSUB 5450
1610 END
1620 REM * * * * * * * * * * * * * * * * * * * * * * * * * * * * * * * * * *
1630 REM *                                                                 *
1640 REM *    SUBROUTINE TO READ DATA FROM THE INPUT FILE                  *
1650 REM *                                                                 *
1660 REM * * * * * * * * * * * * * * * * * * * * * * * * * * * * * * * * * *
1670 REM
1680 INPUT #1, TITLE$
1690 INPUT #1, PTYPE, NCORD, NPROP, NNODE, NMEMB, NREST, NNLOD, NMLOD
1700 IF PTYPE = 4 THEN GOTO 1750
1710 PRINT : PRINT "NOT A GRILLAGE DATA FILE" : END
1720 REM
1730 REM  ---  READ NODAL COORDINATES
1740 REM
1750 FOR I = 1 TO NNODE
1760 INPUT #1, NODE(I,1), NODE(I,2)
1770 NEXT I
1780 REM
1790 REM  ---  READ MEMBER DATA
1800 REM
1810 FOR I = 1 TO NMEMB
1820 INPUT #1, MEMB(I,1), MEMB(I,2), MEMB(I,3), MEMB(I,4), MEMB(I,5), MEMB(I,6)
1830 NEXT I
1840 REM
1850 REM  ---  READ RESTRAINT DATA
1860 REM
1870 FOR I = 1 TO NREST
1880 INPUT #1, REST(I,1), REST(I,2)
1890 REST(I,3) = 0
1900 NEXT I
1910 REM
```

```
1920 REM --- READ NODAL LOADS
1930 REM
1940 FOR I = 1 TO NNLOD
1950 INPUT #1, NLOD(I,1), NLOD(I,2), NLOD(I,3)
1960 NEXT I
1970 REM
1980 REM --- READ MEMBER LOADS
1990 REM
2000 FOR I = 1 TO NMLOD
2010 INPUT #1, MLOD(I,1), MLOD(I,2), MLOD(I,3), MLOD(I,4), MLOD(I,5)
2020 NEXT I
2030 RETURN
2040 REM * * * * * * * * * * * * * * * * * * * * * * * * * * * * * * * * *
2050 REM *                                                                *
2060 REM *      SUBROUTINE TO PRINT DATA                                  *
2070 REM *                                                                *
2080 REM * * * * * * * * * * * * * * * * * * * * * * * * * * * * * * * * *
2090 REM
2100 PRINT
2110 PRINT "+ + + + + + + + +"
2120 PRINT "+   JOB TITLE    +    -    "; TITLE$
2130 PRINT "+ + + + + + + + +"
2140 PRINT
2150 PRINT "+ + + + + + + + + + + +"
2160 PRINT "+   NODAL COORDINATES  +"
2170 PRINT "+ + + + + + + + + + + +"
2180 PRINT
2190 PRINT "NODE              X              Y"
2200 PRINT "NUMBER"
2210 FOR I = 1 TO NNODE
2220 PRINT " "; I, NODE(I,1), NODE(I,2)
2230 NEXT I
2240 PRINT
2250 PRINT "+ + + + + + + + + + + +"
2260 PRINT "+   MEMBER PROPERTIES  +"
2270 PRINT "+ + + + + + + + + + + +"
2280 PRINT
2290 PRINT "MEMBER           I           J           G           E"
2300 FOR I = 1 TO NMEMB
2310 PRINT USING "## ##       "; MEMB(I,1), MEMB(I,2);
2320 FOR J = 3 TO NPROP
2330 PRINT USING "   #.###^^^^"; MEMB(I,J);
2340 NEXT J
2350 PRINT
2360 NEXT I
2370 PRINT
2380 PRINT "+ + + + + + + +"
2390 PRINT "+  RESTRAINTS +"
2400 PRINT "+ + + + + + + +"
2410 PRINT
2420 PRINT "NODE      DIRECTION(S)"
2430 PRINT "NUMBER"
2440 FOR I = 1 TO NREST
2450 PRINT USING " ##         ###"; REST(I,1), REST(I,2)
2460 NEXT I
2470 IF NNLOD = 0 THEN GOTO 2570 ELSE PRINT
2480 PRINT "+ + + + + + + + +"
2490 PRINT "+   NODAL LOADS  +"
2500 PRINT "+ + + + + + + + +"
2510 PRINT
2520 PRINT        "NODE    DIRECTION    VALUE"
2530 PRINT        "NUMBER"
2540 FOR I = 1 TO NNLOD
2550 PRINT USING " ##        #     #.###^^^^"; NLOD(I,1), NLOD(I,2), NLOD(I,3)
2560 NEXT I
2570 IF NMLOD = 0 THEN GOTO 2710
2580 PRINT
2590 PRINT "+ + + + + + + + +"
2600 PRINT "+  MEMBER LOADS  +    -    UDL = 1,    POINT LOAD = 2"
2610 PRINT "+ + + + + + + + +"
2620 PRINT
```

GRILLAGES

```
2630 PRINT "MEMBER     LOAD      MAGNITUDE    DISTANCE FROM"
2640 PRINT "           TYPE                   LOWER NODE"
2650 FOR I = 1 TO NMLOD
2660 PRINT USING "## ##          #"; MLOD(I,1), MLOD(I,2), MLOD(I,3);
2670 PRINT USING "          #.###^^^^"; MLOD(I,4);
2680 IF MLOD(I,3) =  2 THEN PRINT USING "         #.###^^^^"; MLOD(I,5)
2690 IF MLOD(I,3) <> 2 THEN PRINT         "          N/A"
2700 NEXT I
2710 RETURN
2720 REM * * * * * * * * * * * * * * * * * * * * * * * * * * * * * * *
2730 REM *                                                             *
2740 REM *     SUBROUTINE TO GENERATE THE FREEDOM VECTOR               *
2750 REM *                                                             *
2760 REM * * * * * * * * * * * * * * * * * * * * * * * * * * * * * * *
2770 REM
2780 REM --- ZERO THE FREEDOM VECTOR
2790 REM
2800 FOR I    = 1 TO 3*NNODE
2810 FREE(I)  = 0
2820 NEXT I
2830 FOR I    = 1 TO NREST
2840 K        = REST(I,2)
2850 REM
2860 REM --- EVALUATE THE FREEDOM TO BE RESTRAINED FROM RIGHT-MOST DIGIT OF K
2870 REM
2880 J        = 3*REST(I,1) - 3 + (K MOD 10)
2890 FREE(J)  = 1
2900 K        = INT(K/10)
2910 IF K     > 0 THEN GOTO 2880
2920 NEXT I
2930 REM
2940 REM --- NUMBER THE FREEDOMS (RESTRAINTS SET TO ZERO)
2950 REM
2960 FOR I    = 1 TO 3*NNODE
2970 IF FREE(I) = 1 THEN GOTO 3010
2980 NDOF     = NDOF + 1
2990 FREE(I)  = NDOF
3000 GOTO 3020
3010 FREE(I) = 0
3020 NEXT I
3030 RETURN
3040 REM * * * * * * * * * * * * * * * * * * * * * * * * * * * * * * *
3050 REM *                                                             *
3060 REM *     SUBROUTINE TO GENERATE THE STRUCTURE STIFFNESS MATRIX   *
3070 REM *                                                             *
3080 REM * * * * * * * * * * * * * * * * * * * * * * * * * * * * * * *
3090 REM
3100 REM --- ZERO THE STIFFNESS ARRAY AND LOADING VECTOR
3110 REM
3120 FOR I    = 1 TO NDOF
3130 FOR J    = 1 TO 20
3140 FSTIFF(I,J) = 0
3150 NEXT J
3160 LOD(I) = 0
3170 NEXT I
3180 REM
3190 REM --- LOOP FOR EACH ELEMENT
3200 REM
3210 BAND = 0
3220 PRINT "ADDING ELEMENT :- ";
3230 FOR K = 1 TO NMEMB
3240 PRINT K;
3250 IN = MEMB(K,1)
3260 JN = MEMB(K,2)
3270 REM
3280 REM --- CALCULATE ELEMENT LENGTH (STORE IN MEMB(K,7))
3290 REM
3300 MEMB(K,7) = SQR((NODE(JN,1)-NODE(IN,1))^2 + (NODE(JN,2)-NODE(IN,2))^2)
3310 REM
```

```
3320 REM ---  CALCULATE THE COSINE AND SINE OF THE ANGLE THE MEMBER MAKES
3330 REM ---  WITH THE COORDINATE X-AXIS (STORE IN MEMB(K,8) AND MEMB(K,9))
3340 REM
3350 MEMB(K,8) = (NODE(JN,1) - NODE(IN,1)) / MEMB(K,7)
3360 MEMB(K,9) = (NODE(JN,2) - NODE(IN,2)) / MEMB(K,7)
3370 REM
3380 REM ---  GENERATE THE UPPER TRIANGLE OF THE ELEMENT STIFFNESS MATRIX
3390 REM
3400 A2 = MEMB(K,3)
3410 J2 = MEMB(K,4)
3420 G  = MEMB(K,5)
3430 E  = MEMB(K,6)
3440 L  = MEMB(K,7)
3450 C  = MEMB(K,8)
3460 S  = MEMB(K,9)
3470 ESTIFF(1,1) =     12 * E * A2 / L^3
3480 ESTIFF(1,2) =      6 * E * A2 * S / L^2
3490 ESTIFF(1,3) = -    6 * E * A2 * C / L^2
3500 ESTIFF(1,4) = -  ESTIFF(1,1)
3510 ESTIFF(1,5) =    ESTIFF(1,2)
3520 ESTIFF(1,6) =    ESTIFF(1,3)
3530 ESTIFF(2,2) =      4 * E * A2 * S * S / L  +  G * J2 * C * C / L
3540 ESTIFF(2,3) = -  (4 * E * A2  -  G * J2) * S * C / L
3550 ESTIFF(2,4) = -  ESTIFF(1,2)
3560 ESTIFF(2,5) =    (2 * E * A2 * S * S  -  G * J2 * C * C) / L
3570 ESTIFF(2,6) = -  (2 * E * A2  +  G * J2) * S * C / L
3580 ESTIFF(3,3) =    (4 * E * A2 * C * C  +  G * J2 * S * S) / L
3590 ESTIFF(3,4) = -  ESTIFF(1,3)
3600 ESTIFF(3,5) =    ESTIFF(2,6)
3610 ESTIFF(3,6) =    (2 * E * A2 * C * C  -  G * J2 * S * S) / L
3620 ESTIFF(4,4) =    ESTIFF(1,1)
3630 ESTIFF(4,5) = -  ESTIFF(1,2)
3640 ESTIFF(4,6) = -  ESTIFF(1,3)
3650 ESTIFF(5,5) =    ESTIFF(2,2)
3660 ESTIFF(5,6) =    ESTIFF(2,3)
3670 ESTIFF(6,6) =    ESTIFF(3,3)
3680 REM
3690 REM ---  SET UP THE ELEMENT CODE NUMBER
3700 REM
3710 EFREE(1) = FREE(3*IN-2)
3720 EFREE(2) = FREE(3*IN-1)
3730 EFREE(3) = FREE(3*IN)
3740 EFREE(4) = FREE(3*JN-2)
3750 EFREE(5) = FREE(3*JN-1)
3760 EFREE(6) = FREE(3*JN)
3770 REM
3780 REM ---  ADD THE ELEMENT STIFFNESS TERMS
3790 REM
3800 FOR I      = 1 TO 6
3810 IF EFREE(I) = 0 THEN GOTO 3890
3820 FOR J      = 1 TO 6
3830 IF EFREE(J) = 0 THEN GOTO 3880
3840 L          = EFREE(I)
3850 M          = EFREE(J) - L + 1
3860 IF M       > BAND THEN BAND = M
3870 FSTIFF(L,M) = FSTIFF(L,M) + ESTIFF(I,J)
3880 NEXT J
3890 NEXT I
3900 NEXT K
3910 PRINT
3920 RETURN
3930 REM * * * * * * * * * * * * * * * * * * * * * * * * * * * * * * *
3940 REM *                                                             *
3950 REM *     SUBROUTINE TO SET UP THE LOADING VECTOR                 *
3960 REM *                                                             *
3970 REM * * * * * * * * * * * * * * * * * * * * * * * * * * * * * * *
3980 REM
3990 REM ---  ADD NODAL LOADS TO THE LOADING VECTOR
4000 REM
4010 IF NNLOD = 0 THEN GOTO 4100
```

GRILLAGES

```
4020 FOR I    = 1 TO NNLOD
4030 J        = 3*NLOD(I,1) - 3 + NLOD(I,2)
4040 J        = FREE(J)
4050 LOD(J) = NLOD(I,3)
4060 NEXT I
4070 REM
4080 REM  --- ADD MEMBER LOADS TO THE LOADING VECTOR
4090 REM
4100 IF NMLOD = 0 THEN GOTO 4590
4110 FOR I = 1 TO NMLOD
4120 IN    = MLOD(I,1)
4130 JN    = MLOD(I,2)
4140 FOR K          = 1 TO NMEMB
4150 IF MEMB(K,1) <> IN THEN GOTO 4170
4160 IF MEMB(K,2) =  JN THEN GOTO 4180
4170 NEXT K
4180 W = MLOD(I,4)
4190 L = MEMB(K,7)
4200 X = MLOD(I,5)
4210 Y = L - X
4220 C = MEMB(K,8)
4230 S = MEMB(K,9)
4240 REM
4250 REM  --- CALC. AND STORE F.E.F.
4260 REM
4270 ON MLOD(I,3) GOTO 4310, 4390
4280 REM
4290 REM  --- UNIFORMLY DISTRIBUTED LOAD
4300 REM
4310 MLOD(I,6) = - W * L / 2
4320 MLOD(I,7) =   W * L^2 / 12
4330 MLOD(I,8) =   MLOD(I,6)
4340 MLOD(I,9) = - MLOD(I,7)
4350 GOTO 4460
4360 REM
4370 REM  --- POINT LOAD
4380 REM
4390 MLOD(I,6) = - W * (Y/L)^2 * (1 + 2*X/L)
4400 MLOD(I,7) =   W * X * Y^2 / L^2
4410 MLOD(I,8) = - W * (X/L)^2 * (1 + 2*Y/L)
4420 MLOD(I,9) = - W * X^2 * Y / L^2
4430 REM
4440 REM  --- ADD EQUIVALENT JOINT FORCES TO THE LOADING VECTOR
4450 REM
4460 J = FREE(3*IN - 2)
4470 IF J <> 0 THEN LOD(J) = LOD(J) - MLOD(I,6)
4480 J = FREE(3*IN - 1)
4490 IF J <> 0 THEN LOD(J) = LOD(J) + MLOD(I,7)*S
4500 J = FREE(3*IN)
4510 IF J <> 0 THEN LOD(J) = LOD(J) - MLOD(I,7)*C
4520 J = FREE(3*JN - 2)
4530 IF J <> 0 THEN LOD(J) = LOD(J) - MLOD(I,8)
4540 J = FREE(3*JN - 1)
4550 IF J <> 0 THEN LOD(J) = LOD(J) + MLOD(I,9)*S
4560 J = FREE(3*JN)
4570 IF J <> 0 THEN LOD(J) = LOD(J) - MLOD(I,9)*C
4580 NEXT I
4590 RETURN
4600 REM  * * * * * * * * * * * * * * * * * * * * * * * * * * * * * * *
4610 REM  *                                                            *
4620 REM  *   SUBROUTINE TO SOLVE LINEAR SIMULTANEOUS EQUATIONS BY     *
4630 REM  *   GAUSS ELIMINATION WITH NO ROW INTERCHANGE                *
4640 REM  *                                                            *
4650 REM  * * * * * * * * * * * * * * * * * * * * * * * * * * * * * * *
4660 REM
4670 REM  --- LOOP FOR ALL PIVOTS
4680 REM
4690 BB       = BAND
4700 PRINT "USING PIVOT    :- ";
4710 FOR I    = 1 TO NDOF
4720 PRINT I;
4730 REM
```

```
4740 REM --- CHECK IF IN THE UNUSED TRIANGLE
4750 REM
4760 IF I > NDOF-BAND+1 THEN BB = NDOF-I+1
4770 PIVOT   = FSTIFF(I,1)
4780 REM
4790 REM --- NORMALISE
4800 REM
4810 FOR J = 1 TO BB
4820 FSTIFF(I,J) = FSTIFF(I,J) / PIVOT
4830 NEXT J
4840 LOD(I) = LOD(I) / PIVOT
4850 REM
4860 REM --- CHECK IF LAST ROW
4870 REM
4880 IF BB = 1 THEN GOTO 5070
4890 REM
4900 REM --- ELIMINATE (WITHIN BAND) FOR ALL ROWS ABOVE PIVOT
4910 REM
4920 FOR K = 2 TO BB
4930 REM
4940 REM --- CALCULATE ROW NUMBER THEN EVALUATE MULTIPLIER
4950 REM
4960 L       = I + K - 1
4970 MULT    = FSTIFF(I,K) * PIVOT
4980 REM
4990 REM --- LOOP FOR ELEMENTS IN THE ELIMINATION ROW
5000 REM
5010 FOR J = K TO BB
5020 M       = J - K + 1
5030 FSTIFF(L,M) = FSTIFF(L,M) - MULT * FSTIFF(I,J)
5040 NEXT J
5050 LOD(L) = LOD(L) - MULT*LOD(I)
5060 NEXT K
5070 NEXT I
5080 PRINT
5090 REM
5100 REM --- BACK SUBSTITUTE
5110 REM
5120 FOR I = 1 TO NDOF-1
5130 BB      = I
5140 IF I > BAND-1 THEN BB = BAND-1
5150 FOR J = 1 TO BB
5160 LOD(NDOF-I) = LOD(NDOF-I) - FSTIFF(NDOF-I,J+1)*LOD(NDOF-I+J)
5170 NEXT J
5180 NEXT I
5190 RETURN
5200 REM * * * * * * * * * * * * * * * * * * * * * * * * * * * * * *
5210 REM *                                                         *
5220 REM *    SUBROUTINE TO OUTPUT THE DISPLACEMENTS               *
5230 REM *                                                         *
5240 REM * * * * * * * * * * * * * * * * * * * * * * * * * * * * * *
5250 REM
5260 PRINT : PRINT
5270 PRINT "* * * * * * * * * * *"
5280 PRINT "*   DISPLACEMENTS   *         * :- RESTRAINT"
5290 PRINT "* * * * * * * * * * *"
5300 PRINT
5310 PRINT "NODE   Z-DISPLACEMENT    X-ROTATION    Y-ROTATION"
5320 PRINT
5330 FOR I = 1 TO NNODE
5340 PRINT USING "###    "; I;
5350 FOR J = 1 TO 3
5360 K       = FREE(3*I+J-3)
5370 IF K = 0 THEN GOTO 5400
5380 PRINT USING "     #.####^^^^"; LOD(K);
5390 GOTO 5410
5400 PRINT "           *        ";
5410 NEXT J
5420 PRINT
5430 NEXT I
5440 RETURN
```

GRILLAGES

```
5450 REM  * * * * * * * * * * * * * * * * * * * * * * * * * * * * * * *
5460 REM  *                                                             *
5470 REM  *    SUBROUTINE TO CALCULATE AND OUTPUT THE MEMBER END FORCES *
5480 REM  *    AND REACTIONS                                            *
5490 REM  *                                                             *
5500 REM  * * * * * * * * * * * * * * * * * * * * * * * * * * * * * * *
5510 REM
5520 FOR I     = 1 TO NREST
5530 REST(I,3) = 0
5540 NEXT I
5550 FOR K = 1 TO NMEMB
5560 A2 = MEMB(K,3)
5570 J2 = MEMB(K,4)
5580 G  = MEMB(K,5)
5590 E  = MEMB(K,6)
5600 L  = MEMB(K,7)
5610 C  = MEMB(K,8)
5620 S  = MEMB(K,9)
5630 I  = 3*MEMB(K,1) - 3
5640 REM
5650 REM ---  EXTRACT THE NODAL DISPLACEMENTS - STORE TEMP(1) - (6)
5660 REM
5670 FOR M = 1 TO 6
5680 J     = FREE(I+M)
5690 IF J = 0 THEN TEMP(M) = 0 ELSE TEMP(M) = LOD(J)
5700 IF M = 3 THEN I = 3*MEMB(K,2) - 6
5710 NEXT M
5720 REM
5730 REM ---  CALC. MEMBER END FORCES - STORE TEMP(7) - (12)
5740 REM
5750 TEMP(7)  =    TEMP(2)*S - TEMP(3)*C + TEMP(5)*S - TEMP(6)*C
5760 TEMP(7)  =    6 * E * A2 * (2*(TEMP(1)-TEMP(4))/L + TEMP(7)) / L^2
5770 TEMP(8)  =    G * J2 * (TEMP(2)*C + TEMP(3)*S - TEMP(5)*C - TEMP(6)*S) / L
5780 TEMP(9)  = -  2*TEMP(2)*S + 2*TEMP(3)*C - TEMP(5)*S + TEMP(6)*C
5790 TEMP(9)  =    2 * E * A2 * (3*(TEMP(4) - TEMP(1))/L + TEMP(9)) / L
5800 TEMP(10) = -  TEMP(7)
5810 TEMP(11) = -  TEMP(8)
5820 TEMP(12) = -  L*TEMP(7) - TEMP(9)
5830 REM
5840 REM ---  ADD FIXED END FORCES
5850 REM
5860 IF NMLOD = 0 THEN GOTO 5950
5870 FOR I     = 1 TO NMLOD
5880 IF MEMB(K,1) <> MLOD(I,1) THEN GOTO 5940
5890 IF MEMB(K,2) <> MLOD(I,2) THEN GOTO 5940
5900 TEMP(7)  = TEMP(7)  + MLOD(I,6)
5910 TEMP(9)  = TEMP(9)  + MLOD(I,7)
5920 TEMP(10) = TEMP(10) + MLOD(I,8)
5930 TEMP(12) = TEMP(12) + MLOD(I,9)
5940 NEXT I
5950 PRINT : PRINT
5960 PRINT        "* * * * * * * * * *"
5970 PRINT USING "*    ELEMENT ## ##    *"; MEMB(K,1), MEMB(K,2)
5980 PRINT        "* * * * * * * * * *"
5990 PRINT
6000 PRINT "           SHEAR        TORSION        BENDING"
6010 PRINT "           FORCE        MOMENT         MOMENT"
6020 PRINT USING "NODE ##"; MEMB(K,1);
6030 PRINT USING "     #.####^^^^"; TEMP(7), TEMP(8), TEMP(9)
6040 PRINT USING "NODE ##"; MEMB(K,2);
6050 PRINT USING "     #.####^^^^"; TEMP(10), TEMP(11), TEMP(12)
6060 REM
6070 REM ---  ADD ANY CONTRIBUTION FROM THE MEMBER FORCES TO THE REACTIONS
6080 REM
6090 FOR I = 1 TO NREST
6100 J     = 7
6110 IF MEMB(K,1) = REST(I,1) THEN GOTO 6140
6120 J     = 10
6130 IF MEMB(K,2) <> REST(I,1) THEN GOTO 6170
6140 REST(I,3) = REST(I,3) + TEMP(J)
```

```
6150 REST(I,4) = REST(I,4) + C*TEMP(J+1) - S*TEMP(J+2)
6160 REST(I,5) = REST(I,5) + S*TEMP(J+1) + C*TEMP(J+2)
6170 NEXT I
6180 NEXT K
6190 REM
6200 REM  ---  ADD ANY APPLIED LOADS
6210 REM
6220 FOR I = 1 TO NREST
6230 FOR J = 1 TO NNLOD
6240 IF REST(I,1) <> NLOD(J,1) THEN GOTO 6270
6250 K        = NLOD(J,2)
6260 REST(I,2+K) = REST(I,2+K) - NLOD(J,3)
6270 NEXT J
6280 NEXT I
6290 REM
6300 REM  ---  OUTPUT THE REACTIONS
6310 REM
6320 PRINT : PRINT
6330 PRINT "* * * * * * * * *"
6340 PRINT "*   REACTIONS   *"
6350 PRINT "* * * * * * * * *"
6360 PRINT
6370 PRINT "               Z-FORCE    MOMENT ABOUT    MOMENT ABOUT"
6380 PRINT "                          GLOBAL X-AXIS   GLOBAL Y-AXIS"
6390 PRINT
6400 FOR I   = 1 TO NREST
6410 PRINT USING "NODE ##";   REST(I,1);
6420 PRINT USING "     #.####^^^^"; REST(I,3), REST(I,4), REST(I,5)
6430 NEXT I
6440 RETURN
```

Sample Run The following sample run shows the program GRID.GD being used to analyse the structure from example 9.1

```
------------------------------------------------------
PROGRAM    G R I D . G D    FOR THE AUTOMATIC ANALYSIS
OF GRILLAGES
------------------------------------------------------

NAME OF THE DATA FILE = DATA.GD

READING DATA

+ + + + + + + + +
+   JOB TITLE   +     -    EXAMPLE 9.1 - FILE DATA.GD
+ + + + + + + + +

+ + + + + + + + + + + + +
+   NODAL COORDINATES   +
+ + + + + + + + + + + + +

NODE            X           Y
NUMBER
  1           9000       -10000
  2           9000           0
  3              0           0
  4           9000       12000

+ + + + + + + + + + + + +
+   MEMBER PROPERTIES   +
+ + + + + + + + + + + + +

MEMBER        I           J           G           E
 1  2      0.800E+11   0.400E+11   0.600E+01   0.150E+02
 2  3      0.800E+11   0.400E+11   0.600E+01   0.150E+02
 2  4      0.800E+11   0.400E+11   0.600E+01   0.150E+02
```

GRILLAGES

```
+ + + + + + + + +
+   RESTRAINTS  +
+ + + + + + + + +

NODE       DIRECTION(S)
NUMBER
   1         123
   3         123
   4         123


+ + + + + + + + + +
+  MEMBER LOADS  +    -   UDL = 1,   POINT LOAD = 2
+ + + + + + + + + +

MEMBER    LOAD      MAGNITUDE     DISTANCE FROM
          TYPE                    LOWER NODE
 2  3      1         -.800E-01      N/A

GENERATING THE FREEDOM VECTOR

GENERATING THE STRUCTURE STIFFNESS MATRIX
ADDING ELEMENT :-  1  2  3

GENERATING THE LOADING VECTOR

SOLVING EQUATIONS  -  NO. OF EQUATIONS  =  3
SEMI-BANDWIDTH                          =  3
USING PIVOT    :-  1  2  3


* * * * * * * * * * *
*   DISPLACEMENTS   *            * :-  RESTRAINT
* * * * * * * * * * *

NODE      Z-DISPLACEMENT      X-ROTATION        Y-ROTATION

  1            *                  *                 *
  2          -.9794E+01        -.2377E-03         0.5727E-03
  3            *                  *                 *
  4            *                  *                 *

* * * * * * * * * * *
*   ELEMENT  1  2   *
* * * * * * * * * * *

             SHEAR           TORSION          BENDING
             FORCE           MOMENT           MOMENT
NODE  1    0.1239E+03      -.1374E+05       -.6482E+06
NODE  2    -.1239E+03      0.1374E+05       -.5911E+06


* * * * * * * * * * *
*   ELEMENT  2  3   *
* * * * * * * * * * *

             SHEAR           TORSION          BENDING
             FORCE           MOMENT           MOMENT
NODE  2    0.2174E+03      0.6338E+04       0.2520E+05
NODE  3    0.5026E+03      -.6338E+04       0.1258E+07


* * * * * * * * * * *
*   ELEMENT  2  4   *
* * * * * * * * * * *

             SHEAR           TORSION          BENDING
             FORCE           MOMENT           MOMENT
NODE  2    -.9350E+02      0.1145E+05       0.5848E+06
NODE  4    0.9350E+02      -.1145E+05       0.5373E+06
```

```
* * * * * * * * *
*   REACTIONS   *
* * * * * * * * *
                Z-FORCE     MOMENT ABOUT    MOMENT ABOUT
                            GLOBAL X-AXIS   GLOBAL Y-AXIS

NODE  1        0.1239E+03    0.6482E+06     -.1374E+05
NODE  3        0.5026E+03    0.6338E+04     -.1258E+07
NODE  4        0.9350E+02   -.5373E+06      -.1145E+05
```

10 Space Frames

10.1 INTRODUCTION

In practice structural frameworks are almost always three dimensional. If, however, the structural action is predominantly two-dimensional then structural analysis using a two-dimensional mathematical model, such as a plane frame or a grillage, is usually sufficiently accurate. If, on the other hand, the behaviour of the structure is essentially three-dimensional then structural analysis should be conducted using a three-dimensional mathematical model. This chapter is concerned with the most general type of skeletal structure, the three-dimensional rigidly jointed framework.

A computer program capable of dealing with space frames can, of course, be used for the analysis of plane frames and grillages. Usually this will prove much less efficient than using a program specifically written for the type of structure under consideration, and should be avoided where possible.

10.2 THE ELEMENT STIFFNESS MATRIX IN THE MEMBER AXES SYSTEM

The displacement of each node of a space frame is described by three translational and three rotational components of displacement. Hence the degree of freedom of each unrestrained node is six.

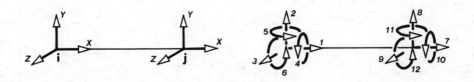

FIGURE 10.1 MEMBER AXES AND FREEDOMS FOR A SPACE FRAME ELEMENT

COMPUTER ANALYSIS OF STRUCTURAL FRAMEWORKS

Figure 10.1 shows the member axes and the member freedoms adopted in this text. Previous chapters have shown that for any type of skeletal structure the member end forces are related to the member end displacements by the following matrix equations:

$$f_{ij} = k_{ii}^{j} \delta_{ij} + k_{ij} \delta_{ji}$$

$$f_{ji} = k_{ji} \delta_{ij} + k_{jj}^{i} \delta_{ji}$$

At this point the reader should note that every stiffness coefficient associated with a space frame element has already been encountered when dealing with plane frames and grillages. Hence there is little difficulty in showing that the above equations take the following form when applied to a space frame member:

$$\begin{bmatrix} f_{ijx} \\ f_{ijy} \\ f_{ijz} \\ m_{ijx} \\ m_{ijy} \\ m_{ijz} \end{bmatrix} = \begin{bmatrix} EA/L & 0 & 0 & 0 & 0 & 0 \\ 0 & 12EI_z/L^3 & 0 & 0 & 0 & 6EI_z/L^2 \\ 0 & 0 & 12EI_y/L^3 & 0 & -6EI_y/L^2 & 0 \\ 0 & 0 & 0 & GJ/L & 0 & 0 \\ 0 & 0 & -6EI_y/L^2 & 0 & 4EI_y/L & 0 \\ 0 & 6EI_z/L^2 & 0 & 0 & 0 & 4EI_z/L \end{bmatrix} \begin{bmatrix} \delta_{ijx} \\ \delta_{ijy} \\ \delta_{ijz} \\ \theta_{ijx} \\ \theta_{ijy} \\ \theta_{ijz} \end{bmatrix}$$

$$+ \begin{bmatrix} -EA/L & 0 & 0 & 0 & 0 & 0 \\ 0 & -12EI_z/L^3 & 0 & 0 & 0 & 6EI_z/L^2 \\ 0 & 0 & -12EI_y/L^3 & 0 & -6EI_y/L^2 & 0 \\ 0 & 0 & 0 & -GJ/L & 0 & 0 \\ 0 & 0 & 6EI_y/L^2 & 0 & 2EI_y/L & 0 \\ 0 & -6EI_z/L^2 & 0 & 0 & 0 & 2EI_z/L \end{bmatrix} \begin{bmatrix} \delta_{jix} \\ \delta_{jiy} \\ \delta_{jiz} \\ \theta_{jix} \\ \theta_{jiy} \\ \theta_{jiz} \end{bmatrix}$$

$$\begin{bmatrix} f_{jix} \\ f_{jiy} \\ f_{jiz} \\ m_{jix} \\ m_{jiy} \\ m_{jiz} \end{bmatrix} = \begin{bmatrix} -EA/L & 0 & 0 & 0 & 0 & 0 \\ 0 & -12EI_z/L^3 & 0 & 0 & 0 & -6EI_z/L^2 \\ 0 & 0 & -12EI_y/L^3 & 0 & 6EI_y/L^2 & 0 \\ 0 & 0 & 0 & -GJ/L & 0 & 0 \\ 0 & 0 & -6EI_y/L^2 & 0 & 2EI_y/L & 0 \\ 0 & 6EI_z/L^2 & 0 & 0 & 0 & 2EI_z/L \end{bmatrix} \begin{bmatrix} \delta_{ijx} \\ \delta_{ijy} \\ \delta_{ijz} \\ \theta_{ijx} \\ \theta_{ijy} \\ \theta_{ijz} \end{bmatrix}$$

$$+ \begin{bmatrix} EA/L & 0 & 0 & 0 & 0 & 0 \\ 0 & 12EI_z/L^3 & 0 & 0 & 0 & -6EI_z/L^2 \\ 0 & 0 & 12EI_y/L^3 & 0 & 6EI_y/L^2 & 0 \\ 0 & 0 & 0 & GJ/L & 0 & 0 \\ 0 & 0 & 6EI_y/L^2 & 0 & 4EI_y/L & 0 \\ 0 & -6EI_z/L^2 & 0 & 0 & 0 & 4EI_z/L \end{bmatrix} \begin{bmatrix} \delta_{jix} \\ \delta_{jiy} \\ \delta_{jiz} \\ \theta_{jix} \\ \theta_{jiy} \\ \theta_{jiz} \end{bmatrix}$$

SPACE FRAMES

10.3 TRANSFORMATION OF FORCE AND DISPLACEMENT

When dealing with space trusses (Chapter 6), the matrix required to transform a vector of linear nodal displacements in three-dimensional space from one cartesian coordinate system to another sharing a common origin was shown to be

$$\begin{bmatrix} \delta_{ijx} \\ \delta_{ijy} \\ \delta_{ijz} \end{bmatrix} = \begin{bmatrix} l_x & m_x & n_x \\ l_y & m_y & n_y \\ l_z & m_z & n_z \end{bmatrix} \begin{bmatrix} \Delta_{ix} \\ \Delta_{iy} \\ \Delta_{iz} \end{bmatrix}$$

In the previous chapter the transformation of rotational displacements was seen to be identical to the transformation of linear displacements, hence

$$\begin{bmatrix} \Theta_{ijx} \\ \Theta_{ijy} \\ \Theta_{ijz} \end{bmatrix} = \begin{bmatrix} l_x & m_x & n_x \\ l_y & m_y & n_y \\ l_z & m_z & n_z \end{bmatrix} \begin{bmatrix} \Theta_{ix} \\ \Theta_{iy} \\ \Theta_{iz} \end{bmatrix}$$

As linear displacements are independent of rotational displacements, the transformation of a nodal displacement vector for a space frame, from the global to the member axes system is as follows:

$$\begin{bmatrix} \delta_{ijx} \\ \delta_{ijy} \\ \delta_{ijz} \\ \Theta_{ijx} \\ \Theta_{ijy} \\ \Theta_{ijz} \end{bmatrix} = \begin{bmatrix} l_x & m_x & n_x & 0 & 0 & 0 \\ l_y & m_y & n_y & 0 & 0 & 0 \\ l_z & m_z & n_z & 0 & 0 & 0 \\ 0 & 0 & 0 & l_x & m_x & n_x \\ 0 & 0 & 0 & l_y & m_y & n_y \\ 0 & 0 & 0 & l_z & m_z & n_z \end{bmatrix} \begin{bmatrix} \Delta_{ijx} \\ \Delta_{ijy} \\ \Delta_{ijx} \\ \Theta_{ijx} \\ \Theta_{ijy} \\ \Theta_{ijz} \end{bmatrix}$$

i.e.
$$\delta_{ij} = T_{ij} \Delta_i$$

The same matrix will also transform vectors of nodal force, i.e.

$$f_{ij} = T_{ij} F_{ij}$$

l_x, m_x, and n_x are, as defined in Chapter 6, the cosines of the angles that the member "x" axis makes with the global "x", "y", and "z" axes respectively (i.e. the direction cosines of the member "x" axis).

i.e.
$$l_x = (x_j - x_i)/L_{ij}$$

$$m_x = (y_j - y_i)/L_{ij}$$
$$n_x = (z_j - z_i)/L_{ij}$$

where (x_i, y_i, z_i) and (x_j, y_j, z_j) are the coordinates of nodes "i" and "j" (i.e. the coordinates of the ends of member "i,j"), and L_{ij} is the length of member "i,j" (see fig 10.2).

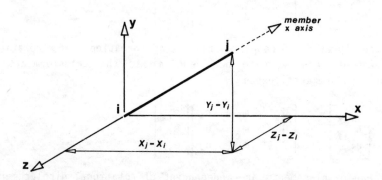

FIGURE 10.2   CALCULATION OF THE DIRECTION COSINES FOR THE MEMBER "X" AXIS

The problem here is to find the remaining elements of T_{ij}, as l_x, m_x, and n_x define only the orientation of the member "x" axis (i.e. the member "y" and "z" axes can be rotated about the member "x" axis to take up any orientation). To proceed the orientation of one of these axes must be defined. In this text the member "y" axis will be orientated to lie in the global "x,y" plane.

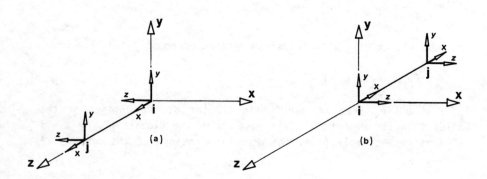

FIGURE 10.3   MEMBER LYING PARALLEL TO THE GLOBAL "Z" AXIS

SPACE FRAMES

If, however, the member "x" axis lies along the global "z" axis then the condition that the member "y" axis must lie in the global "x,y" plane is automatically satisfied, and a further condition is necessary. In this text the member "y" axis of a member lying along the global "z" axis will be chosen to coincide with the global "y" axis. There are two possible cases, and these are illustrated in fig 10.3. The complete set of direction cosines can be found by inspection.

Case (a)

$$l_x = 0 \qquad m_x = 0 \qquad n_x = 1$$
$$l_y = 0 \qquad m_y = 1 \qquad n_y = 0$$
$$l_z = -1 \qquad m_z = 0 \qquad n_z = 0$$

Case (b)

$$l_x = 0 \qquad m_x = 0 \qquad n_x = -1$$
$$l_y = 0 \qquad m_y = 1 \qquad n_y = 0$$
$$l_z = 1 \qquad m_z = 0 \qquad n_z = 0$$

For members not lying along the global "z" axis the evaluation of the direction cosines of the member "y" and "z" axes is less straightforward. The problem is best tackled using vector cross products. The cross product, $\vec{C}$, of two vectors, $\vec{A}$ and $\vec{B}$, is defined as follows:

$$\vec{A} = \vec{i} x_a + \vec{j} y_a + \vec{k} z_a$$

and

$$\vec{B} = \vec{i} x_b + \vec{j} y_b + \vec{k} z_b$$

$$\vec{C} = \vec{A} \times \vec{B} = \begin{vmatrix} \vec{i} & \vec{j} & \vec{k} \\ x_a & y_a & z_a \\ x_b & y_b & z_b \end{vmatrix}$$

where

$\vec{C}$ is a third vector normal to the plane of $\vec{A}$ and $\vec{B}$ and is directed such that $\vec{A}$, $\vec{B}$, and $\vec{C}$ form a right-hand system. The length of the $\vec{C}$ is $|\vec{A}||\vec{B}|\sin(\vec{A},\vec{B})|$.

$\sin(\vec{A},\vec{B})$ is the absolute value of the sine of the angle between $\vec{A}$ and $\vec{B}$.

$\vec{i},\vec{j},\vec{k}$ are unit vectors in the direction of the "x", "y", and "z" axes respectively.

x_a, y_a, z_a are the magnitudes of the components of vector $\vec{A}$ in the directions of the "x", "y", and "z" axes respectively.

x_b, y_b, z_b are the magnitudes of the components of vector $\vec{B}$ in the directions of the "x", "y", and "z" axes respectively.

l_x, m_x, and n_x are, in effect, the components in the directions of the global "x", "y", and "z" axes of a unit vector lying along the member "x" axis. The member "y" axis is perpendicular to the member "x" axis and, if the member "y" axis lies in the global "x,y" plane, it must also be perpendicular to the global "z" axis. Hence the cross product of unit vectors lying along the global "z" and the member "x" axes must result in a vector, $\vec{Y}$, in the direction of the member "y" axis (note that the unit vectors are expressed in terms of the global axes system, and that the components of the unit vector on the member "x" axis are equal to the direction cosines for that axis).

$$\vec{Y} = \begin{vmatrix} \vec{i} & \vec{j} & \vec{k} \\ 0 & 0 & 1 \\ l_x & m_x & n_x \end{vmatrix} = -\vec{i}\, m_x + \vec{j}\, l_x + \vec{k}\, 0$$

The length of $\vec{Y}$ is $||1|| \, ||1|| \, \sqrt{(1 - n_x^2)}|$, as $\sin\theta = \sqrt{(1 - \cos^2\theta)}$

l_y, m_y, and n_y are the components in the global axes system of a unit vector lying on the member "y" axis. Hence the direction cosines of the member "y" axis are found by scaling $\vec{Y}$ to make a unit vector:

$$\vec{Y}/\sqrt{(1 - n_x^2)} = -\vec{i}\, m_x/\sqrt{(1 - n_x^2)} + \vec{j}\, l_x/\sqrt{(1 - n_x^2)} + \vec{k}\, 0$$

i.e.
$$l_y = -m_x/\sqrt{(1 - n_x^2)}$$
$$m_y = l_x/\sqrt{(1 - n_x^2)}$$
$$n_y = 0$$

A vector in the direction of the member "z" axis is generated by taking the cross product of unit vectors lying along the member "x" and "y" axes (note that the length of the resulting vector is unity).

$$\vec{Y} = \begin{vmatrix} \vec{i} & \vec{j} & \vec{k} \\ l_x & m_x & n_x \\ l_y & m_y & n_y \end{vmatrix}$$

$$= \vec{i}(m_x n_y - m_y n_x) - \vec{j}(l_x n_y - l_y n_x) + \vec{k}(l_x m_y - l_y m_x)$$

$$= -\vec{i}(l_x n_x/\sqrt{(1 - n_x^2)}) - \vec{j}(m_x n_x/\sqrt{(1 - n_x^2)}) + \vec{k}((l_x^2 + m_x^2)/\sqrt{(1 - n_x^2)})$$

hence

$$l_z = -l_x n_x/\sqrt{(1 - n_x^2)}$$
$$m_z = -m_x n_x/\sqrt{(1 - n_x^2)}$$
$$n_z = \sqrt{(1 - n_x^2)}$$

As all of the direction cosines have now been expressed in terms of l_x, m_x, and n_x the subscript "x" is no longer necessary and the transformation matrix becomes

$$T_{ij} = \begin{bmatrix} 1 & m & n & 0 & 0 & 0 \\ -m/D & 1/D & 0 & 0 & 0 & 0 \\ -ln/D & -mn/D & D & 0 & 0 & 0 \\ 0 & 0 & 0 & 1 & m & n \\ 0 & 0 & 0 & -m/D & 1/D & 0 \\ 0 & 0 & 0 & -ln/D & -mn/D & D \end{bmatrix} \quad (10.1)$$

where

$$D = \sqrt{(1 - n^2)}$$

The transformation matrix shown in equation (10.1) tranforms vectors of nodal force and nodal displacement from the global axes system to the member axes system for space frame member "i,j". The inverse of that matrix, shown in equation (10.2), effects the reverse transformation (i.e. from the member to the global axes system):

$$T_{ij}^{-1} = \begin{bmatrix} 1 & -m/D & -ln/D & 0 & 0 & 0 \\ m & 1/D & -mn/D & 0 & 0 & 0 \\ n & 0 & D & 0 & 0 & 0 \\ 0 & 0 & 0 & 1 & -m/D & -ln/D \\ 0 & 0 & 0 & m & 1/D & -mn/D \\ 0 & 0 & 0 & n & 0 & D \end{bmatrix} \quad (10.2)$$

10.4 THE ELEMENT STIFFNESS MATRIX IN THE GLOBAL AXES SYSTEM

As seen in previous chapters the member stiffness submatrices in the global axes system are obtained by the following triple matrix multiplications:

$$K_{ii}^j = T_{ij}^{-1} k_{ii}^j T_{ij}$$
$$K_{ij} = T_{ij}^{-1} k_{ij} T_{ij}$$
$$K_{ji} = T_{ij}^{-1} k_{ji} T_{ij}$$
$$K_{jj}^i = T_{ij}^{-1} k_{jj}^i T_{ij}$$

At this point it should be noted that the relationship between the member end forces and the member end displacements assumes that the member "y" and "z" axes coincide with the principal axes of the section. However, earlier in the chapter the member "y" axis was forced to lie in the global "x,y" plane, and this restriction means that the member "y" and "z" axes will not necessarily coincide with the principal axes (y^*, z^*) of the section. The general case is illustrated in fig 10.4. Note that fig 10.4 looks down member "i,j" from end "i" to end "j". Hence β is the *anticlockwise* rotation of the member (when viewed from end "i") that will make the principal axes coincide with the member axes. Note also that the principal "x" axis and the member "x" axis are always coincident.

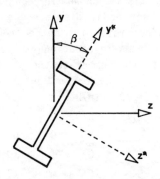

FIGURE 10.4   PRINCIPAL AXES

The reader should recognise that the required transformation is identical to the transformation from the global to the member axis system for a grillage member (see Chapter 9). Hence the transformation of member end displacements and rotations from the member axes system to the principal axes system is effected by the following matrix:

$$T^*_{ij} = \begin{bmatrix} 1 & 0 & 0 & 0 & 0 & 0 \\ 0 & \cos\beta & \sin\beta & 0 & 0 & 0 \\ 0 & -\sin\beta & \cos\beta & 0 & 0 & 0 \\ 0 & 0 & 0 & 1 & 0 & 0 \\ 0 & 0 & 0 & 0 & \cos\beta & \sin\beta \\ 0 & 0 & 0 & 0 & -\sin\beta & \cos\beta \end{bmatrix}$$

The reverse transformation is effected by

$$T^{*-1}_{ij} = \begin{bmatrix} 1 & 0 & 0 & 0 & 0 & 0 \\ 0 & \cos\beta & -\sin\beta & 0 & 0 & 0 \\ 0 & \sin\beta & \cos\beta & 0 & 0 & 0 \\ 0 & 0 & 0 & 1 & 0 & 0 \\ 0 & 0 & 0 & 0 & \cos\beta & -\sin\beta \\ 0 & 0 & 0 & 0 & \sin\beta & \cos\beta \end{bmatrix}$$

hence

$$\begin{bmatrix} \delta^*_{ijx} \\ \delta^*_{ijy} \\ \delta^*_{ijz} \\ \theta^*_{ijx} \\ \theta^*_{ijy} \\ \theta^*_{ijz} \end{bmatrix} = \begin{bmatrix} 1 & 0 & 0 & 0 & 0 & 0 \\ 0 & \cos\beta & \sin\beta & 0 & 0 & 0 \\ 0 & -\sin\beta & \cos\beta & 0 & 0 & 0 \\ 0 & 0 & 0 & 1 & 0 & 0 \\ 0 & 0 & 0 & 0 & \cos\beta & \sin\beta \\ 0 & 0 & 0 & 0 & -\sin\beta & \cos\beta \end{bmatrix} \begin{bmatrix} \delta_{ijx} \\ \delta_{ijy} \\ \delta_{ijz} \\ \theta_{ijx} \\ \theta_{ijy} \\ \theta_{ijz} \end{bmatrix}$$

As stated previously the member stiffness relationship is only valid in terms of the principal axes, i.e.

$$f^*_{ij} = k^{j*}_{ii} \delta^*_{ij} + k^*_{ij} \delta^*_{ji}$$

but

$$f_{ij} = T^{*-1}_{ij} f^*_{ij} \quad \text{and} \quad \delta^*_{ij} = T^*_{ij} \delta_{ij}$$

hence

$$f_{ij} = T^{*-1}_{ij} k^{j*}_{ii} T^*_{ij} \delta_{ij} + T^{*-1}_{ij} k^*_{ij} T^*_{ij} \delta_{ji}$$

The above equation gives the relationship between the member end forces and the member end displacements in terms of the member axes system, but using stiffness submatrices in the principal axes system (the stars denote matrices that are associated with the principal axes system). To express forces and displacements in the global axes system the transformation matrices shown in equations (10.1) and (10.2) are used:

$$F_{ij} = T_{ij}^{-1} f_{ij} \quad \text{and} \quad \delta_{ij} = T_{ij} \Delta_i$$

hence

$$F_{ij} = T_{ij}^{-1} T_{ij}^{*-1} k_{ii}^{j*} T_{ij}^{*} T_{ij} \Delta_{ij} + T_{ij}^{-1} T_{ij}^{*-1} k_{ij} T_{ij}^{*} T_{ij} \Delta_{ji} \tag{10.3}$$

The equation of equilibrium for node "i" of a framed structure has been shown in previous chapters to be

$$P_i = F_{ia} + F_{ib} + F_{ic} + \ldots\ldots + F_{in}$$

where P_i is the vector of external forces at node "i" (note that this vector may contain both applied loads and reactive forces), and F_{ij} is the vector of member end forces, in the global axes system, at end "i" of member "i,j".

Using equation (10.3) to substitute for the member end forces gives

$$\begin{aligned} P_i &= K_{ii}^a \Delta_i + K_{ia} \Delta_a + K_{ii}^b \Delta_i + K_{ib} \Delta_b + \\ & \quad K_{ii}^c \Delta_i + K_{ic} \Delta_c + \ldots\ldots + K_{ii}^n \Delta_i + K_{in} \Delta_n \\ &= K_{ii} \Delta_i + K_{ia} \Delta_a + K_{ib} \Delta_b \ldots\ldots + K_{in} \Delta_n \end{aligned}$$

where

$$K_{ii} = \sum K_{ii}^j = \sum T_{ij}^{-1} T_{ij}^{*-1} k_{ii}^{j*} T_{ij}^{*} T_{ij} \tag{10.4}$$

and

$$K_{ij} = T_{ij}^{-1} T_{ij}^{*-1} k_{ij}^{*} T_{ij}^{*} T_{ij} \tag{10.5}$$

The summation is for all members connected to node "i".

From here the final structure stiffness equation is developed using the same procedures as were used for all other types of frameworks.

Third Node Method - Program   TNODE.SF

The rotation that will make the principal axes coincide with the member axis cannot always be assessed by inspection. Take for example the member shown in fig 10.5(a). Figure 10.5(b) shows the same member viewed looking down the global "y" axis. The problem in hand is to calculate the angle through which the member must be rotated in order to make the principal axes coincide with the member axes (note that anticlockwise rotation of the member when viewed looking from end "i" to end "j" is positive). A convenient way

SPACE FRAMES

forward is to locate a point "p" that lies in the principal "x,y" plane (but not on the principal "x" axis). This point is known as the *third node* and nodes "i", "j" and "p" define the principal "x,y" plane. In the case of the member shown in fig 10.5 the global "y" axis lies in the principal "x,y" plane of the member. Thus any point, except the node "i", on the global "y" axis will suffice for "p".

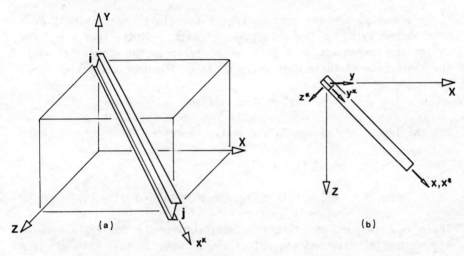

FIGURE 10.5   PRINCIPAL AND MEMBER AXES

The cross product of two independent vectors lying on the principal "x,y" plane will result in a vector normal to that plane (i.e. in the direction of the principal "z" axis). The nodes "i" and "j" provide one vector, and the nodes "i" and "p" provide the other, hence:

$$\vec{Z}^* = \begin{vmatrix} \vec{i} & \vec{j} & \vec{k} \\ (x_j - x_i) & (y_j - y_i) & (z_j - z_i) \\ (x_p - x_i) & (y_p - y_i) & (z_p - z_i) \end{vmatrix}$$

$$= \vec{i}\,[(y_j - y_i)(z_p - z_i) - (y_p - y_i)(z_j - z_i)] -$$
$$\vec{j}\,[(x_j - x_i)(z_p - z_i) - (x_p - x_i)(z_j - z_i)] +$$
$$\vec{k}\,[(x_j - x_i)(y_p - y_i) - (x_p - x_i)(y_j - y_i)]$$

$$= \vec{i}\,X_z^* + \vec{j}\,Y_z^* + \vec{k}\,Z_z^*$$

The length of this vector is found by Pythagoras' Theorem, and is then used to find the direction cosines of the principal "z" axis.

$$L = \sqrt{(X_z^{*2} + Y_z^{*2} + Z_z^{*2})}$$

hence

$$l_z^* = X_z^*/L$$
$$m_z^* = Y_z^*/L$$
$$n_z^* = Z_z^*/L$$

l_z^*, m_z^*, and n_z^* are the components (in the global axes system) of a unit vector lying on the principal "z" axis. Similarly l_z, m_z, and n_z are the components of a unit vector lying on the member "z" axis. The angle between these unit vectors can be found by the cosine rule:

$$\beta = \cos^{-1}((a^2 + b^2 - c^2)/2ab)$$

but the length of "a" and "b" is unity, therefore

$$\beta = \cos^{-1}(1 - c^2/2)$$

where

$$c^2 = (l_z^* - l_z)^2 + (m_z^* - m_z)^2 + (n_z^* - n_z)^2$$

There is, however, one further complication. If the angle between the member "y" axis and the principal "z" axis is less than 90° then the wrong angle will be found as illustrated in fig 10.6, and β must be set equal to (360 - β)° (note that the angle calculated in these cases is, in fact, the anticlockwise rotation necessary to make the principal axes coincide with the member axes). The direction cosines of the principal "z" axis are already known and the direction cosines of the member "y" axis are readily available, allowing the angle between these axes to be calculated using the cosine rule.

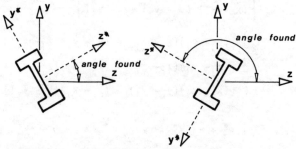

FIGURE 10.6   CASES WHERE THE WRONG ANGLE IS EVALUATED

Note that this complication arises because the vector cross product always produces a right hand system.

SPACE FRAMES

Example 10.1

Find β for the member shown in the following diagram if the coordinates of "i" and "j" are (0,10,0) and (10,0,10) respectively.

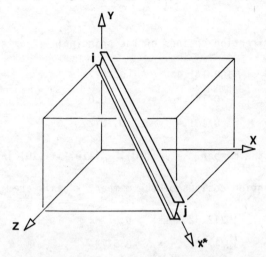

"p" can be located anywhere on the principal "x,y" plane (the principal "x" axis excepted). Here "p" will be located at (10,10,10) (i.e. directly "above" node "j"). Taking node "i" as the origin, the coordinates of nodes "j" and "p" become (10,-10,10) and (10,0,10) respectively. The following vector cross product generates a vector in the direction of the principal "z" axis:

$$\vec{z}^* = \begin{vmatrix} \vec{i} & \vec{j} & \vec{k} \\ 10 & -10 & 10 \\ 10 & 0 & 10 \end{vmatrix} = -100\vec{i} - 0\vec{j} + 100\vec{k}$$

Calculate the length of the vector

$$L = \sqrt{(100^2 + 100^2)} = 141.4$$

As the cross product of vectors produces a right-hand system, and the principal axes are themselves a right-hand system, the axes will point in the directions indicated in the preceding diagram which shows the member viewed looking down the global "y" axis towards the origin.

Evaluate the direction cosines of the principal "z" axis.

$$l_z^* = -100/141.4 = -0.7072$$
$$m_z^* = -0/141.4 = 0.0$$
$$n_z^* = 100/141.4 = 0.7072$$

The length of the member $= \sqrt{(10)^2 + (-10)^2 + (10)^2} = 17.32$

Hence the direction cosines of the member "x" axis are

$$l_x = 10/17.32 = 0.5774$$
$$m_x = -10/17.32 = -0.5774$$
$$n_x = 10/17.32 = 0.5774$$

The direction cosines of the member "z" axis can now be calculated.

$$D = \sqrt{(1 - n_x^2)} = 0.8165$$
$$l_z = -l_x n_x / D = -0.4083$$
$$m_z = -m_x n_x / D = 0.4083$$
$$n_z = D = 0.8165$$

Now calculate β.

$$\beta = \cos^{-1}(1 - c^2/2)$$

where

$$c^2 = (l_z^* - l_z)^2 + (m_z^* - m_z)^2 + (n_z^* - n_z)^2$$
$$= (-0.7072 + 0.4083)^2 + (0 - 0.4083)^2 + (0.7072 - 0.8165)^2$$
$$= 0.268$$

Hence

$$\beta = 30°$$

Check the angle between the member "y" axis and the principal "z" axis.

$$l_y = -m_x/D = 0.7072$$
$$m_y = l_x/D = 0.7072$$
$$n_y = 0$$
$$\cos\alpha = (1 - c^2/2)$$

where
$$c^2 = (l_z^* - l_y)^2 + (m_z^* - m_y)^2 + (n_z^* - n_y)^2$$
$$= (-0.7072 - 0.7072)^2 + (0 - 0.7072)^2 + (0.7072 - 0)^2$$
$$= 3.000$$

hence
$$\cos\alpha = -0.5 \text{ (negative, therefore } \alpha > 90°)$$

Therefore the true value of β is 30°.

Note that if cosα is positive then the true value of β is (360 - β)

Example 10.2

Calculate β for the member considered in example 10.1, but choose "p" to be the point (10,-10,10) (i.e. directly "below" node "j").

Make node "i" the origin then generate a vector in the direction of the principal "z" axis.

$$\vec{z}^* = \begin{vmatrix} \vec{i} & \vec{j} & \vec{k} \\ 10 & -10 & 10 \\ 10 & -20 & 10 \end{vmatrix}$$
$$= 100\vec{i} - 0\vec{j} - 100\vec{k}$$

The length of vector $\vec{z}^*$ is

$$L = \sqrt{(100^2 + 100^2)} = 141.4$$

As the cross product of vectors produces a right-hand system, and the principal axes are themselves a right-hand system, the axes will point in the directions indicated in the following diagram which shows the member viewed looking down the global "y" axis towards the origin.

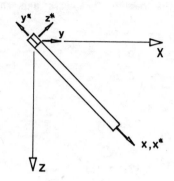

Evaluate the direction cosines of the principal "z" axis.

$$l_z^* = 100/141.4 = 0.7072$$
$$m_z^* = -0/141.4 = 0.0$$
$$n_z^* = -100/141.4 = -0.7072$$

The length of the member and the direction cosines of the member "x", "y", and "z" axes were found in the previous example and are copied here for convenience.

$$L = 17.32$$
$$l_x = 0.5774 \quad l_y = 0.7072 \quad l_z = -0.4083$$
$$m_x = -0.5774 \quad m_y = 0.7072 \quad m_z = 0.4083$$
$$n_x = 0.5774 \quad n_y = 0.0 \quad n_z = 0.8165$$

Calculate β.

$$\beta = \cos^{-1}(1 - c^2/2)$$

where

$$c^2 = (l_z^* - l_z)^2 + (m_z^* - m_z)^2 + (n_z^* - n_z)^2$$
$$= (0.7072 + 0.4083)^2 + (0 - 0.4083)^2 + (-0.7072 - 0.8165)^2$$
$$= 3.733$$

hence

$$\beta = 150°$$

SPACE FRAMES

Check the angle between the member "y" axis and the principal "z" axis (direction cosines of the member "y" axis are unchanged from the previous example).

where
$$\cos\alpha = (1 - c^2/2)$$
$$c^2 = (l_z^* - l_y)^2 + (m_z^* - m_y)^2 + (n_z^* - n_y)^2$$
$$= (0.7072 - 0.7072)^2 + (0 - 0.7072)^2 + (-0.7072 - 0)^2$$
$$= 1.000$$

hence
$$\cos\alpha = 0.5 \text{ (positive, i.e } \alpha < 90°)$$

Therefore the true value of β is (360-30)° (i.e. 330°). Note that the examples 10.1 and 10.2 have considered the same member and that the angles found from the two choices of "p" are 180° different from each other.

Program TNODE.SF

There follows the listing of a program that utilises the theory presented in this section to calculate β for a space frame member using the third node method.

```
1000 REM ******************************************************************
1010 REM *                                                                *
1020 REM *    PROGRAM   T N O D E . S F    TO CALCULATE THE ANGLE         *
1030 REM *    BETWEEN THE PRINCIPAL Y-AXIS AND THE MEMBER Y-AXIS          *
1040 REM *    FOR A SPACE FRAME MEMBER                                    *
1050 REM *                                                                *
1060 REM *    J.A.D.BALFOUR                                               *
1070 REM *                                                                *
1080 REM ******************************************************************
1090 REM
1100 REM  OPTION BASE 1
1110 REM
1120 REM  --- NOTE THAT VARIABLES ARE DEFINED IN APPENDIX A
1130 REM
1140 PRINT : PRINT
1150 PRINT "-------------------------------------------------------"
1160 PRINT
1170 PRINT "PROGRAM    T N O D E . S F    TO CALCULATE BETA FOR"
1180 PRINT "SPACE FRAME MEMBERS"
1190 PRINT
1200 PRINT "-------------------------------------------------------"
1210 PRINT
1220 INPUT "NUMBER OF MEMBERS = ", NMEMB
1230 PRINT
1240 FOR I = 1 TO NMEMB
1250 PRINT
1260 PRINT "MEMBER "; I
1270 PRINT
1280 INPUT "X-COORDINATE LOWER  NODE = ", XI
1290 INPUT "Y-COORDINATE LOWER  NODE = ", YI
1300 INPUT "Z-COORDINATE LOWER  NODE = ", ZI
1310 PRINT
```

```
1320 INPUT "X-COORDINATE HIGHER NODE  =  ", XJ
1330 INPUT "Y-COORDINATE HIGHER NODE  =  ", YJ
1340 INPUT "Z-COORDINATE HIGHER NODE  =  ", ZJ
1350 PRINT
1360 INPUT "X-COORDINATE THIRD  NODE  =  ", XP
1370 INPUT "Y-COORDINATE THIRD  NODE  =  ", YP
1380 INPUT "Z-COORDINATE THIRD  NODE  =  ", ZP
1390 REM
1400 REM   ---   MOVE ORIGIN TO NODE I
1410 REM
1420 XJ = XJ - XI
1430 YJ = YJ - YI
1440 ZJ = ZJ - ZI
1450 XP = XP - XI
1460 YP = YP - YI
1470 ZP = ZP - ZI
1480 REM
1490 REM   ---   CALCULATE THE DIRECTION COSINES OF THE PRINCIPAL Z-AXIS
1500 REM
1510 LZP = YJ*ZP - YP*ZJ
1520 MZP = XP*ZJ - XJ*ZP
1530 NZP = XJ*YP - XP*YJ
1540 L   = SQR(LZP^2 + MZP^2 + NZP^2)
1550 LZP = LZP / L
1560 MZP = MZP / L
1570 NZP = NZP / L
1580 REM
1590 REM   ---   CALCULATE THE DIRECTION COSINES OF THE MEMBER X-AXIS
1600 REM
1610 L   = SQR(XJ^2 + YJ^2 + ZJ^2)
1620 LX = XJ / L
1630 MX = YJ / L
1640 NX = ZJ / L
1650 REM
1660 REM   ---   CALCULATE THE DIRECTION COSINES OF THE MEMBER Z-AXIS
1670 REM
1680 IF ABS(NX) > .9999 THEN GOTO 1730
1690 NZ  =  SQR(1 - NX^2)
1700 LZ = -LX * NX / NZ
1710 MZ = -MX * NX / NZ
1720 GOTO 1790
1730 LZ = -NX
1740 MZ = 0
1750 NZ = 0
1760 REM
1770 REM   ---   CALCULATE BETA
1780 REM
1790 X    = 1 - ((LZP-LZ)^2+(MZP-MZ)^2+(NZP-NZ)^2)/2
1800 IF ABS(X) > .9999 THEN BETA = 0
1810 IF ABS(X) > .9999 THEN GOTO 1910
1820 BETA = (180 * (1.5708-ATN(X/SQR(1-X*X)))) / (4 * ATN(1))
1830 REM
1840 REM   ---     CHECK ANGLE BETWEEN THE MEMBER Y-AXIS AND THE PRINCIPAL Z-AXIS
1850 REM
1860 LY  = -MX/NZ
1870 MY  =  LX/NZ
1880 CYZ = (1 - ((LZP-LY)^2 + (MZP-MY)^2 + (NZP)^2)/2)
1890 IF CYZ > 0 THEN BETA = 360 - BETA
1900 PRINT
1910 PRINT "BETA  =  "; BETA; " DEGREES"
1920 NEXT I
1930 END
```

SPACE FRAMES

Sample Run There follows a sample run which shows program TNODE.SF being used to solve examples 10.1 and 10.2. Input from the keyboard is shown in bold typeface.

```
--------------------------------------------------
PROGRAM    T N O D E . S F    TO CALCULATE BETA FOR
SPACE FRAME MEMBERS
--------------------------------------------------

NUMBER OF MEMBERS = 2

MEMBER 1

X-COORDINATE LOWER  NODE = 0
Y-COORDINATE LOWER  NODE = 10
Z-COORDINATE LOWER  NODE = 0

X-COORDINATE HIGHER NODE = 10
Y-COORDINATE HIGHER NODE = 0
Z-COORDINATE HIGHER NODE = 10

X-COORDINATE THIRD  NODE = 10
Y-COORDINATE THIRD  NODE = 10
Z-COORDINATE THIRD  NODE = 10

BETA =   30.0002 DEGREES
MEMBER 2

X-COORDINATE LOWER  NODE = 0
Y-COORDINATE LOWER  NODE = 10
Z-COORDINATE LOWER  NODE = 0

X-COORDINATE HIGHER NODE = 10
Y-COORDINATE HIGHER NODE = 0
Z-COORDINATE HIGHER NODE = 10

X-COORDINATE THIRD  NODE = 10
Y-COORDINATE THIRD  NODE = -10
Z-COORDINATE THIRD  NODE = 10

BETA =   210 DEGREES
```

10.5 MEMBER END FORCES AND REACTIONS

The relationship between the forces at the ends of member "i,j" due to displacements at "i" and "j", in the principal axes system, has already been seen to be

$$f^*_{ij} = k^{j*}_{ii} \delta^*_{ij} + k^*_{ij} \delta^*_{ji}$$

but

$$\delta^*_{ij} = T^*_{ij} \delta_{ij} \qquad \delta^*_{ji} = T^*_{ij} \delta_{ji}$$

$$\delta_{ij} = T_{ij} \Delta_i \qquad \delta_{ji} = T_{ij} \Delta_j$$

hence

$$f^*_{ij} = k^{j*}_{ii} T^*_{ij} T_{ij} \Delta_i + k^*_{ij} T^*_{ij} T_{ij} \Delta_j \qquad (10.6)$$

Once the final stiffness equation has been solved the nodal displacements are used to find member end forces at end "i" of each member using equation (10.6). An equation similar to equation (10.6) could be used to find the forces at end "j", but it is more efficient to consider the equilibrium of the element which shows that

$$f^*_{jix} = -f^*_{ijx}$$
$$f^*_{jiy} = -f^*_{ijy}$$
$$f^*_{jiz} = -f^*_{ijz}$$
$$m^*_{jix} = -m^*_{ijx}$$
$$m^*_{jiy} = -m^*_{ijy} - L_{ij} f^*_{ijz}$$
$$m^*_{jiz} = -m^*_{ijz} + L_{ij} f^*_{ijy}$$

If equivalent joint forces have been used to cater for settlement, lack of fit, etc., as described in section 8.2, then the final member end forces are found by superimposing the fixed end force system on the forces due to nodal displacements.

10.6 SPACE FRAME PROGRAM - SFRAME.SF

This section presents the program SFRAME.SF for the automatic analysis of space frames. The program is similar in structure to the grillage analysis program GRID.GD, except that the facility to deal with member loads has been omitted.

Input The only piece of information the user supplies to the program from the keyboard is the name of the data file to be read by the program. This data file can be constructed using either a text editor or the data preprocessor program PRE.MA/PRE.01/PRE.02 described in section 5.11. If using a text editor the data file should be constructed to the following format:

<job title (not to exceed one line)>
5,3,9,<no. of nodes>,<no. of members>,<no. of restrained nodes>,
<no. of nodal loads>
<x-coordinate node 1>,<y-coordinate node 1>,<z-coordinate node 1>
<x-coordinate node 2>,<y-coordinate node 2>,<z-coordinate node 2>
<x-coordinate node 3>,<y-coordinate node 3>,<z-coordinate node 3>
         .   .   .
         .   .   .
<x-coordinate node n>,<y-coordinate node n>,<z-coordinate node n>

SPACE FRAMES

<node "i" member 1>,<node "j" member 1>,<area member 1>,
<I_y member 1>,<I_z member 1>,<J member 1>,<β member 1>,<G member 1>,
<E member 1>
<node "i" member 2>,<node "j" member 2>,<area member 2>,
<I_y member 2>,<I_z member 2>,<J member 2>,<β member 2>,<G member 2>,
<E member 2>
         .   .   .
           .   .   .
  Note - "i" must be less then   "j"
           .   .   .
         .   .   .
<node "i" member n>,<node "j" member n>,<area member n>,
<I_y member n>,<I_z member n>,<J member n>,<β member n>,<G member n>,
<E member n>
<restrained node no. 1 node number>,<direction(s)>
<restrained node no. 2 node number>,<direction(s)>
         .   .   .
           .   .   .
  Note - restraint directions are as follows:
         restraint in the global "x" direction = 1
         restraint in the global "y" direction = 2
         restraint in the global "z" direction = 3
         rotational restraint about "x" axis = 4
         rotational restraint about "y" axis = 5
         rotational restraint about "z" axis = 6
           .   .   .
         Input restraints as a composite number
           .   .   .
         .   .   .
<restrained node no. n node number>,<direction(s)>
<nodal load no. 1 node number>,<direction>,<magnitude>
<nodal load no. 2 node number>,<direction>,<magnitude>
           .   .   .
         .   .   .
  Note - load directions are as follows:
         load in the global "x" direction = 1
         load in the global "y" direction = 2
         load in the global "z" direction = 2
         moment about global "x" axis   = 4
         moment about global "y" axis   = 5
         moment about global "z" axis   = 6
           .   .   .
         .   .   .
<nodal load no. n node number>,<direction>,<magnitude>

COMPUTER ANALYSIS OF STRUCTURAL FRAMEWORKS

Comments on the Algorithm The length of each member and the direction cosines for the member "x" axis are calculated from the member node numbers and the nodal coordinates. The following equations were derived in section 10.3 and show how the element stiffness submatrices in the global axes system can be obtained from the element stiffness submatrices in the principal axes system using transformation matrices:

and
$$K_{ii}^j = T_{ij}^{-1} T_{ij}^{*-1} k_{ii}^{j*} T_{ij}^{*} T_{ij}$$
$$K_{ij} = T_{ij}^{-1} T_{ij}^{*-1} k_{ij}^{*} T_{ij}^{*} T_{ij}$$

From the above equations it can be seen that to evaluate one global element stiffness submatrix involves multiplying five 6 x 6 matrices together. Using standard matrix multiplication this requires 864 multiplications and 720 additions. This volume of arithmetic is unmanageable by hand. To set up the matrices and conduct the matrix multiplication on a computer is straightforward. However, setting up the matrices takes a fair amount of coding, and if a relatively slow computer is used then the time taken to evaluate the global element stiffness submatrices can become unacceptably long. Significant savings of computer time and memory can be made by expanding the matrix multiplication as follows:

$$T_{ij}^{-1} T_{ij}^{*-1} = \begin{bmatrix} T_1 & T_2 & T_3 & 0 & 0 & 0 \\ T_4 & T_5 & T_6 & 0 & 0 & 0 \\ T_7 & T_8 & T_9 & 0 & 0 & 0 \\ 0 & 0 & 0 & T_1 & T_2 & T_3 \\ 0 & 0 & 0 & T_4 & T_5 & T_6 \\ 0 & 0 & 0 & T_7 & T_8 & T_9 \end{bmatrix}$$

where

T_1 = 1 (0)
T_2 = $-(m \cos\beta + \ln \sin\beta)/D$ $(-n \sin\beta)$
T_3 = $(m \sin\beta - \ln \cos\beta)/D$ $(-n \cos\beta)$
T_4 = m (0)
T_5 = $(l \cos\beta - mn \sin\beta)/D$ (c)
T_6 = $-(l \sin\beta + mn \cos\beta)/D$ (-s)
T_7 = n (n)
T_8 = $D \sin\beta$ (0)
T_9 = $D \cos\beta$ (0)

(the expressions in brackets are those assumed for the "special" case where the member lies parallel to the coordinate "z" axis).

and

$$T^*_{ij} T_{ij} = [T^{-1}_{ij} T^{*-1}_{ij}]^T$$

$$= \begin{bmatrix} T_1 & T_4 & T_7 & 0 & 0 & 0 \\ T_2 & T_5 & T_8 & 0 & 0 & 0 \\ T_3 & T_6 & T_9 & 0 & 0 & 0 \\ 0 & 0 & 0 & T_1 & T_4 & T_7 \\ 0 & 0 & 0 & T_2 & T_5 & T_8 \\ 0 & 0 & 0 & T_3 & T_6 & T_9 \end{bmatrix}$$

Writing the k^{j*}_{ii}, k^*_{ij}, k^*_{ji}, and k^{j*}_{jj} submatrices in terms of "K" factors allows the global stiffness submatrices to be evaluated in terms of the "T" and "K" factors as follows:

$$k^{j*}_{ii} = \begin{bmatrix} K_1 & 0 & 0 & 0 & 0 & 0 \\ 0 & K_2 & 0 & 0 & 0 & K_7 \\ 0 & 0 & K_3 & 0 & K_8 & 0 \\ 0 & 0 & 0 & K_4 & 0 & 0 \\ 0 & 0 & K_9 & 0 & K_5 & 0 \\ 0 & K_{10} & 0 & 0 & 0 & K_6 \end{bmatrix}$$

where

K_1	=	EA/L	K_2 =	$12EI_z/L^3$
K_3	=	$12EI_y/L^3$	K_4 =	GJ/L
K_5	=	$4EI_y/L$	K_6 =	$4EI_z/L$
K_7	=	$6EI_z/L^2$	K_8 =	$-6EI_y/L^2$
K_9	=	$-6EI_y/L^2$	K_{10} =	$6EI_z/L^2$

$$k_{ij} = \begin{bmatrix} -K_1 & 0 & 0 & 0 & 0 & 0 \\ 0 & -K_2 & 0 & 0 & 0 & K_7 \\ 0 & 0 & -K_3 & 0 & K_8 & 0 \\ 0 & 0 & 0 & -K_4 & 0 & 0 \\ 0 & 0 & -K_9 & 0 & -K_5/2 & 0 \\ 0 & -K_{10} & 0 & 0 & 0 & -K_6/2 \end{bmatrix}$$

$$K^j_{ii} = T^{-1}_{ij} T^{*-1}_{ij} k^{j*}_{ii} T^*_{ij} T_{ij} = \begin{bmatrix} A & B \\ C & D \end{bmatrix}$$

where A =

$$\begin{bmatrix} (K_1T_1^2+K_2T_2^2+K_3T_3^2) & (K_1T_1T_4+K_2T_2T_5+K_3T_3T_6) & (K_1T_1T_7+K_2T_2T_8+K_3T_3T_9) \\ (K_1T_4T_1+K_2T_5T_2+K_3T_6T_3) & (K_1T_4^2+K_2T_5^2+K_3T_6^2) & (K_1T_4T_7+K_2T_5T_8+K_3T_6T_9) \\ (K_1T_7T_1+K_2T_8T_2+K_3T_9T_3) & (K_1T_7T_4+K_2T_8T_5+K_3T_9T_6) & (K_1T_7^2+K_2T_8^2+K_3T_9^2) \end{bmatrix}$$

$$B = \begin{bmatrix} (K_7T_2T_3+K_8T_3T_2) & (K_7T_2T_6+K_8T_3T_5) & (K_7T_2T_9+K_8T_3T_8) \\ (K_7T_5T_3+K_8T_6T_2) & (K_7T_5T_6+K_8T_6T_5) & (K_7T_5T_9+K_8T_6T_8) \\ (K_7T_8T_3+K_8T_9T_2) & (K_7T_8T_6+K_8T_9T_5) & (K_7T_8T_9+K_8T_9T_8) \end{bmatrix}$$

$$C = \begin{bmatrix} (K_9T_2T_3+K_{10}T_3T_2) & (K_9T_2T_6+K_{10}T_3T_5) & (K_9T_2T_9+K_{10}T_3T_8) \\ (K_9T_5T_3+K_{10}T_6T_2) & (K_9T_5T_6+K_{10}T_6T_5) & (K_9T_5T_9+K_{10}T_6T_8) \\ (K_9T_8T_3+K_{10}T_9T_2) & (K_9T_8T_6+K_{10}T_9T_5) & (K_9T_8T_9+K_{10}T_9T_8) \end{bmatrix}$$

$$D = \begin{bmatrix} (K_4T_1^2+K_5T_2^2+K_6T_3^2) & (K_4T_1T_4+K_5T_2T_5+K_6T_3T_6) & (K_4T_1T_7+K_5T_2T_8+K_6T_3T_9) \\ (K_4T_4T_1+K_5T_5T_2+K_6T_6T_3) & (K_4T_4^2+K_5T_5^2+K_6T_6^2) & (K_4T_4T_7+K_5T_5T_8+K_6T_6T_9) \\ (K_4T_7T_1+K_5T_8T_2+K_6T_9T_3) & (K_4T_7T_4+K_5T_8T_5+K_6T_9T_6) & (K_4T_7^2+K_5T_8^2+K_6T_9^2) \end{bmatrix}$$

To evaluate one global element substiffness matrix using the above expansions reduces the arthmetic to 180 multiplications and 54 additions. Further savings can be made by taking advantage of the symmetry of the global element stiffness matrix and the similarity of the global element stiffness submatrices. In SFRAME.SF to generate the upper triangle of the global element stiffness submatrix requires only 138 multiplications/divisions and 39 additions/subtractions. Contrast this with the 3456 multiplications and 2880 additions required to evaluate the global element stiffness submatrices by matrix multiplication of 5 number 6 x 6 matrices.

Using code numbers the global element stiffness matrix for each element in turn is added to the upper triangle of the final structure stiffness matrix. The loading vector is then generated and the final stiffness equation is solved using the Gaussian elimination subroutine from Chapter 4. Once the unknown nodal displacements have been found they are used to calculate the member forces. Note that the "T" and "K" factors that were used to evaluate the member stiffness are also used to simplify the calculation of the member forces. The reactions are evaluated by summing the reaction contributions from the members connected to the restrained nodes. If node "i" is restrained then the contribution, R_{ij}, from member "i,j" to the reactions at node "i" is (in the global axes system) is as follows:

$$\begin{bmatrix} R_{ijx} \\ R_{ijy} \\ R_{ijz} \\ RM_{ijx} \\ RM_{ijy} \\ RM_{ijz} \end{bmatrix} = \begin{bmatrix} T_1 & T_2 & T_3 & 0 & 0 & 0 \\ T_4 & T_5 & T_6 & 0 & 0 & 0 \\ T_7 & T_8 & T_9 & 0 & 0 & 0 \\ 0 & 0 & 0 & T_1 & T_2 & T_3 \\ 0 & 0 & 0 & T_4 & T_5 & T_6 \\ 0 & 0 & 0 & T_7 & T_8 & T_9 \end{bmatrix} \begin{bmatrix} f^*_{ijx} \\ f^*_{ijy} \\ f^*_{ijz} \\ m^*_{ijx} \\ m^*_{ijy} \\ m^*_{ijz} \end{bmatrix}$$

$$= \begin{bmatrix} T_1 f^*_{ijx} + T_2 f^*_{ijy} + T_3 f^*_{ijz} \\ T_4 f^*_{ijx} + T_5 f^*_{ijy} + T_6 f^*_{ijz} \\ T_7 f^*_{ijx} + T_8 f^*_{ijy} + T_9 f^*_{ijz} \\ T_1 m^*_{ijx} + T_2 m^*_{ijy} + T_3 m^*_{ijz} \\ T_4 m^*_{ijx} + T_5 m^*_{ijy} + T_6 m^*_{ijz} \\ T_7 m^*_{ijx} + T_8 m^*_{ijy} + T_9 m^*_{ijz} \end{bmatrix}$$

where

R_{ijx} is the reactive component of force in the global "x" direction.

R_{ijy} is the reactive component of force in the global "y" direction.

R_{ijz} is the reactive component of force in the global "z" direction.

RM_{ijx} is the reactive component of moment about the global "x" axis.

RM_{ijy} is the reactive component of moment about the global "y" axis.

RM_{ijz} is the reactive component of moment about the global "z" axis.

The final reaction at node "i" is given by the following equation:

$$R_i = \sum R_{ij} - P_i$$

where the summation is for all of the members attached to node "i".

Output The data from the data file is echoed and messages are output to indicate the beginning of each solution phase. After solution of the stiffness equation nodal displacements are output. The axial force, shear forces (in two planes), bending moments (also in two planes), and torsional moment at each end of each member are then calculated and output. Finally the components of reaction (in the global axes system) at each restrained node are output.

Listing

```
1000 REM ***********************************************************************
1010 REM *                                                                     *
1020 REM *      PROGRAM    S F R A M E . S F     FOR THE AUTOMATIC ANALYSIS    *
1030 REM *      OF SPACE FRAMES                                                *
1040 REM *                                                                     *
1050 REM *      DATA FROM DATA FILE GENERATED BY A TEXT EDITOR OR BY THE       *
1060 REM *      PREPROCESSOR PROGRAM P R E . M A / P R E . 0 1 / P R E . 0 2   *
1070 REM *                                                                     *
1080 REM *      J.A.D.BALFOUR                                                  *
1090 REM *                                                                     *
1100 REM ***********************************************************************
1110 REM
1120 REM ---   NOTE THAT VARIABLES ARE DEFINED IN APPENDIX A
1130 REM
1140 OPTION BASE 1
1150 REM
1160 REM ---   THE FOLLOWING DIMENSION STATEMENT SETS THE PROBLEM SIZE
1170 REM
1180 COMMON FILE$
1190 DIM NODE(20,3), MEMB(30,13), REST(20,8), NLOD(40,3),    FSTIFF(40,20)
1200 DIM LOD(40),     FREE(120),    EFREE(12),   ESTIFF(12,12), U(24)
1210 PRINT : PRINT
1220 PRINT "--------------------------------------------------------"
1230 PRINT
1240 PRINT "PROGRAM    S F R A M E . S F    FOR THE AUTOMATIC ANALYSIS"
1250 PRINT "OF SPACE FRAMES"
1260 PRINT
1270 PRINT "--------------------------------------------------------"
1280 REM
1290 REM ---   CHECK IF THIS PROGRAM CHAINED FROM PRE.MA/PRE.01/PRE.02
1300 REM
1310 IF LEN(FILE$) <> 0 THEN GOTO 1340
1320 PRINT
1330 INPUT          "NAME OF THE DATA FILE = ", FILE$
1340 OPEN "I", #1, FILE$
1350 PRINT
1360 PRINT          "READING DATA"
1370 GOSUB 1610
1380 REM
1390 REM           PRINT DATA
1400 GOSUB 2000
1410 PRINT
1420 PRINT          "GENERATING THE FREEDOM VECTOR"
1430 GOSUB 2570
1440 PRINT
1450 PRINT          "GENERATING THE STRUCTURE STIFFNESS MATRIX"
1460 GOSUB 2900
1470 PRINT
1480 PRINT          "GENERATING THE LOADING VECTOR"
1490 GOSUB 4020
1500 PRINT
1510 PRINT          "SOLVING EQUATIONS  -  NO. OF EQUATIONS = "; NDOF
1520 PRINT          "SEMI-BANDWIDTH                         = "; BAND
1530 GOSUB 4160
1540 REM
1550 REM           OUTPUT NODAL DISPLACEMENTS
1560 GOSUB 4760
1570 REM
1580 REM           OUTPUT MEMBER FORCES
1590 GOSUB 5020
1600 END
1610 REM * * * * * * * * * * * * * * * * * * * * * * * * * * * * * * * *
1620 REM *                                                             *
1630 REM *     SUBROUTINE TO READ DATA FROM THE INPUT FILE             *
1640 REM *                                                             *
1650 REM * * * * * * * * * * * * * * * * * * * * * * * * * * * * * * * *
1660 REM
```

```
1670 REM  ---  READ TITLE AND BASIC PROBLEM DATA
1680 REM
1690 INPUT #1, TITLE$
1700 INPUT #1, PTYPE, NCORD, NPROP, NNODE, NMEMB, NREST, NNLOD
1710 IF PTYPE = 5 THEN GOTO 1760
1720 PRINT : PRINT "NOT A SPACE FRAME DATA FILE" : END
1730 REM
1740 REM  ---  READ NODAL COORDINATES
1750 REM
1760 FOR I = 1 TO NNODE
1770 INPUT #1, NODE(I,1), NODE(I,2), NODE(I,3)
1780 NEXT I
1790 REM
1800 REM  ---  READ MEMBER DATA
1810 REM
1820 FOR I = 1 TO NMEMB
1830 INPUT #1, MEMB(I,1), MEMB(I,2), MEMB(I,3), MEMB(I,4), MEMB(I,5), MEMB(I,6)
1840 INPUT #1, MEMB(I,7), MEMB(I,8), MEMB(I,9)
1850 NEXT I
1860 REM
1870 REM  ---  READ RESTRAINT DATA
1880 REM
1890 FOR I = 1 TO NREST
1900 INPUT #1, REST(I,1), REST(I,2)
1910 REST(I,3) = 0
1920 NEXT I
1930 REM
1940 REM  ---  READ LOADING DATA
1950 REM
1960 FOR I = 1 TO NNLOD
1970 INPUT #1, NLOD(I,1), NLOD(I,2), NLOD(I,3)
1980 NEXT I
1990 RETURN
2000 REM * * * * * * * * * * * * * * * * * * * * * * * * * * * * * * *
2010 REM *                                                            *
2020 REM *     SUBROUTINE TO PRINT THE INPUT DATA                     *
2030 REM *                                                            *
2040 REM * * * * * * * * * * * * * * * * * * * * * * * * * * * * * * *
2050 REM
2060 PRINT
2070 PRINT "+ + + + + + + + +"
2080 PRINT "+    JOB TITLE   +   -   "; TITLE$
2090 PRINT "+ + + + + + + + +"
2100 PRINT
2110 PRINT "+ + + + + + + + + + + + +"
2120 PRINT "+   NODAL COORDINATES   +"
2130 PRINT "+ + + + + + + + + + + + +"
2140 PRINT
2150 PRINT "NODE            X               Y               Z"
2160 PRINT "NUMBER"
2170 FOR I = 1 TO NNODE
2180 PRINT " "; I, NODE(I,1), NODE(I,2), NODE(I,3)
2190 NEXT I
2200 PRINT
2210 PRINT "+ + + + + + + + + + + + +"
2220 PRINT "+   MEMBER PROPERTIES   +"
2230 PRINT "+ + + + + + + + + + + + +"
2240 PRINT
2250 PRINT "MEMBER          A          IYY       IZZ        J       BETA";
2260 PRINT "        G         E"
2270 PRINT "NUMBER"
2280 FOR I = 1 TO NMEMB
2290 PRINT USING "## ##      "; MEMB(I,1), MEMB(I,2);
2300 FOR J = 3 TO NPROP
2310 PRINT USING " #.###^^^^"; MEMB(I,J);
2320 NEXT J
2330 PRINT
2340 NEXT I
2350 PRINT
```

```
2360 PRINT "+ + + + + + + + +"
2370 PRINT "+    RESTRAINTS   +"
2380 PRINT "+ + + + + + + + +"
2390 PRINT
2400 PRINT "NODE      DIRECTION(S)"
2410 PRINT "NUMBER"
2420 FOR I = 1 TO NREST
2430 PRINT USING "  ##         ######"; REST(I,1); REST(I,2)
2440 NEXT I
2450 PRINT
2460 PRINT "+ + + + + + + + + +"
2470 PRINT "+    NODAL LOADS    +"
2480 PRINT "+ + + + + + + + + +"
2490 PRINT
2500 PRINT "NODE     DIRECTION    VALUE"
2510 PRINT "NUMBER"
2520 FOR I = 1 TO NNLOD
2530 PRINT USING "  ##         #      #.###^^^^"; NLOD(I,1), NLOD(I,2), NLOD(I,3)
2540 NEXT I
2550 PRINT
2560 RETURN
2570 REM  * * * * * * * * * * * * * * * * * * * * * * * * * * * * * * * *
2580 REM  *                                                              *
2590 REM  *     SUBROUTINE TO GENERATE THE FREEDOM VECTOR                *
2600 REM  *                                                              *
2610 REM  * * * * * * * * * * * * * * * * * * * * * * * * * * * * * * * *
2620 REM
2630 REM  --- ZERO THE FREEDOM VECTOR
2640 REM
2650 FOR I    = 1 TO 6*NNODE
2660 FREE(I) = 0
2670 NEXT I
2680 FOR I    = 1 TO NREST
2690 REM
2700 REM  --- LOOP FOR ALL DIGITS IN THE RESTRAINT NUMBER
2710 REM
2720 FOR J    = 1 TO LEN(STR$(REST(I,2)))
2730 K        = VAL(MID$(STR$(REST(I,2)), J, 1))
2740 IF K<1 OR K>6 THEN GOTO 2770
2750 K        = 6 * (REST(I,1)-1) + VAL(MID$(STR$(REST(I,2)), J, 1))
2760 FREE(K) = 1
2770 NEXT J
2780 NEXT I
2790 REM
2800 REM  --- NUMBER THE FREEDOMS (RESTRAINTS SET TO ZERO)
2810 REM
2820 FOR I    = 1 TO 6*NNODE
2830 IF FREE(I) = 1 THEN GOTO 2870
2840 NDOF     = NDOF + 1
2850 FREE(I) = NDOF
2860 GOTO 2880
2870 FREE(I) = 0
2880 NEXT I
2890 RETURN
2900 REM  * * * * * * * * * * * * * * * * * * * * * * * * * * * * * * * *
2910 REM  *                                                              *
2920 REM  *     SUBROUTINE TO GENERATE THE STRUCTURE STIFFNESS MATRIX    *
2930 REM  *                                                              *
2940 REM  * * * * * * * * * * * * * * * * * * * * * * * * * * * * * * * *
2950 REM
2960 REM  --- ZERO THE STIFFNESS MATRIX AND LOADING VECTOR
2970 REM
2980 FOR I    = 1 TO NDOF
2990 FOR J    = 1 TO 20
3000 FSTIFF(I,J) = 0
3010 NEXT J
3020 LOD(I) = 0
3030 NEXT I
3040 REM
```

SPACE FRAMES

```
3050 REM    ---  LOOP FOR EACH ELEMENT
3060 REM
3070 BAND = 0
3080 PRINT "ADDING ELEMENT :- ";
3090 FOR K = 1 TO NMEMB
3100 PRINT K;
3110 IN = MEMB(K,1)
3120 JN = MEMB(K,2)
3130 REM
3140 REM    ---  CALCULATE ELEMENT LENGTH (STORE IN MEMB(K,10))
3150 REM
3160 MEMB(K,10) = (NODE(JN,1)-NODE(IN,1))^2 + (NODE(JN,2)-NODE(IN,2))^2
3170 MEMB(K,10) = SQR(MEMB(K,10) + (NODE(JN,3)-NODE(IN,3))^2)
3180 REM
3190 REM    ---  CALCULATE THE DIRECTION COSINES OF THE ANGLES THE MEMBER MAKES
3200 REM    ---  WITH THE COORDINATE X-AXIS (STORE IN MEMB(K,11) TO MEMB(K,13))
3210 REM
3220 MEMB(K,11) = (NODE(JN,1) - NODE(IN,1)) / MEMB(K,10)
3230 MEMB(K,12) = (NODE(JN,2) - NODE(IN,2)) / MEMB(K,10)
3240 MEMB(K,13) = (NODE(JN,3) - NODE(IN,3)) / MEMB(K,10)
3250 REM
3260 REM    ---  GENERATE THE UPPER TRIANGLE OF THE ELEMENT STIFFNESS MATRIX
3270 REM
3280 GOSUB 6040
3290 ESTIFF(1,1)   =   K1*T1*T1 + K2*T2*T2 + K3*T3*T3
3300 ESTIFF(1,2)   =   K1*T1*T4 + K2*T2*T5 + K3*T3*T6
3310 ESTIFF(1,3)   =   K1*T1*T7 + K2*T2*T8 + K3*T3*T9
3320 ESTIFF(1,4)   =   K7*T2*T3 + K8*T3*T2
3330 ESTIFF(1,5)   =   K7*T2*T6 + K8*T3*T5
3340 ESTIFF(1,6)   =   K7*T2*T9 + K8*T3*T8
3350 ESTIFF(2,2)   =   K1*T4*T4 + K2*T5*T5 + K3*T6*T6
3360 ESTIFF(2,3)   =   K1*T4*T7 + K2*T5*T8 + K3*T6*T9
3370 ESTIFF(2,4)   =   K7*T5*T3 + K8*T6*T2
3380 ESTIFF(2,5)   =   K7*T5*T6 + K8*T6*T5
3390 ESTIFF(2,6)   =   K7*T5*T9 + K8*T6*T8
3400 ESTIFF(3,3)   =   K1*T7*T7 + K2*T8*T8 + K3*T9*T9
3410 ESTIFF(3,4)   =   K7*T8*T3 + K8*T9*T2
3420 ESTIFF(3,5)   =   K7*T8*T6 + K8*T9*T5
3430 ESTIFF(3,6)   =   K7*T8*T9 + K8*T9*T8
3440 ESTIFF(4,4)   =   K4*T1*T1 + K5*T2*T2 + K6*T3*T3
3450 ESTIFF(4,5)   =   K4*T1*T4 + K5*T2*T5 + K6*T3*T6
3460 ESTIFF(4,6)   =   K4*T1*T7 + K5*T2*T8 + K6*T3*T9
3470 ESTIFF(4,10)  = (ESTIFF(4,4) - 3*K4*T1*T1) / 2
3480 ESTIFF(4,11)  = (ESTIFF(4,5) - 3*K4*T1*T4) / 2
3490 ESTIFF(4,12)  = (ESTIFF(4,6) - 3*K4*T1*T7) / 2
3500 ESTIFF(5,5)   =   K4*T4*T4 + K5*T5*T5 + K6*T6*T6
3510 ESTIFF(5,6)   =   K4*T4*T7 + K5*T5*T8 + K6*T6*T9
3520 ESTIFF(5,10)  =   ESTIFF(4,11)
3530 ESTIFF(5,11)  = (ESTIFF(5,5) - 3*K4*T4*T4) / 2
3540 ESTIFF(5,12)  = (ESTIFF(5,6) - 3*K4*T4*T7) / 2
3550 ESTIFF(6,6)   =   K4*T7*T7 + K5*T8*T8 + K6*T9*T9
3560 ESTIFF(6,10)  =   ESTIFF(4,12)
3570 ESTIFF(6,11)  =   ESTIFF(5,12)
3580 ESTIFF(6,12)  = (ESTIFF(6,6) - 3*K4*T7*T7) / 2
3590 FOR I = 1 TO 3
3600 FOR J = 1 TO 3
3610 ESTIFF(I,J+9)    =     ESTIFF(I,J+3)
3620 IF I <= J THEN    ESTIFF(I,J+6) = - ESTIFF(I,J)
3630 IF J < I THEN     ESTIFF(I,J+6) = - ESTIFF(J,I)
3640 ESTIFF(I+3,J+6) = - ESTIFF(J,I+3)
3650 ESTIFF(I+6,J+9) = - ESTIFF(I,J+3)
3660 IF J < I THEN GOTO 3690
3670 ESTIFF(I+6,J+6) =   ESTIFF(I,J)
3680 ESTIFF(I+9,J+9) =   ESTIFF(I+3,J+3)
3690 NEXT J
3700 NEXT I
3710 REM
3720 REM    ---  SET UP THE ELEMENT CODE NUMBER
3730 REM
```

```
3740 EFREE(1)    = FREE(6*IN-5)
3750 EFREE(2)    = FREE(6*IN-4)
3760 EFREE(3)    = FREE(6*IN-3)
3770 EFREE(4)    = FREE(6*IN-2)
3780 EFREE(5)    = FREE(6*IN-1)
3790 EFREE(6)    = FREE(6*IN)
3800 EFREE(7)    = FREE(6*JN-5)
3810 EFREE(8)    = FREE(6*JN-4)
3820 EFREE(9)    = FREE(6*JN-3)
3830 EFREE(10)   = FREE(6*JN-2)
3840 EFREE(11)   = FREE(6*JN-1)
3850 EFREE(12)   = FREE(6*JN)
3860 REM
3870 REM   ---   ADD THE ELEMENT STIFFNESS
3880 REM
3890 FOR I       = 1 TO 12
3900 IF EFREE(I) = 0 THEN GOTO 3980
3910 FOR J       = 1 TO 12
3920 IF EFREE(J) = 0 THEN GOTO 3970
3930 L           = EFREE(I)
3940 M           = EFREE(J) - L + 1
3950 IF M        > BAND THEN BAND = M
3960 FSTIFF(L,M) = FSTIFF(L,M) + ESTIFF(I,J)
3970 NEXT J
3980 NEXT I
3990 NEXT K
4000 PRINT
4010 RETURN
4020 REM   * * * * * * * * * * * * * * * * * * * * * * * * * * * * * * * *
4030 REM   *                                                              *
4040 REM   *     SUBROUTINE TO SET UP THE LOADING VECTOR                  *
4050 REM   *                                                              *
4060 REM   * * * * * * * * * * * * * * * * * * * * * * * * * * * * * * * *
4070 REM
4080 REM   ---   ADD NODAL LOADS TO THE LOADING VECTOR
4090 REM
4100 FOR I       = 1 TO NNLOD
4110 J           = 6*NLOD(I,1) - 6 + NLOD(I,2)
4120 J           = FREE(J)
4130 LOD(J)      = NLOD(I,3)
4140 NEXT I
4150 RETURN
4160 REM   * * * * * * * * * * * * * * * * * * * * * * * * * * * * * * * *
4170 REM   *                                                              *
4180 REM   *     SUBROUTINE TO SOLVE LINEAR SIMULTANEOUS EQUATIONS BY     *
4190 REM   *     GAUSS ELIMINATION WITH NO ROW INTERCHANGE                *
4200 REM   *                                                              *
4210 REM   * * * * * * * * * * * * * * * * * * * * * * * * * * * * * * * *
4220 REM
4230 REM   ---   LOOP FOR ALL PIVOTS
4240 REM
4250 BB          = BAND
4260 PRINT "USING PIVOT    :- ";
4270 FOR I  = 1 TO NDOF
4280 PRINT I;
4290 REM
4300 REM   ---   CHECK IF IN THE UNUSED TRIANGLE
4310 REM
4320 IF I        > NDOF-BAND+1 THEN BB = NDOF-I+1
4330 PIVOT       = FSTIFF(I,1)
4340 REM
4350 REM   ---   NORMALISE
4360 REM
4370 FOR J       = 1 TO BB
4380 FSTIFF(I,J) = FSTIFF(I,J) / PIVOT
4390 NEXT J
4400 LOD(I) = LOD(I) / PIVOT
4410 REM
4420 REM   ---   CHECK IF FIRST ROW
4430 REM
4440 IF BB  = 1 THEN GOTO 4630
4450 REM
```

```
4460 REM  ---  ELIMINATE (WITHIN BAND) FOR ALL ROWS ABOVE PIVOT
4470 REM
4480 FOR K    = 2 TO BB
4490 REM
4500 REM  ---  CALCULATE ROW NUMBER THEN EVALUATE MULTIPLIER
4510 REM
4520 L        = I + K - 1
4530 MULT     = FSTIFF(I,K) * PIVOT
4540 REM
4550 REM  ---  LOOP FOR ELEMENTS IN THE ELIMINATION ROW
4560 REM
4570 FOR J    = K TO BB
4580 M        = J - K + 1
4590 FSTIFF(L,M) = FSTIFF(L,M) - MULT * FSTIFF(I,J)
4600 NEXT J
4610 LOD(L) = LOD(L) - MULT*LOD(I)
4620 NEXT K
4630 NEXT I
4640 PRINT
4650 REM
4660 REM  ---  BACK SUBSTITUTE
4670 REM
4680 FOR I    = 1 TO NDOF-1
4690 BB       = I
4700 IF I > BAND-1 THEN BB = BAND-1
4710 FOR J    = 1 TO BB
4720 LOD(NDOF-I) = LOD(NDOF-I) - FSTIFF(NDOF-I,J+1)*LOD(NDOF-I+J)
4730 NEXT J
4740 NEXT I
4750 RETURN
4760 REM  * * * * * * * * * * * * * * * * * * * * * * * * * * * * * *
4770 REM  *                                                          *
4780 REM  *     SUBROUTINE TO OUTPUT THE DISPLACEMENTS               *
4790 REM  *                                                          *
4800 REM  * * * * * * * * * * * * * * * * * * * * * * * * * * * * * *
4810 REM
4820 PRINT : PRINT
4830 PRINT "* * * * * * * * * * *"
4840 PRINT "*   DISPLACEMENTS   *            * :- RESTRAINT"
4850 PRINT "* * * * * * * * * * *"
4860 PRINT
4870 PRINT "NODE       X-DISP.      Y-DISP.      Z-DISP.";
4880 PRINT "    X-ROTN.      Y-ROTN.      Z-ROTN."
4890 PRINT
4900 FOR I = 1 TO NNODE
4910 PRINT USING "###    "; I;
4920 FOR J = 1 TO 6
4930 K       = FREE(6*I+J-6)
4940 IF K = 0 THEN GOTO 4970
4950 PRINT USING " #.####^^^^"; LOD(K);
4960 GOTO 4980
4970 PRINT          "        *       ";
4980 NEXT J
4990 PRINT
5000 NEXT I
5010 RETURN
5020 REM  * * * * * * * * * * * * * * * * * * * * * * * * * * * * * *
5030 REM  *                                                          *
5040 REM  *     SUBROUTINE TO CALCULATE AND OUTPUT THE MEMBER END FORCES  *
5050 REM  *     AND REACTIONS                                        *
5060 REM  *                                                          *
5070 REM  * * * * * * * * * * * * * * * * * * * * * * * * * * * * * *
5080 REM
5090 REM  ---  ZERO REACTION LOCATIONS IN THE RESTRAINT ARRAY
5100 REM
5110 FOR I    = 1 TO NREST
5120 FOR J    = 3 TO 8
5130 REST(I,J) = 0
5140 NEXT J
5150 NEXT I
5160 REM
```

COMPUTER ANALYSIS OF STRUCTURAL FRAMEWORKS

```
5170 REM  ---  LOOP FOR ALL MEMBERS
5180 REM
5190 FOR K = 1 TO NMEMB
5200 I     = 6*MEMB(K,1) - 6
5210 REM
5220 REM  ---  EXTRACT THE NODAL DISPLACEMENTS - STORE U(1) - (12)
5230 REM
5240 FOR M = 1 TO 12
5250 J     = FREE(I+M)
5260 IF J  = 0 THEN U(M) = 0 ELSE U(M) = LOD(J)
5270 IF M  = 6 THEN I = 6*MEMB(K,2) - 12
5280 NEXT M
5290 REM
5300 REM  ---  CALC. MEMBER END FORCES - STORE U(13) - (24)
5310 REM
5320 GOSUB 6040
5330 U(13) =          K1*((U(1)-U(7))*T1   + (U(2)-U(8))*T4  + (U(3)-U(9))*T7)
5340 U(14) =          K2*((U(1)-U(7))*T2   + (U(2)-U(8))*T5  + (U(3)-U(9))*T8)
5350 U(14) = U(14) + K7*((U(4)+U(10))*T3   + (U(5)+U(11))*T6 + (U(6)+U(12))*T9)
5360 U(15) =          K3*((U(1)-U(7))*T3   + (U(2)-U(8))*T6  + (U(3)-U(9))*T9)
5370 U(15) = U(15) + K8*((U(4)+U(10))*T2   + (U(5)+U(11))*T5 + (U(6)+U(12))*T8)
5380 U(16) =          K4*((U(4)-U(10))*T1  + (U(5)-U(11))*T4 + (U(6)-U(12))*T7)
5390 U(17) =          K5*((U(4)+U(10)/2)*T2+ (U(5)+U(11)/2)*T5+(U(6)+U(12)/2)*T8)
5400 U(17) = U(17) + K8*((U(1)-U(7))*T3    + (U(2)-U(8))*T6  + (U(3)-U(9))*T9)
5410 U(18) =          K6*((U(4)+U(10)/2)*T3+ (U(5)+U(11)/2)*T6+(U(6)+U(12)/2)*T9)
5420 U(18) = U(18) + K7*((U(1)-U(7))*T2    + (U(2)-U(8))*T5  + (U(3)-U(9))*T8)
5430 U(19) =-U(13)
5440 U(20) =-U(14)
5450 U(21) =-U(15)
5460 U(22) =-U(16)
5470 U(23) =-U(17) - U(15)*MEMB(K,10)
5480 U(24) =-U(18) + U(14)*MEMB(K,10)
5490 PRINT : PRINT
5500 PRINT        "* * * * * * * * * * *"
5510 PRINT USING "*     ELEMENT ## ##     *"; MEMB(K,1), MEMB(K,2)
5520 PRINT        "* * * * * * * * * * *"
5530 PRINT
5540 PRINT "                AXIAL       SHEAR         SHEAR";
5550 PRINT "      TORSION    BENDING      BENDING"
5560 PRINT "                FORCE       FORCE Y       FORCE Z";
5570 PRINT "      MOMENT     MOMENT Y     MOMENT Z"
5580 PRINT USING "NODE ##"; MEMB(K,1);
5590 PRINT USING "  #.####^^^^"; U(13), U(14), U(15), U(16), U(17), U(18)
5600 PRINT USING "NODE ##"; MEMB(K,2);
5610 PRINT USING "  #.####^^^^"; U(19), U(20), U(21), U(22), U(23), U(24)
5620 REM
5630 REM  ---  ADD ANY CONTRIBUTION FROM THE MEMBER FORCES TO THE REACTIONS
5640 REM
5650 FOR I = 1 TO NREST
5660 J     = 13
5670 IF MEMB(K,1) =  REST(I,1) THEN GOTO 5700
5680 J     = 19
5690 IF MEMB(K,2) <> REST(I,1) THEN GOTO 5760
5700 REST(I,3) = REST(I,3) + T1*U(J)    + T2*U(J+1) + T3*U(J+2)
5710 REST(I,4) = REST(I,4) + T4*U(J)    + T5*U(J+1) + T6*U(J+2)
5720 REST(I,5) = REST(I,5) + T7*U(J)    + T8*U(J+1) + T9*U(J+2)
5730 REST(I,6) = REST(I,6) + T1*U(J+3)  + T2*U(J+4) + T3*U(J+5)
5740 REST(I,7) = REST(I,7) + T4*U(J+3)  + T5*U(J+4) + T6*U(J+5)
5750 REST(I,8) = REST(I,8) + T7*U(J+3)  + T8*U(J+4) + T9*U(J+5)
5760 NEXT I
5770 NEXT K
5780 REM
5790 REM  ---  ADD ANY APPLIED LOADS
5800 REM
5810 FOR I = 1 TO NREST
5820 FOR J = 1 TO NNLOD
5830 IF REST(I,1) <> NLOD(J,1) THEN GOTO 5860
5840 K     = NLOD(J,2)
5850 REST(I,2+K) = REST(I,2+K) - NLOD(J,3)
5860 NEXT J
5870 NEXT I
5880 REM
```

SPACE FRAMES

```
5890 REM  ---  OUTPUT THE REACTIONS
5900 REM
5910 PRINT : PRINT
5920 PRINT "* * * * * * * * *"
5930 PRINT "*   REACTIONS   *    -    GLOBAL AXES"
5940 PRINT "* * * * * * * * *"
5950 PRINT
5960 PRINT "                  X-FORCE      Y-FORCE     Z-FORCE";
5970 PRINT "     X-MOMENT     Y-MOMENT    Z-MOMENT"
5980 FOR I = 1 TO NREST
5990 PRINT USING "NODE ##"; REST(I,1);
6000 PRINT USING "  #.####^^^^"; REST(I,3), REST(I,4), REST(I,5), REST(I,6),
6010 PRINT USING "  #.####^^^^"; REST(I,7), REST(I,8)
6020 NEXT I
6030 RETURN
6040 REM  * * * * * * * * * * * * * * * * * * * * * * * * * * * * * * * * *
6050 REM  *                                                                *
6060 REM  *    SUBROUTINE TO EVALUATE THE STIFFNESS AND TRANSFORMATION     *
6070 REM  *    COEFFICIENTS  K1 - K8  AND  T1 - T9                         *
6080 REM  *                                                                *
6090 REM  * * * * * * * * * * * * * * * * * * * * * * * * * * * * * * * * *
6100 REM
6110 REM  ---  STIFFNESS COEFFICIENTS
6120 REM
6130 K1 =      MEMB(K,9) * MEMB(K,3) / MEMB(K,10)
6140 K2 = 12 * MEMB(K,9) * MEMB(K,5) / MEMB(K,10)^3
6150 K3 = 12 * MEMB(K,9) * MEMB(K,4) / MEMB(K,10)^3
6160 K4 =      MEMB(K,8) * MEMB(K,6) / MEMB(K,10)
6170 K5 =  4 * MEMB(K,9) * MEMB(K,4) / MEMB(K,10)
6180 K6 =  4 * MEMB(K,9) * MEMB(K,5) / MEMB(K,10)
6190 K7 =  6 * MEMB(K,9) * MEMB(K,5) / MEMB(K,10)^2
6200 K8 = -6 * MEMB(K,9) * MEMB(K,4) / MEMB(K,10)^2
6210 REM
6220 REM  ---  TRANSFORMATION COEFFICIENTS
6230 REM
6240 AR = ATN(1)*MEMB(K,7)/45
6250 C  = COS(AR)
6260 S  = SIN(AR)
6270 REM
6280 REM  ---  CHECK IF THE ELEMENT LIES PARALLEL TO THE COORDINATE Z-AXIS
6290 REM
6300 IF ABS(MEMB(K,13)) > .999 THEN GOTO 6420
6310 D  = SQR(1 - MEMB(K,13)^2)
6320 T1 =      MEMB(K,11)
6330 T2 = (-MEMB(K,12)*C - MEMB(K,11)*MEMB(K,13)*S) / D
6340 T3 = ( MEMB(K,12)*S - MEMB(K,11)*MEMB(K,13)*C) / D
6350 T4 =      MEMB(K,12)
6360 T5 = ( MEMB(K,11)*C - MEMB(K,12)*MEMB(K,13)*S) / D
6370 T6 = (-MEMB(K,11)*S - MEMB(K,12)*MEMB(K,13)*C) / D
6380 T7 =      MEMB(K,13)
6390 T8 =  S * D
6400 T9 =  C * D
6410 GOTO 6510
6420 T1 =   0
6430 T2 = - MEMB(K,13)*S
6440 T3 = - MEMB(K,13)*C
6450 T4 =   0
6460 T5 =   C
6470 T6 = - S
6480 T7 =   MEMB(K,13)
6490 T8 =   0
6500 T9 =   0
6510 RETURN
```

Sample Run The following sample run shows program SFRAME.SF being used to analyse the cranked cantilever shown below.

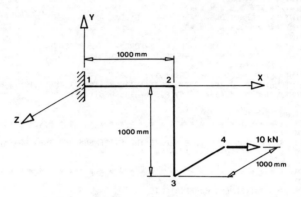

The structure is constructed from solid 10 mm x 20 mm steel sections. The following diagram shows how the members are connected. Also shown are the principal axes and the member axes systems. Using these principal axes results in the following values of ß (ß being the angle through which the member must be rotated about its "x" axis to make the principal axes coincide with the member axes). Clockwise rotation is positive when the member is viewed looking from end "i" to end "j".

$$\beta_{12} = 0°$$
$$\beta_{23} = 0°$$
$$\beta_{34} = 90°$$

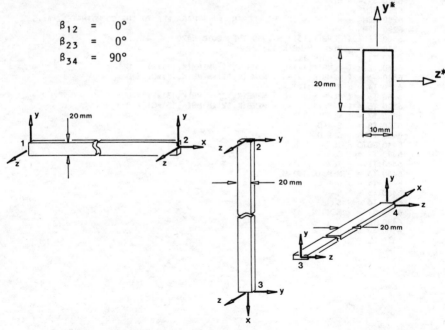

SPACE FRAMES

PROGRAM   S F R A M E . S F   FOR THE AUTOMATIC ANALYSIS
OF SPACE FRAMES

NAME OF THE DATA FILE = **DATA.SF**

READING DATA

```
+ + + + + + + +
+   JOB TITLE   +    -   CRANKED CANTILEVER - FILE DATA.SF
+ + + + + + + +
```

```
+ + + + + + + + + + + +
+   NODAL COORDINATES   +
+ + + + + + + + + + + +
```

NODE NUMBER	X	Y	Z
1	0	0	0
2	1000	0	0
3	1000	-1000	0
4	1000	-1000	-1000

```
+ + + + + + + + + + + +
+   MEMBER PROPERTIES   +
+ + + + + + + + + + + +
```

MEMBER NUMBER	A	IYY	IZZ	J	BETA	G	E
1 2	0.200E+03	0.167E+04	0.667E+04	0.667E+04	0.000E+00	0.800E+02	0.209E+03
2 3	0.200E+03	0.167E+04	0.667E+04	0.667E+04	0.000E+00	0.800E+02	0.209E+03
3 4	0.200E+03	0.167E+04	0.667E+04	0.667E+04	0.900E+02	0.800E+02	0.209E+03

```
+ + + + + + + + +
+   RESTRAINTS   +
+ + + + + + + + +
```

NODE NUMBER	DIRECTION(S)
1	123456

```
+ + + + + + + + + +
+   NODAL LOADS   +
+ + + + + + + + + +
```

NODE NUMBER	DIRECTION	VALUE
4	1	0.100E-01

GENERATING THE FREEDOM VECTOR

GENERATING THE STRUCTURE STIFFNESS MATRIX
ADDING ELEMENT :- 1 2 3

GENERATING THE LOADING VECTOR

SOLVING EQUATIONS - NO. OF EQUATIONS = 18
SEMI-BANDWIDTH   = 12
USING PIVOT   :- 1 2 3 4 5 6 7 8 9 10 11 12 13 14 15 16 17 18

```
* * * * * * * * * *
*   DISPLACEMENTS   *           * :- RESTRAINT
* * * * * * * * * *
```

NODE	X-DISP.	Y-DISP.	Z-DISP.	X-ROTN.	Y-ROTN.	Z-ROTN.
1	*	*	*	*	*	*
2	0.2392E-03	0.3590E+01	0.1434E+02	0.2572E-04	-.2869E-01	0.7179E-02
3	0.9572E+01	0.3590E+01	0.1430E+02	0.4542E-04	-.4744E-01	0.1077E-01
4	0.5940E+02	0.3635E+01	0.1430E+02	0.4544E-04	-.5103E-01	0.1077E-01

COMPUTER ANALYSIS OF STRUCTURAL FRAMEWORKS

```
* * * * * * * * * * *
*   ELEMENT  1  2   *
* * * * * * * * * * *
            AXIAL      SHEAR       SHEAR      TORSION     BENDING     BENDING
            FORCE      FORCE Y     FORCE Z    MOMENT      MOMENT Y    MOMENT Z
NODE  1   -.1000E-01  -.6754E-05  0.1371E-04  -.1372E-01  0.9988E+01  -.1001E+02
NODE  2   0.1000E-01  0.6754E-05  -.1371E-04  0.1372E-01  -.1000E+02  0.1000E+02


* * * * * * * * * * *
*   ELEMENT  2  3   *
* * * * * * * * * * *
            AXIAL      SHEAR       SHEAR      TORSION     BENDING     BENDING
            FORCE      FORCE Y     FORCE Z    MOMENT      MOMENT Y    MOMENT Z
NODE  2   0.9966E-05  -.1000E-01  0.1371E-04  -.1000E+02  -.1372E-01  -.1000E+02
NODE  3   -.9966E-05  0.1000E-01  -.1371E-04  0.1000E+02  0.7041E-05  0.2174E-03


* * * * * * * * * * *
*   ELEMENT  3  4   *
* * * * * * * * * * *
            AXIAL      SHEAR       SHEAR      TORSION     BENDING     BENDING
            FORCE      FORCE Y     FORCE Z    MOMENT      MOMENT Y    MOMENT Z
NODE  3   0.0000E+00 -1.0000E-02  0.9721E-08  -.4967E-06  -.1111E-04  -.1000E+02
NODE  4   0.0000E+00  1.0000E-02  -.9721E-08  0.4967E-06  0.1388E-05  0.1200E-02


* * * * * * * * *
*  REACTIONS    *    -   GLOBAL AXES
* * * * * * * * *
            X-FORCE    Y-FORCE     Z-FORCE    X-MOMENT    Y-MOMENT    Z-MOMENT
NODE  1   -.1000E-01  -.6754E-05  0.1371E-04  -.1372E-01  0.9988E+01  -.1001E+02
```

11 Buying and Using Commercial Frame Analysis Programs

11.1 INTRODUCTION

This book is intended to give engineering students and practising engineers an understanding of the theory behind commercial frame analysis programs. It is also intended give some insight into how that theory is turned into useful computer programs.

The user of commercial frame analysis progams need know little about the program to obtain results. There is general agreement, however, that better and safer use of the program will come from the user having some understanding of the structural theory underlying the program and an appreciation of how that theory might have been implemented.

Not so many years ago, civil engineers made very limited use of computers for structural analysis. In those days few engineering concerns owned a computer, being forced to use large structural analysis programs supported by computer bureaux. These programs were sometimes unnecessarily complicated to use. Often the engineers using them had little experience of computing, and hence an underlying distrust of computers. On top of this computing costs were invariably high.

This situation has now dramatically changed. Today's engineers are much more computer-conscious: powerful, low cost microcomputers are widely available, and using computers to analyse structures has become the norm rather than the exception. There are a number of reasons behind the growth in popularity of computer based structural analysis. Firstly (and perhaps most importantly), computer-based structural analysis is cost-effective. Secondly, computer methods often yield "better" results because they demand less idealisation of the true structural behaviour. Finally it can be argued that computer analysis is less error-prone than traditional structural analysis because the arithmetic is done by the computer, rather than the engineer.

11.2 BUYING A FRAME ANALYSIS COMPUTER PROGRAM

Program Language

Frame analysis programs are usually written in either BASIC or FORTRAN. BASIC is a compact, interpretive language that is widely used on microcomputers. The early versions of BASIC were heavily criticised for their lack of structure which tended to encourage poor programming habits. Later (extended) versions of BASIC are much better structured. The disadvantages of using extended versions of BASIC are that they require more memory and they may run more slowly. On some machines the BASIC interpreter is stored in read only memory (ROM). On others it is loaded into random access memory (RAM) from disk. The principal advantages of BASIC are its low cost, wide availability, and simplicity. Its principal disadvantages are its slowness and the number of variants of the language.

FORTRAN was one of the world's first high-level computing languages. It is a compiled language that is widely used by engineers and scientists. FORTRAN compilers require large amounts of memory and this restricts FORTRAN to the more powerful computers. Compiled code runs many times faster than code that has to be interpreted - giving FORTRAN a distinct advantage over BASIC for the solution of large structural problems. Another advantage of FORTRAN is that it is a highly standardised language. This allows the same program to be run on different machines with little or no modification. The speed and portability of FORTRAN ensure that it will continue to be an important language in scientific computing. Finally, the reader should note that compilers are available for some versions of BASIC, and that using compiled BASIC can result in substantial reductions in execution times.

Matching Machine, Language, and Program

The engineer faced with the task of buying a frame analysis program and a computer system to run it on will find a bewildering variety of choice. One of his prime considerations will almost certainly be cost. At one end of the spectrum are finite element packages (which invariably support line elements) requiring powerful computers, where the cost of hardware plus software will run into tens of thousands of pounds. At the other end of the spectrum are simple frame analysis programs designed to run on inexpensive microcomputers, where the total cost of the system will be measured in hundreds of pounds.

BUYING AND USING COMMERCIAL FRAME ANALYSIS PROGRAMS

The first step in deciding which type of system to buy is to decide what it will be used for. What types of structure will be analysed? What is a typical job size? What is the size of the biggest job likely to be? How often will the program be used and how many engineers will use it?

If the program will be frequently used to analyse large frameworks then the short execution times offered by compiled FORTRAN may be a deciding factor. On the other hand if large problems are never likely to arise then a frame analysis program written in BASIC intended to run on a microcomputer may be the cost-effective solution. One point to note about FORTRAN is that some software vendors sell only compiled code. This is done for security reasons. It precludes the purchaser from modifying the code (e.g. the format of the output), and the program is machine-specific (i.e. compiled code has no portability). If large jobs arise only infrequently, then one solution is to purchase a system that is suitable for all but the large jobs (these would be run externally at a computer bureau).

If many engineers are likely to need simultaneous access to the program then there are many ways to meet this requirement. These range from giving each engineer (or small group of engineers) a stand alone system, through networked stand alone systems, to terminals into a single, centralised mainframe computer.

There are always likely to be a number of alternatives and the program and machine purchased will depend upon the budget, the existing facilities, the hardware required by the program, the need for data sharing, the number of engineers involved, and the likely program usage. Each situation must be individually assessed because the computing requirements of each engineering concern are unique, and the cost and nature of the available hardware and software vary continuously.

The Scope of the Program

The facilities offered by different frame analysis programs vary widely. Not all programs deal with all types of structural frameworks. The maximum problem size that the program will deal with is usually related to the size of the computer's memory. Some frame analysis programs allow large problems to be solved by using equation solving techniques that require only part of the structure stiffness matrix to be held in the computer's memory at any one time, with the rest being stored on disk. Such equation solving techniques are quite complex and require a fair amount of coding. As they make heavy use of secondary storage they tend to be much slower than techniques that don't use secondary storage. Member loads, local axes, elastic supports, and members with pins are facilities that are offered by some programs, but not others. The engineer should always check that any program considered offers the facilities that are required.

The User Interface

The only interaction that the engineer has with automatic frame analysis programs is in the preparation of the input data and in the interpretation of the results. Consequently the user's opinion of the program will be largely based upon the ease of data preparation, and the readability of the output. This being the case, no program should be purchased without first considering the way in which data is supplied to the program and the format of the output.

Graphical output is enormously valuable. If the program has facilities to draw the structure from the input data then input errors involving the structure's geometry are quickly detected. Even more useful is the facility to draw the deformed shape of the structure from the results. This gives the engineer a feel for the structure's behaviour, and may help detect erroneous input data. Input and output from frame analysis programs is usually available as arrays. This facilitates the implementation of graphical output.

11.3 USING FRAME ANALYSIS PROGRAMS

A scale drawing of the structure showing node numbers, dimensions, loads, and restraints aids the preparation of input data. Often data can be given directly to the computer from such a drawing.

Node Numbers

Some programs internally renumber the nodes to minimise the bandwidth of the structure stiffness matrix. Most do not and it is good practice to number the nodes such that the maximum difference between the node numbers at the ends of any element is kept to a minimum (or near minimum).

Initial Member Sizing

One of the problems facing the designer of a structural framework is the initial choice of member sizes. Once these have been chosen then the designer iterates around the loop shown in fig 11.1 until a satisfactory structure is found.

BUYING AND USING COMMERCIAL FRAME ANALYSIS PROGRAMS

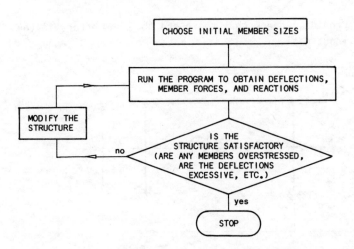

FIGURE 11.1   THE DESIGN CYCLE

Initial member sizes can be selected in many ways. Using member sizes from a similar structure can prove efficient. Length to breadth (or span to depth) ratios can be used as the length to breadth of structural framework members tends to be in the range 10 to 20. A preliminary analysis of the structure by hand can be used to determine initial member sizes - as can pure guesswork!

In the opinion of the author there are two good reasons in favour of spending some time over the initial sizing of members. Firstly, the number of times the program is run before a satisfactory solution is found will usually be reduced. Secondly, and more importantly, the engineer will gain a feel for how the structure will perform. Should the structure behave differently from the engineer's expectations when it is subsequently analysed by the computer, then either the computer analysis is wrong due to erroneous input data, or the engineer has misunderstood how the structure will behave. In either case the source of the problem must be identified, understood, and corrected if necessary. To illustrate, consider the problem of choosing initial member sizes for the structure shown in fig 11.2.

There follows an example of how the bending moments and sway deflections might be estimated (note that the initial member sizes can be calculated from the estimated bending moments). As the computer will later be used to analyse the structure, the preliminary calculations only serve to:

(i)   enable reasonable initial member sizes to be selected

(ii) give an insight into the behaviour of the structure

(iii) provide a check that there are no gross errors in the input data.

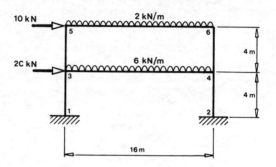

FIGURE 11.2   TWO STOREY PORTAL FRAME

The emphasis is therefore on simplicity, while still retaining the essential characteristics of the true structural behaviour. Provided that the calculations yield results that serve the objectives listed above then the method of calculation is immaterial. Hence the calculations that follow are only one example of how the problem might be tackled.

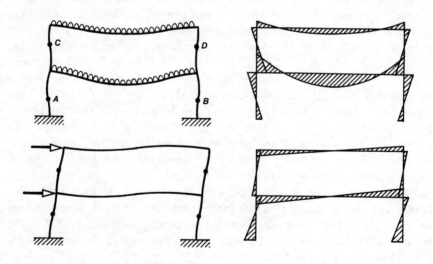

FIGURE 11.3   DEFORMED SHAPES AND BENDING MOMENT DIAGRAMS

For structures such as the portal shown in fig 11.2, axial deformations can be safely neglected. It is now helpful to decide the relative flexural stiffness (EI/L) of the members. For simplicity all members will be assumed to have the same flexural stiffness. If the effects of the vertical and the horizontal loads are separated then the general shape of the deformed structure and of the bending moment diagrams can be sketched as shown in fig 11.3.

Inspection of the deformations caused by the vertical loads indicates that there are points of contraflexure (zero curvature, therefore zero moment) at A, B, C, and D. The points of contraflexure at A and B occur at one-third of the column height. Consideration of the loading and the relative rotational stiffness of the joints indicates that joints 3 and 4 will rotate more than joints 5 and 6. It is, therefore, reasonable to assume that the points of contraflexure at C and D occur two-thirds of the way up the columns. Further, if the points of contraflexure are assumed to undergo no translation then the part of the structure bounded by points A, B, C, and D is as shown in fig 11.4.

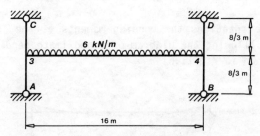

FIGURE 11.4   SUBSTRUCTURE FORMED BY A,B,C,D

The bending moments for the substructure shown in fig 11.4 are most easily calculated by moment distribution. As moment distribution is not within the scope of this book the bending moments will be evaluated using the stiffness method.

By treating the columns as members with pins the substructure shown in fig 11.4 has two freedoms as illustrated in fig 11.5.

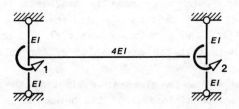

FIGURE 11.5   FREEDOMS AND RELATIVE STIFFNESS VALUES

Figure 11.5 also shows the relative EI values of the members (EI/L was previously assumed to be equal for all members). The coefficients of the stiffness matrix can now be calculated as follows:

$$K_{11} = (2(9EI/8) + 16EI/16) = 13EI/4$$

$$K_{21} = EI/2$$

$$K_{22} = K_{11}$$

The structure stiffness equation is as follows:

$$\begin{bmatrix} 13EI/4 & EI/2 \\ EI/2 & 13EI/4 \end{bmatrix} \begin{bmatrix} \theta_1 \\ \theta_2 \end{bmatrix} = \begin{bmatrix} 128 \\ -128 \end{bmatrix}$$

Noting that the rotations will be equal and opposite simplifies their evaluation, and θ_1 turns out to be $46.5/EI$.

From the rotations the bending moments at the joints are easily calculated, allowing the diagram of approximate bending moments to be drawn as shown in fig 11.6

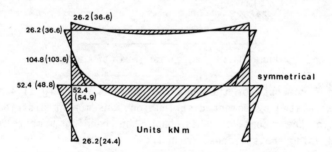

FIGURE 11.6   APPROXIMATE BENDING MOMENTS DUE TO VERTICAL LOAD

The exact values (which are shown in brackets) indicate that, in general, the estimated bending moments are reasonable, and would certainly serve for the purposes of initial member sizing. It is interesting to note that the exact value of the rotation at joint 3 is $48.8/EI$.

Attention will now be given to estimating the bending moments caused by the horizontal loads. If the beams where infinitely stiff compared to the columns then the joints would not rotate and the deformed shape would be as shown in fig 11.7.

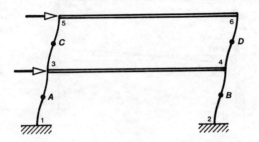

FIGURE 11.7   DEFORMED SHAPE WITH NO JOINT ROTATION

There is a point of contraflexure on each column. If the stiffness of beam 3,4 is now reduced then the joints 3 and 4 will rotate causing points of contraflexure on the upper and lower columns to move together. If the stiffness of beam 5,6 is reduced (beam 3,4 remaining infinitely stiff) then the points of contraflexure on the upper columns will migrate towards the tops of these columns. The beams and columns have been assumed to have the same EI/L value for the purposes of these calculations. It is reasonable, therefore, to assume that the points of contraflexure at A and B occur three-quarters of the way up the lower columns, and the points of contraflexure at C and D occur half-way up the upper columns. This leads to the freebody diagrams shown in fig 11.8.

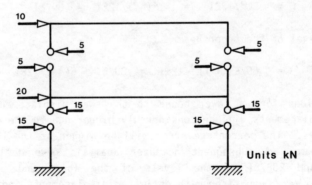

FIGURE 11.8   FREEBODY DIAGRAMS

Hence the diagram of approximate bending moments for horizontal loads can be drawn as shown in fig 11.9. Again the exact values are given in brackets for the purposes of comparison.

371

COMPUTER ANALYSIS OF STRUCTURAL FRAMEWORKS

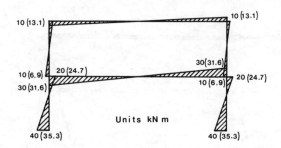

FIGURE 11.9   APPROXIMATE BENDING MOMENT DIAGRAM FOR HORIZONTAL LOADS

Sway Displacements

In this instance the vertical loads cause no sidesway, and therefore do not appear in the calculation of sway displacements. The sway displacements associated with horizontal loads could be easily estimated from the approximate bending moment diagram for horizontal loads using techniques such as the moment-area method. However such methods are not within the scope of this book and here approximate values for the sway displacements will be found by assuming that the beams are infinitely stiff relative to the columns. The resulting deformed shape is shown in fig 11.7, and the sway displacements are:

At the level of the lower beam

$$\delta \;=\; FL^3/12EI \;=\; 15(4^3)/12EI \;=\; 80/EI$$

At the level of the upper beam

$$\delta \;=\; 80/EI + FL^3/12EI \;=\; 80/EI + 5(4^3)/12EI \;=\; 107/EI$$

These values form a lower bound to the sway displacements and the true displacements will be considerably bigger due to the rotation of the joints. The percentage error will be bigger at the level of the upper beam. A subsequent computer analysis gave deflections of 122/EI and 208/EI at the levels of the lower and upper beams respectively. Comparison with estimated displacements indicates that no gross error was present in the input data.

Member Properties

The initial member sizes are calculated from the estimated member forces. The initial member sizes allow the section properties of the members to be evaluated.

For trusses and plane frames this process is usually straightforward. It is worth noting, however, that in the early stages of analysis it is not efficient to design the reinforcement for reinforced concrete members as sufficiently accurate results can be obtained by assuming the members to consist of plane concrete that won't crack in tension.

Grillages and space frames require the torsional constants for the members. Often approximate formulae are employed. In some cases a parametric study of the sensitivity of the problem to the torsional stiffness of the members is worth while. The engineer estimates maximum and minimum values of the torsional stiffness and compares results using these values.

In Chapter 10 the relationship between the member axes and the principal axes for a space frame element was seen to be at the discretion of the program writer. The program user should therefore consult the user manual to make sure that the data supplied to the program accurately describes the problem in hand.

Restraints

In section 2.6 boundary conditions were discussed in some detail. When preparing restraint data for a frame analysis program the engineer must consider how the boundary conditions that are input to the program will be achieved in practice. For instance, if a connection to a footing is assumed to be fully fixed then the engineer must ensure that the actual footing is big enough, and the connection is rigid enough for this assumption to be valid. Again parametric studies can be worth while and the facility to include elastic supports is particularly useful in this instance.

Local Axes

Not all frame analysis programs allow the use of local axes. The way in which local axes are defined is at the discretion of the program writer. Hence the program user must pay careful attention to the user manual when preparing data for problems involving the use of local axes.

Member Loads

Most frame analysis programs allow the use of a variety of member loads. The user manual will describe the load types that are available and will define how they should be input to the computer.

Units

Any unit system can be used to prepare the input data for a frame analysis program. The only restriction is that the units must be *consistent*. For instance, if Newtons are used for force and centimetres for length then nodal coordinates must be given in cm, the elastic constants in N/cm^2, the cross sectional areas in cm^2, the torsional constants in cm^4, etc.

Results

Results will be in the same units as the input data. For example, if the unit of force is the kgf and the unit of length is the metre, then displacements will be in m, rotations will be radians (always), forces will be in kgf, moments will be in kgf-m, etc. Nodal displacements will be output in the nodal axes systems (see section 3.6), member end forces (stress resultants) will be in the principal axes of the member, and reactions are normally output in nodal axes systems for the restrained nodes.

Appendix A
Definition of Program Variables

VARIABLE	DESCRIPTION
A	Cross sectional area of the member.
A(I,J)	GAUSSJ.EQ - A(I,J) contains the augmented matrix. Later overwritten by the decomposed matrix. GAUSS.EQ and UDU.EQ - A(I,J) contains the upper semi-band of the coefficient matrix, stored as shown in fig 3.15. Overwritten by the decomposed matrix. INVERT.EQ - A(I,J) initially contains the coefficient matrix. Overwritten by the inverse, or the inverse of the inverse, of the coefficient matrix.
A2	Second moment of area.
ALPHA	Anticlockwise rotation of the element from the global "x" axis in degrees.
ANS	Numeric reply to a prompt from the computer.
ANS$	String reply to a prompt from the computer.
AR	Value of ALPHA in radians.
B(I,J)	Used to store the right hand sides, column by column (Overwritten by the solution vectors).
BAND BND	Semi-bandwidth of the coefficient matrix.
BB	Width applicable at the current row.
BETA	Anticlockwise rotation of a space frame member about its "x" axis that will make the principal axes coincide with the member axes (rotation viewed from end "i").
C	Cosine of ALPHA.
C(I,J)	Used to preserve the coefficient matrix.
CO	Coefficient of thermal expansion.
COUNT	Counter when sorting data.
CYZ	Cosine of the angle between the member "y" axis and the

	principal "z" axis.
D(I,J)	Inverse of the coefficient matrix copied to this array while calculating the inverse of the inverse of the coefficient matrix.
DIRN	Directions of restraint.
E	Elastic constant (Young's modulus).
EFREE(I)	Element code number.
EJF(I)	Vector of equivalent joint forces.
ESTIFF(I,J)	Element stiffness matrix.
FEF(I)	Vector of fixed end forces.
FILE$	Name of data file.
FLAG	Flag, used when sorting data.
FORCE	Axial force in a truss member.
FREE(I)	Freedom vector.
FRI	moment at end "i" of element "i,j".
FRJ	moment at end "j" of element "i,j".
FSTIFF(I,J)	Final structure stiffness matrix.
FXI	"x" force at end "i" of element "i,j".
FXJ	"x" force at end "j" of element "i,j".
FYI	"y" force at end "i" of element "i,j".
FYJ	"y" force at end "j" of element "i,j".
G	Modulus of rigidity.
I	Loop counter or array subscript.
IN	Number of node "i".
ISTIFF(I,J)	Initial structure stiffness matrix.
J	Loop counter or array subscript.
J2	Torsional constant.
JN	Number of node "j".
K	Loop counter or array subscript.
K1-K8	Stiffness coefficients.
L	Member length (also occasionally used as an array subscript).
LOD(I)	Load vector.
LTYPE	Load type.
LX,MX,NX	Direction cosines of the member "x" axis.
LXP,MXP,NXP	Direction cosines of the principal "z" axis.
LZ,MZ,NZ	Direction cosines of the member "z" axis.
M	Array subscript.
MEMB(I,J)	Row "i" contains data for element "i", stored as follows.

PROGRAM	PTRUSS.PT	STRUSS.ST	PFRAME.PF	GRID.GD	SFRAME.SF
(I,1)	I	I	I	I	I
(I,2)	J	J	J	J	J
(I,3)	A	A	A	I2	A

APPENDIX A

(I,4)	E	E	I2	J2	IY
(I,5)	L	L	PTYPE	G	IZ
(I,6)	C	LX	E	E	J2
(I,7)	S	MX	L	L	BETA
(I,8)		NX	C	C	G
(I,9)			S	S	E
(I,10)					L
(I,11)					LX
(I,12)					MX
(I,13)					NX

Where

	I	node "i".
	J	node "j".
	A	cross sectional area.
	E	elastic constant (Young's modulus).
	I2	second moment of area.
	J2	torsional constant.
	PTYPE	element type.
	G	modulus of rigidity.
	IY	second moment of area about principal "y" axis.
	IZ	second moment of area about principal "z" axis.
	BETA	anticlockwise rotation of element that will make the principal axes coincide with the member axes.
	L	element length.
	C	cosine of the angle the element makes with the global "x" axis.
	S	sine of the angle the element makes with the global "x" axis.
	LX,MX,NX	direction cosines of the member "x" axis.
MLOD(I,J)		Row "i" contains details of member load "i".
	(I,1)	Lower node number of loaded element.
	(I,2)	Higher node number of loaded element.
	(I,3)	Load type, 1 for a UDL, 2 for a point load.
	(I,4)	Magnitude of the load. Note that a positive member load acts in the direction of the member "y" axis.
	(I,5)	Distance of load from lower node (only meaningful for point loads).

	(I,6)-(I,9) Fixed end forces (plane frame and grillage elements only).
MTYPE	Member type 1 - fixed/fixed
	2 - fixed/pinned
	3 - pinned/fixed
	4 - pinned/pinned
MULT	Multiplier for the pivotal row.
NCØRD	Number of coordinate components (2 or 3).
NDOF	Degree of freedom of the structure (either initial or final).
NEQ	Number of equations to be solved.
NLOAD	Number of load cases.
NLOD(I,J)	Row "i" contains details of nodal load "i".
	(I,1) number of loaded node.
	(I,2) direction of the load.
	(I,3) magnitude of the load.
NMEMB	Number of members.
NMLOD	Number of member loads.
NNLOD	Number of nodal loads.
NNODE	Number of nodes.
NODE(I,J)	Nodal coordinates.
	(I,1) "x" coordinate of node "i".
	(I,2) "y" coordinate of node "i".
	(I,3) "z" coordinate of node "i" (3 dimensional structures only).
NPROP	Number of member properties (used to read and write array MEMB(I,J)).
NR	Number of right hand sides.
NREST	Number of restrained nodes.
NS	Node where settlement occurs.
P	Magnitude of point load.
PTYPE	Problem type 1 - Plane truss
	2 - Space truss
	3 - Plane frame
	4 - Grillage
	5 - Space frame
R	Rotational settlement.
REST(I,J)	Row "i" contains details of restrained node "i".
	(I,1) number of restrained node.
	(I,2) composite restraint number.
	(I,3)-(I,8) used to store reaction components.
S	Sine of ALPHA.
SUM	Summation used in forward substitution.
T$	Temporary string variable.

APPENDIX A

T1-T9	Transformation coefficients.
TEMP	Temporary variable.
TEMP(I)	Temporary array.
TITLE$	Job title.
U(I)	Used to store member end displacements and member end forces.
W	Intensity of UDL.
XI	"x" displacement or "x" coordinate of node "i".
XJ	"x" displacement or "x" coordinate of node "j".
XP	"x" coordinate of node "p".
YI	"y" displacement or "y" coordinate of node "i".
YJ	"y" displacement or "y" coordinate of node "j".
YP	"y" coordinate of node "p".
ZI	"z" displacement or "z" coordinate of node "i".
ZJ	"z" displacement or "z" coordinate of node "j".
ZP	"z" coordinate of node "p".

Appendix B
Summary of Key Formulae and Matrices

MEMBER AXES SYSTEMS

These are as described in section 3.6 (see fig 3.12).

ELEMENT STIFFNESS MATRICES

The relationship between the member end displacements and the member end forces (expressed in the member axes system) is given by the following equation:

$$\begin{bmatrix} f_{ij} \\ \hline f_{ji} \end{bmatrix} = \begin{bmatrix} k_{ii}^j & | & k_{ij} \\ \hline k_{ji} & | & k_{jj}^i \end{bmatrix} \begin{bmatrix} \delta_{ij} \\ \hline \delta_{ji} \end{bmatrix} \qquad (B.1)$$

Plane Truss

$$\begin{bmatrix} f_{ijx} \\ f_{ijy} \\ \hline f_{jix} \\ f_{jiy} \end{bmatrix} = \begin{bmatrix} EA/L & 0 & | & -EA/L & 0 \\ 0 & 0 & | & 0 & 0 \\ \hline -EA/L & 0 & | & EA/L & 0 \\ 0 & 0 & | & 0 & 0 \end{bmatrix} \begin{bmatrix} \delta_{ijx} \\ \delta_{ijy} \\ \delta_{jix} \\ \delta_{jiy} \end{bmatrix}$$

Space Truss

$$\begin{bmatrix} f_{ijx} \\ f_{ijy} \\ f_{ijz} \\ \hline f_{jix} \\ f_{jiy} \\ f_{jiz} \end{bmatrix} = \begin{bmatrix} EA/L & 0 & 0 & | & -EA/L & 0 & 0 \\ 0 & 0 & 0 & | & 0 & 0 & 0 \\ 0 & 0 & 0 & | & 0 & 0 & 0 \\ \hline -EA/L & 0 & 0 & | & EA/L & 0 & 0 \\ 0 & 0 & 0 & | & 0 & 0 & 0 \\ 0 & 0 & 0 & | & 0 & 0 & 0 \end{bmatrix} \begin{bmatrix} \delta_{ijx} \\ \delta_{ijy} \\ \delta_{ijz} \\ \delta_{jix} \\ \delta_{jiy} \\ \delta_{jiz} \end{bmatrix}$$

APPENDIX B

Plane Frames - Various end connections

Members containing pins are dealt with by using the variables $K_1 - K_8$ in the element stiffness matrix as shown in the following equation:

$$\begin{bmatrix} f_{ijx} \\ f_{ijy} \\ m_{ij} \\ \hline f_{jix} \\ f_{jiy} \\ m_{ji} \end{bmatrix} = \begin{bmatrix} EA/L & 0 & 0 & | & -EA/L & 0 & 0 \\ & K_1EI/L^3 & K_2EI/L^2 & | & 0 & -K_1EI/L^3 & K_3EI/L^2 \\ & & K_4EI/L & | & 0 & K_5EI/L^2 & K_6EI/L \\ \hline & \text{symmetrical} & & | & EA/L & 0 & 0 \\ & & & | & & K_1EI/L^3 & K_7EI/L^2 \\ & & & | & & & K_8EI/L \end{bmatrix} \begin{bmatrix} \delta_{ijx} \\ \delta_{ijy} \\ \theta_{ij} \\ \hline \delta_{jix} \\ \delta_{jiy} \\ \theta_{ji} \end{bmatrix}$$

MEMBER TYPE	NODE "i"	NODE "j"	K_1	K_2	K_3	K_4	K_5	K_6	K_7	K_8
1	Fixed	Fixed	12	6	6	4	-6	2	-6	4
2	Fixed	Pinned	3	3	0	3	-3	0	0	0
3	Pinned	Fixed	3	0	3	0	0	0	-3	3
4	Pinned	Pinned	0	0	0	0	0	0	0	0

Grillages

$$\begin{bmatrix} f_{ijz} \\ m_{ijx} \\ m_{ijy} \\ \hline f_{jiz} \\ m_{jix} \\ m_{jiy} \end{bmatrix} = \begin{bmatrix} 12EI/L^3 & 0 & -6EI/L^2 & | & -12EI/L^3 & 0 & -6EI/L^2 \\ 0 & GJ/L & 0 & | & 0 & -GJ/L & 0 \\ -6EI/L^2 & 0 & 4EI/L & | & 6EI/L^2 & 0 & 2EI/L \\ \hline -12EI/L^3 & 0 & 6EI/L^2 & | & 12EI/L^3 & 0 & 6EI/L^2 \\ 0 & -GJ/L & 0 & | & 0 & GJ/L & 0 \\ -6EI/L^2 & 0 & 2EI/L & | & 6EI/L^2 & 0 & 4EI/L \end{bmatrix} \begin{bmatrix} \delta_{ijz} \\ \theta_{ijx} \\ \theta_{ijy} \\ \hline \delta_{jiz} \\ \theta_{jix} \\ \theta_{jiy} \end{bmatrix}$$

Space Frames

From equation (B.1) the forces at end "i" are found from the following equation:

$$f_{ij} = k_{ii}^j \delta_{ij} + k_{ij} \delta_{ji}$$

i.e.

$$\begin{bmatrix} f_{ijx} \\ f_{ijy} \\ f_{ijz} \\ m_{ijx} \\ m_{ijy} \\ m_{ijz} \end{bmatrix} = \begin{bmatrix} EA/L & 0 & 0 & 0 & 0 & 0 \\ 0 & 12EI_z/L^3 & 0 & 0 & 0 & 6EI_z/L^2 \\ 0 & 0 & 12EI_y/L^3 & 0 & -6EI_y/L^2 & 0 \\ 0 & 0 & 0 & GJ/L & 0 & 0 \\ 0 & 0 & -6EI_y/L^2 & 0 & 4EI_y/L & 0 \\ 0 & 6EI_z/L^2 & 0 & 0 & 0 & 4EI_z/L \end{bmatrix} \begin{bmatrix} \delta_{ijx} \\ \delta_{ijy} \\ \delta_{ijz} \\ \theta_{ijx} \\ \theta_{ijy} \\ \theta_{ijz} \end{bmatrix}$$

$$+ \begin{bmatrix} -EA/L & 0 & 0 & 0 & 0 & 0 \\ 0 & -12EI_z/L^3 & 0 & 0 & 0 & 6EI_z/L^2 \\ 0 & 0 & -12EI_y/L^3 & 0 & -6EI_y/L^2 & 0 \\ 0 & 0 & 0 & -GJ/L & 0 & 0 \\ 0 & 0 & 6EI_y/L^2 & 0 & 2EI_y/L & 0 \\ 0 & -6EI_z/L^2 & 0 & 0 & 0 & 2EI_z/L \end{bmatrix} \begin{bmatrix} \delta_{jix} \\ \delta_{jiy} \\ \delta_{jiz} \\ \theta_{jix} \\ \theta_{jiy} \\ \theta_{jiz} \end{bmatrix}$$

Forces at end "j" are found from using the following equation (see equation (B.1)):

$$f_{ji} = k_{ji} \delta_{ij} + k^i_{jj} \delta_{ji}$$

$$\begin{bmatrix} f_{jix} \\ f_{jiy} \\ f_{jiz} \\ m_{jix} \\ m_{jiy} \\ m_{jiz} \end{bmatrix} = \begin{bmatrix} -EA/L & 0 & 0 & 0 & 0 & 0 \\ 0 & -12EI_z/L^3 & 0 & 0 & 0 & -6EI_z/L^2 \\ 0 & 0 & -12EI_y/L^3 & 0 & 6EI_y/L^2 & 0 \\ 0 & 0 & 0 & -GJ/L & 0 & 0 \\ 0 & 0 & -6EI_y/L^2 & 0 & 2EI_y/L & 0 \\ 0 & 6EI_z/L^2 & 0 & 0 & 0 & 2EI_z/L \end{bmatrix} \begin{bmatrix} \delta_{ijx} \\ \delta_{ijy} \\ \delta_{ijz} \\ \theta_{ijx} \\ \theta_{ijy} \\ \theta_{ijz} \end{bmatrix}$$

$$+ \begin{bmatrix} EA/L & 0 & 0 & 0 & 0 & 0 \\ 0 & 12EI_z/L^3 & 0 & 0 & 0 & -6EI_z/L^2 \\ 0 & 0 & 12EI_y/L^3 & 0 & 6EI_y/L^2 & 0 \\ 0 & 0 & 0 & GJ/L & 0 & 0 \\ 0 & 0 & 6EI_y/L^2 & 0 & 4EI_y/L & 0 \\ 0 & -6EI_z/L^2 & 0 & 0 & 0 & 4EI_z/L \end{bmatrix} \begin{bmatrix} \delta_{jix} \\ \delta_{jiy} \\ \delta_{jiz} \\ \theta_{jix} \\ \theta_{jiy} \\ \theta_{jiz} \end{bmatrix}$$

TRANSFORMATION MATRICES

The following transformation matrices transform nodal load and displacement vectors from the global axes system to the member axes system.

The transformation matrix T_{ij} transforms force and displacement vectors at nodes "i" and "j" from *the global axes system to the member axes system* (for member "i,j").

Plane Truss

$$T_{ij} = \begin{bmatrix} \cos\alpha & \sin\alpha \\ -\sin\alpha & \cos\alpha \end{bmatrix}$$

APPENDIX B

where α is the clockwise rotation of member "i,j" about "i" that will make the member axes coincide with the global axes.

Space Truss

$$T_{ij} = \begin{bmatrix} l_x & m_x & n_x \\ l_y & m_y & n_y \\ l_z & m_z & n_z \end{bmatrix}$$

l_x, m_x, n_x, are the direction cosines of the member "x" axis etc.

Plane Frame

$$T_{ij} = \begin{bmatrix} \cos\alpha & \sin\alpha & 0 \\ -\sin\alpha & \cos\alpha & 0 \\ 0 & 0 & 1 \end{bmatrix}$$

Grillage

$$T_{ij} = \begin{bmatrix} 1 & 0 & 0 \\ 0 & \cos\alpha & \sin\alpha \\ 0 & -\sin\alpha & \cos\alpha \end{bmatrix}$$

Space Frame

$$T_{ij} = \begin{bmatrix} l & m & n & 0 & 0 & 0 \\ -m/D & l/D & 0 & 0 & 0 & 0 \\ -ln/D & -mn/D & D & 0 & 0 & 0 \\ 0 & 0 & 0 & l & m & n \\ 0 & 0 & 0 & -m/D & l/D & 0 \\ 0 & 0 & 0 & -ln/D & -mn/D & D \end{bmatrix}$$

l, m, n, are the direction cosines of the member "x" axis, and $D = \sqrt{(1 - n^2)}$.

In the case of a space frame member the member axes system does not necessarily coincide with the principal axes of the member. The following matrix transforms nodal force and displacement vectors from the member axes system to the principal axes system.

$$T^*_{ij} = \begin{bmatrix} 1 & 0 & 0 & 0 & 0 & 0 \\ 0 & \cos\beta & \sin\beta & 0 & 0 & 0 \\ 0 & -\sin\beta & \cos\beta & 0 & 0 & 0 \\ 0 & 0 & 0 & 1 & 0 & 0 \\ 0 & 0 & 0 & 0 & \cos\beta & \sin\beta \\ 0 & 0 & 0 & 0 & -\sin\beta & \cos\beta \end{bmatrix}$$

where β is the anticlockwise rotation of member "i,j" (viewed from end "i") that will make the principal axes coincide with the global axes.

GLOBAL ELEMENT STIFFNESS SUBMATRICES

$$K^j_{ii} = T^{-1}_{ij} k^j_{ii} T_{ij}$$
$$K_{ij} = T^{-1}_{ij} k_{ij} T_{ij}$$
$$K_{ji} = T^{-1}_{ij} k_{ji} T_{ij} = K^T_{ij}$$
$$K^i_{jj} = T^{-1}_{ij} k^i_{jj} T_{ij}$$

Plane Truss

$$K^j_{ii} = \begin{bmatrix} EA/L \cos^2\alpha & EA/L \cos\alpha \sin\alpha \\ EA/L \cos\alpha \sin\alpha & EA/L \sin^2\alpha \end{bmatrix} = K^i_{jj}$$

Find K_{ij} and K_{ji} by multiplying corresponding elements of K^j_{ii} by $\begin{bmatrix} -1 & -1 \\ -1 & -1 \end{bmatrix}$

Space Truss

$$K^j_{ii} = EA/L \begin{bmatrix} l_x^2 & l_x m_x & l_x n_x \\ l_x m_x & m_x^2 & m_x n_x \\ l_x n_x & m_x n_x & n_x^2 \end{bmatrix} = K^{jj}_i$$

Find K_{ij} and K_{ji} by multiplying corresponding elements of K^j_{ii} by $\begin{bmatrix} -1 & -1 & -1 \\ -1 & -1 & -1 \\ -1 & -1 & -1 \end{bmatrix}$

APPENDIX B

Plane Frames

Using the plane frame stiffness factors K_1-K_8, defined previously, allows the global element stiffness submatrices for plane frame elements having a variety of end connections to be developed as follows:

$$K_{ii}^j = T_{ij}^{-1} k_{ii}^j T_{ij}$$

$$= \begin{bmatrix} (EA\cos^2\alpha/L + K_1 EI\sin^2\alpha/L^3) & (EA/L - K_1 EI/L^3)\sin\alpha\cos\alpha & -K_2 EI\sin\alpha/L^2 \\ (EA/L - K_1 EI/L^3)\sin\alpha\cos\alpha & (EA\sin^2\alpha/L + K_1 EI\cos^2\alpha/L^3) & K_2 EI\cos\alpha/L^2 \\ -K_2 EI\sin\alpha/L^2 & K_2 EI\cos\alpha/L^2 & K_4 EI/L \end{bmatrix}$$

$$K_{ij} = T_{ij}^{-1} k_{ij} T_{ij}$$

$$= \begin{bmatrix} -(EA\cos^2\alpha/L + K_1 EI\sin^2\alpha/L^3) & -(EA/L - K_1 EI/L^3)\sin\alpha\cos\alpha & -K_3 EI\sin\alpha/L^2 \\ -(EA/L - K_1 EI/L^3)\sin\alpha\cos\alpha & -(EA\sin^2\alpha/L + K_1 EI\cos^2\alpha/L^3) & K_3 EI\cos\alpha/L^2 \\ -K_5 EI\sin\alpha/L^2 & K_5 EI\cos\alpha/L^2 & K_6 EI/L \end{bmatrix}$$

$$K_{ji} = K_{ij}^T$$

$$K_{jj}^i = T_{ij}^{-1} k_{jj}^i T_{ij}$$

$$= \begin{bmatrix} (EA\cos^2\alpha/L + K_1 EI\sin^2\alpha/L^3) & (EA/L - K_1 EI/L^3)\sin\alpha\cos\alpha & -K_7 EI\sin\alpha/L^2 \\ (EA/L - K_1 EI/L^3)\sin\alpha\cos\alpha & (EA\sin^2\alpha/L + K_1 EI\cos^2\alpha/L^3) & K_7 EI\cos\alpha/L^2 \\ -K_7 EI\sin\alpha/L^2 & K_7 EI\cos\alpha/L^2 & K_8 EI/L \end{bmatrix}$$

Space Frame

The principal axes of plane frame and grillage members are assumed to coincide with the member axes. For space frame members this is not necessarily the case. Hence, the determination of the global element stiffness submatrices requires an additional transformation, as shown in the following equation:

and
$$K_{ii}^j = T_{ij}^{-1} T_{ij}^{*-1} k_{ii}^{j*} T_{ij}^* T_{ij}$$

$$K_{ij} = T_{ij}^{-1} T_{ij}^{*-1} k_{ij}^* T_{ij}^* T_{ij}$$

Defining transformation factors T_1-T_9, and stiffness factors K_1-K_{10}, as follows, allows the global elements stiffness matrix to be expressed in terms of these factors (see section 10.6).

$$
\begin{aligned}
T_1 &= 1 & (0) \\
T_2 &= -(m\cos\beta + \ln\sin\beta)/D & (-n\sin\beta) \\
T_3 &= (m\sin\beta - \ln\cos\beta)/D & (-n\cos\beta) \\
T_4 &= m & (0) \\
T_5 &= (l\cos\beta - mn\sin\beta)/D & (c) \\
T_6 &= -(l\sin\beta + mn\cos\beta)/D & (-s) \\
T_7 &= n & (n) \\
T_8 &= D\sin\beta & (0) \\
T_9 &= D\cos\beta & (0)
\end{aligned}
$$

(the expressions in brackets are those assumed for the "special" case where the member lies parallel to the coordinate "z" axis).

$$
\begin{aligned}
K_1 &= EA/L & K_2 &= 12EI_z/L^3 \\
K_3 &= 12EI_y/L^3 & K_4 &= GJ/L \\
K_5 &= 4EI_y/L & K_6 &= 4EI_z/L \\
K_7 &= 6EI_z/L^2 & K_8 &= -6EI_y/L^2 \\
K_9 &= -6EI_y/L^2 & K_{10} &= 6EI_z/L^2
\end{aligned}
$$

If the global element stiffness submatrix is expressed in terms of the sixteen 3 x 3 global element stiffness submatrices indicated by the broken lines as shown, then there is a similarity between the submatrices which allows the equation to be rewritten as follows:

$$
\begin{bmatrix} F_{ij} \\ M_{ij} \\ F_{ji} \\ M_{ji} \end{bmatrix}
=
\begin{bmatrix}
A & B & -A & B \\
B^T & C & -B & D \\
-A^T & -B^T & A & -B \\
B^T & D^T & -B^T & C
\end{bmatrix}
\begin{bmatrix} \Delta_i \\ \theta_i \\ \Delta_j \\ \theta_j \end{bmatrix}
$$

Hence only four of the sixteen 3 x 3 submatrices need be found, and

$$
A = \begin{bmatrix}
(K_1T_1^2 + K_2T_2^2 + K_3T_3^2) & (K_1T_1T_4 + K_2T_2T_5 + K_3T_3T_6) & (K_1T_1T_7 + K_2T_2T_8 + K_3T_3T_9) \\
(K_1T_4T_1 + K_2T_5T_2 + K_3T_6T_3) & (K_1T_4^2 + K_2T_5^2 + K_3T_6^2) & (K_1T_4T_7 + K_2T_5T_8 + K_3T_6T_9) \\
(K_1T_7T_1 + K_2T_8T_2 + K_3T_9T_3) & (K_1T_7T_4 + K_2T_8T_5 + K_3T_9T_6) & (K_1T_7^2 + K_2T_8^2 + K_3T_9^2)
\end{bmatrix}
$$

$$
B = \begin{bmatrix}
(K_7T_2T_3 + K_8T_3T_2) & (K_7T_2T_6 + K_8T_3T_5) & (K_7T_2T_9 + K_8T_3T_8) \\
(K_7T_5T_3 + K_8T_6T_2) & (K_7T_5T_6 + K_8T_6T_5) & (K_7T_5T_9 + K_8T_6T_8) \\
(K_7T_8T_3 + K_8T_9T_2) & (K_7T_8T_6 + K_8T_9T_5) & (K_7T_8T_9 + K_8T_9T_8)
\end{bmatrix}
$$

APPENDIX B

$$C = \begin{bmatrix} (K_4T_1^2+K_5T_2^2+K_6T_3^2) & (K_4T_1T_4+K_5T_2T_5+K_6T_3T_6) & (K_4T_1T_7+K_5T_2T_8+K_6T_3T_9) \\ (K_4T_4T_1+K_5T_5T_2+K_6T_6T_3) & (K_4T_4^2+K_5T_5^2+K_6T_6^2) & (K_4T_4T_7+K_5T_5T_8+K_6T_6T_9) \\ (K_4T_7T_1+K_5T_8T_2+K_6T_9T_3) & (K_4T_7T_4+K_5T_8T_5+K_6T_9T_6) & (K_4T_7^2+K_5T_8^2+K_6T_9^2) \end{bmatrix}$$

$$D = \begin{bmatrix} (C_{11}-3K_4T_1T_1)/2 & (C_{12}-3K_4T_1T_4)/2 & (C_{13}-3K_4T_1T_7)/2 \\ (C_{21}-3K_4T_4T_1)/2 & (C_{22}-3K_4T_4T_4)/2 & (C_{23}-3K_4T_4T_7)/2 \\ (C_{31}-3K_4T_7T_1)/2 & (C_{32}-3K_4T_7T_4)/2 & (C_{33}-3K_4T_7T_7)/2 \end{bmatrix}$$

hence

$$K_{ii}^j = \begin{bmatrix} A & B \\ B^T & C \end{bmatrix} \qquad K_{ij} = \begin{bmatrix} -A & B \\ -B^T & D \end{bmatrix}$$

$$K_{ji} = K_{ij}^T \qquad K_{jj}^i = \begin{bmatrix} A & -B \\ -B^T & C \end{bmatrix}$$

NODAL FORCE VECTOR

The vector of applied forces acting on a node, in the global axes system, is given by the following equation:

$$P_i = P_i^n + \sum P_{ij}^e$$

It is usually convenient to express the equivalent joint force vectors as transformed fixed end force vectors as follows:

$$P_i = P_i^n + \sum T_{ij}^{-1} f_{ij}^f$$

where f_{ij}^f is the vector of fixed end forces at end "i" of member "i,j" in the member axes system.

EQUATION OF NODAL EQUILIBRIUM

If a framed structure is in a state of static equilibrium then each node must be in equilibrium under the action of the external applied forces (which may include reactive forces) and the member end forces. The equation of nodal equilibrium can be expressed in terms of the global element stiffness submatrices as follows:

$$P_i = K_{ii}\Delta_i + K_{ia}\Delta_a + K_{ib}\Delta_b + \ldots + K_{in}\Delta_n$$

where

$$K_{ii} = \sum K_{ii}^j = \sum T_{ij}^{-1} k_{ii}^j T_{ij}$$

$$K_{ij} = T_{ij}^{-1} k_{ij} T_{ij}$$

The summation is for all elements connected to node "i".

The global element stiffness matrices for a space frame involve an additional transformation as follows:

$$K_{ii}^j = T_{ij}^{-1} T_{ij}^{*-1} k_{ii}^{j*} T_{ij}^{*} T_{ij}$$

and

$$K_{ij} = T_{ij}^{-1} T_{ij}^{*-1} k_{ij}^{*} T_{ij}^{*} T_{ij}$$

MEMBER END FORCES

If the member axes coincide with the principal axes then the element stiffness submatrices in the member axes system can be used to calculate the member end forces from the member end displacements using the following equation:

$$f_{ij} = k_{ii}^j T_{ij} \Delta_i + k_{ij} T_{ij} \Delta_j$$

The element stiffness submatrices and the transformation matrices have already been presented. It is computationally advantageous to expand the above equations as follows:

Plane Truss

$$f_{ijx} = EA/L(\cos\alpha \, \Delta_{ix} + \sin\alpha \, \Delta_{iy} - \cos\alpha \, \Delta_{jx} - \sin\alpha \, \Delta_{jy})$$
$$f_{ijy} = 0$$
$$f_{jix} = -f_{ijx}$$
$$f_{jiy} = 0$$

Space Truss

$$f_{ijx} = EA/L\{l_x(\Delta_{ix} - \Delta_{jx}) + m_x(\Delta_{iy} - \Delta_{jy}) + n_x(\Delta_{iz} - \Delta_{jz})\}$$
$$f_{ijy} = 0$$
$$f_{ijz} = 0$$
$$f_{jix} = -f_{ijx}$$
$$f_{jiy} = 0$$

APPENDIX B

$$f_{jiz} = 0$$

Plane Frame

$$f_{ijx} = EA/L \{(\Delta_{ix} - \Delta_{jx})\cos\alpha + (\Delta_{iy} - \Delta_{jy})\sin\alpha\}$$
$$f_{ijy} = K_1 EI/L^3\{(-\Delta_{ix} + \Delta_{jx})\sin\alpha + (\Delta_{iy} - \Delta_{jy})\cos\alpha\} +$$
$$EI/L^2(K_2\theta_i + K_3\theta_j)$$
$$m_{ij} = EI/L^2\{(-K_2\Delta_{ix} - K_5\Delta_{jx})\sin\alpha + (K_2\Delta_{iy} + K_5\Delta_{jy})\cos\alpha\} +$$
$$EI/L (K_4\theta_i + K_6\theta_j)$$
$$f_{jix} = -f_{ijx}$$
$$f_{jiy} = -f_{ijy}$$
$$m_{ji} = -m_{ij} + L f_{ijy}$$

Grillage

$$f_{ijz} = 6EI\{(2\Delta_{iz} - \Delta_{jz})/L + (\theta_{ix} + \theta_{jx})\sin\alpha - (\theta_{iy} + \theta_{jy})\cos\alpha\}/L^2$$
$$m_{ijx} = GJ \{(\theta_{ix} - \theta_{jx})\cos\alpha + (\theta_{iy} - \theta_{jy})\sin\alpha\}/L$$
$$m_{ijy} = 2EI \{-3(\Delta_{iz} - \Delta_{jz})/L - (2\theta_{ix} + \theta_{jx})\sin\alpha + (2\theta_{iy} + \theta_{jy})\cos\alpha\}/L$$
$$f_{jiz} = -f_{ijz}$$
$$m_{jix} = -m_{ijx}$$
$$m_{jiy} = -m_{ijy} - L f_{ijz}$$

Space Frame

To find the member end forces for a space frame element requires an additional transformation, as shown by the following equation:

$$f_{ij} = k_{ii}^{j*} T_{ij} T_{ij}^* \Delta_i + k_{ij}^* T_{ij} T_{ij}^* \Delta_j$$

Using the "T" and "K" factors defined previously.

$$f_{ijx}^* = K_1(T_1\Delta_{ix} + T_4\Delta_{iy} + T_7\Delta_{iz} - T_1\Delta_{jx} - T_4\Delta_{jy} - T_7\Delta_{jz})$$
$$f_{ijy}^* = K_2(T_2\Delta_{ix} + T_5\Delta_{iy} + T_8\Delta_{iz} - T_2\Delta_{jx} - T_5\Delta_{jy} - T_8\Delta_{jz})$$
$$+ K_7(T_3\theta_{ix} + T_6\theta_{iy} + T_9\theta_{iz} + T_3\theta_{jx} + T_6\theta_{jy} + T_9\theta_{jz})$$
$$f_{ijz}^* = K_3(T_3\Delta_{ix} + T_6\Delta_{iy} + T_9\Delta_{iz} - T_3\Delta_{jx} - T_6\Delta_{jy} - T_9\Delta_{jz})$$
$$+ K_8(T_2\theta_{ix} + T_5\theta_{iy} + T_8\theta_{iz} + T_2\theta_{jx} + T_5\theta_{jy} + T_8\theta_{jz})$$

$$m^*_{ijx} = K_4(T_1\theta_{ix} + T_4\theta_{iy} + T_7\theta_{iz} - T_1\theta_{jx} - T_4\theta_{jy} - T_7\theta_{jz})$$

$$m^*_{ijy} = K_8(T_3\Delta_{ix} + T_6\Delta_{iy} + T_9\Delta_{iz} - T_3\Delta_{jx} - T_6\Delta_{jy} - T_9\Delta_{jz})$$
$$+ K_5\{T_2\theta_{ix} + T_5\theta_{iy} + T_8\theta_{iz}$$
$$+ (T_2\theta_{jx} + T_5\theta_{jy} + T_8\theta_{jz})/2\}$$

$$m^*_{ijz} = K_7(T_2\Delta_{ix} + T_5\Delta_{iy} + T_8\Delta_{iz} - T_2\Delta_{jx} - T_5\Delta_{jy} - T_8\Delta_{jz})$$
$$+ K_6\{T_3\theta_{ix} + T_6\theta_{iy} + T_9\theta_{iz}$$
$$+ (T_3\theta_{jx} + T_6\theta_{jy} + T_9\theta_{jz})/2\}$$

$$f^*_{jix} = -f^*_{ijx}$$
$$f^*_{jiy} = -f^*_{ijy}$$
$$f^*_{jiz} = -f^*_{ijx}$$
$$m^*_{jix} = -m^*_{ijx}$$
$$m^*_{jiy} = -m^*_{ijy} - L f^*_{ijz}$$
$$m^*_{jiz} = -m^*_{ijz} + L f^*_{ijy}$$

Note the the equations presented in this section give the forces associated with the nodal displacements. If member loads are present then the fixed end forces must be superimposed on the forces associated with the nodal displacements.

REACTIONS

In general

$$R_i = \Sigma F_{ij} - P_i$$

where the summation is for all members connected to the restrained node "i". Note that the member end forces will have been found in terms of the member axes system and must be transformed into the global axes system prior to summation. Hence the above equation should be written as follows:

$$R_i = \Sigma T^{-1}_{ij} f_{ij} - P_i$$

Again the principal axes have been assumed to coincide with the member axes. In the case of space frames this is not necessarily the case and an additional transformation is necessary, i.e.

$$R_i = \Sigma T^{-1}_{ij} T^{*-1}_{ij} f_{ij} - P_i$$

Note that it is computationally efficient to expand the above expressions.

Index

Axes systems,
    coordinate, 55
    global, 55-56
    local, 56
    member, 56-57
    nodal, 55-56
Axial deformation, 30-31

Bandwidth, 58-60
Beam behaviour, 30-36
    axial, 30
    flexural, 32
    torsional, 31
Boundary conditions, 12-15, 41
Brick element, 10-11

Castigliano's first theorem, 46
Cholesky decomposition, 75-76
Code number, 124, 175, 217
Coefficient matrix,
    definition of, 62
    inversion of, 83-85
    triangular decomposition of, 75-82
Coefficient of thermal expansion, 235-236
Commercial frame analysis programs,
    data preparation, 366-374
    initial member sizing, 366-372
    language, 364
    machine, 364-365
    scope, 365-366
    the user interface, 366
Cross product of vectors, 331-332
Crout reduction, 75

Data preprocessor program PRE.MA/PRE.01/
    PRE.02, 147-164
Deformation of structures, 23-25
Degree of freedom
    definition of, 38
    examples, 41-42
    nodal, 39
Degree of indeterminacy, 27
Direction cosines, 167
Displacement method, 44
Displacement vector, 44

Elastic supports, 265-269
Element stiffness matrix,
    grillage element, 306-307, 309-310
    plane frame element, 192-194, 197-199
        program ESTIFF.PF, 206-210

plane truss element, 95-97, 100-102
    program ESTIFF.PT, 110-115
space frame element, 327-328, 334-345,
    348-350
space truss element, 165-166, 170-172
Element types, 10
Elements with pins, 270-285
Equilibrium, 15-23
Equivalent joint forces, 232-235
    program FIX.PF, 248-255
    for members with pins, 279-280

Final structure stiffness matrix,
    for a plane frame, 216-223
        program FSTIFF.PF, 219-233
    for a plane truss, 122-131
        program FSTIFF.PT, 127-131
    for a space truss, 173-175
Finite element method, 38
Fixed end forces, 232-234, 238-239
    program FIX.PF, 248-255
Flexibility matrix, 44
Flexibility method, 44
Flexural deformation, 32-36
Forces,
    external, 15
    internal, 15
Frame structures,
    classification of, 37
    definition of, 10
Freebody diagrams, 16-23
Freedoms, 42
Freedom vector, 123, 175

Gauss elimination, 69-74
    program GAUSS.EQ, 71-74
Gauss-Jordan elimination, 63-67
    program GAUSSJ.EQ, 65-67
Generalised force and displacement, 39
Global axes, 55-56
Grids (see grillages)
Grillages, 305-326
    program GRID.GD, 314-326

Idealisation,
    at the structural boundaries, 12-15
    need for, 7-8
    of loading, 8-10
    of materials, 11-12
    of the structure, 2,10-15
Ill-conditioned equations, 85-94

Inclined supports, 255-264
Initial member sizing, 366-372
Initial structure stiffness matrix,
    plane frame, 201-205
        program ISTIFF.PF, 211-214
    plane truss, 105-110
        program ISTIFF.PT, 115-118
    space truss, 173

Lack of fit, 235, 246-247
Limit states,
    serviceability, 12
    ultimate, 12
Line element, 10-11
Load vector, 44
Loads,
    applied, 16
    design, 8
    idealisation of, 8-10
    live, 8
Local axes, 56, 255-264, 373

Materials, idealisation of, 11
Matrix inversion, 83-85
    program INVERT.EQ, 90-94
Matrix methods of structural analysis, 42-44
Mechanism, 7, 27
Member axes, 56-57
    grillage, 306
    plane frame, 193
    plane truss, 96
    space frame, 327
    space truss, 165
Member end forces,
    for members with pins, 279-285
    plane frame, 224-230
        proram MFORCE.PF, 226-230
    plane truss, 131-136
        proram MFORCE.PT, 133-136
    space frame, 345-346
    space truss, 176
Member loads, 231-235, 239-243, 373
Member properties, 372-373
Modulus of rigidity, 31
Multiple right hand sides, 67-69

Node, definition of, 38
Node numbers, 366
Nodal axes, 55-56
Nodal degree of freedom, 39
Nodal displacement vector, 39
Nodal equilibrium, 102-105, 170-171, 199-201, 336
Nodal force vector, 39
Non-prismatic members, 285-287
Non-singular equations, 85-86

Plane frames, 192-304
    program PFRAME.PF, 287-304
Plane trusses, 95-164
    program PTRUSS.PT, 136-147
Plate element, 10-11
Poisson's ratio, ,31
Preliminary design, 1
Principal axes, 334-345
    program TNODE.SF, 343-345
Program languages, 364

Reactions, 15-16
    plane frame, 226
    plane truss, 132-133
    space frame, 350-351

    space truss, 176-177
Restraints, 373
Restraint numbers,
    plane frame, 216
    plane truss, 123
    space truss, 174
Results, 374
Right hand side,
    definition of, 62
    multiple, 67-69

Serviceability limit state, 12
Settlement, 237-238, 245-246
Shear modulus, 31
Shape function, 33-36, 247-248, 275
Simultaneous linear equations, 62-94
Singular equations, 85-86
Skeletal structures, see frame structures
Small deflection theory, 24-25
Space frame, 327-362
    program SFRAME.SF, 346-362
Space trusses, 165-191
    program STRUSS.ST, 177-191
Statically determinate structures, 25
Statically indeterminate structures, 25-29
Stepped member, 285-287
Stiffness coefficients, 44-48
Stiffness matrix, 44
Stiffness method,
    definition of, 44
    direct application of, 49-55
Stress resultants, 16, 57
Structure,
    definition of, 6
    stable, 6
    unstable, 6
Structure stiffness matrix,
    properties of, 57-61
Superposition, 29-30

Tapered members, 285-286
Temperature effects, 235-236, 243-244
Third node method, 336-345
    program TNODE.SF, 343-345
Torsional constant, 31
Torsional deformation, 31, 305
Transformation of force and displacement,
    for grillages, 307-309
    for plane frames, 194-197
    for plane trusses, 97-100
    for space frames, 329-345
        program TNODE.SF, 343-345
    for space trusses, 166-169
Triangular decomposition, 75-82
    program UDU.EQ, 79-82

Ultimate limit state, 12
Units, 110, 374

Vector cross product, 331-332